NEW FRONTIERS IN SUPERCONDUCTIVITY RESEARCH

NEW FRONTIERS IN SUPERCONDUCTIVITY RESEARCH

BARRY P. MARTINS
EDITOR

Nova Science Publishers, Inc.
New York

For permission to use material from this book please contact us:
Telephone 631-231-7269; Fax 631-231-8175
Web Site: http://www.novapublishers.com

Library of Congress Cataloging-in-Publication Data

New frontiers in superconductivity research / Barry P. Martins, editor.
p. cm.
Includes bibliographical references and index.
ISBN 1-59454-850-1
1. Superconductivity--Research. I. Martins, Barry P.
QC611.96.N495 2006
537.6'23072--dc22
2005033124

Published by Nova Science Publishers, Inc. ✦ New York

CONTENTS

PREFACE

Superconductivity is the ability of certain materials to conduct electrical current with no resistance and extremely low losses. High temperature superconductors, such as La2-xSrxCuOx (Tc=40K) and YBa2Cu3O7-x (Tc=90K), were discovered in 1987 and have been actively studied since. In spite of an intense, world-wide, research effort during this time, a complete understanding of the copper oxide (cuprate) materials is still lacking. Many fundamental questions are unanswered, particularly the mechanism by which high-Tc superconductivity occurs. More broadly, the cuprates are in a class of solids with strong electron-electron interactions. An understanding of such "strongly correlated" solids is perhaps the major unsolved problem of condensed matter physics with over ten thousand researchers working on this topic.

The aspects of vortices in high temperature superconductors are interesting. These superconductors (in cuprates, for example) are layered materials due to the alternating CuO plane and charge reservoir layers. They therefore have (highly) anisotropic physical properties along and perpendicular to ab-plane. When the coherence length ξ_c along the c-axis drops below the layer separation d at a particular temperature, the vortex system is considered to go from three-dimensional (3D) to two-dimensional (2D) regime at which the conventional Abrikosov vortices suggestively disappears, forming novel types of vortices for which the vortex supercurrents are no longer necessarily circular or the axes rectilinear. Instead, pancake, tilted, combine and Josephson vortices are formed with the change of orientation (θ) of the magnetic field (**H**) with respect to the ab-plane. The experimental observations of these vortices are discussed in Chapter 1, using such data as vortex scanning, and flux flow resistivity as a function of temperature T (below the superconducting transition temperature T_c), H and θ collected on different cuprate single crystals, whereby lock-in transition angles and hence the θ-T and H-T phase diagrams are extracted. Further, the correlations among the vortex lock-in transition, irreversibility field, and dimensional crossover have also been discussed. In particular, there is difficulty to understand the experimental θ versus H data. This has been attributed to the understanding of these vortices in terms of the superconducting order parameter having s-wave symmetry instead of this parameter with d-wave symmetry pertinent to these layered structures.

The effect of fractal clusters on magnetic and transport properties of percolative superconductors is considered. The superconductor contains percolative superconducting cluster, carrying a transport current, and clusters of a normal phase, which act as pinning centers. A prototype of such a structure is a high-temperature superconducting (HTS) wire.

The superconducting core of first generation HTS wires is the conglomeration of superconducting micro-crystallites containing normal-phase clusters inside. There are the clusters of columnar defects in superconducting layer, which is the current-carrying element of second generation HTS wire. It is found that normal-phase clusters can have essential fractal features that affect the vortex dynamics. The fractal dimension of the boundary of normal-phase clusters in YBCO films is estimated and the cluster statistics is analyzed. Depinning and transport of vortices in fractal superconducting structures are investigated. Transition of the superconductor into a resistive state corresponds to the percolation transition from a pinned vortex state to a resistive state when the vortices are free to move. It is revealed that a mixed state of the vortex glass type is realized in the superconducting system involved. The fractal distribution of critical currents is derived and its peculiarities are studied in Chapter 2. It is found that there is the range of fractal dimension where this distribution has anomalous statistical properties, specifically, its dispersion becomes infinite. The current-voltage characteristics of superconductors containing fractal clusters are obtained. Dependencies of the free vortex density on the fractal dimension as well as the resistance on the transport current are studied. It is found that the fractality of the cluster boundary intensifies pinning and thereby raises the critical current. This feature enables the current-carrying capability of a superconductor to be enhanced without changing of its chemical composition.

The problem of the vortex entry into type-II superconductors (bulk samples, thin films) is analysed theoretically. Two approaches for the problem are discussed in Chapter 3:

i. the Gibbs-London theory, based on the concept of geometrical barrier, controlling vortex entry/exit into/out from superconductor, and
ii. the quantum-mechanical Ginzburg-Landau theory, employing the concept of the wave function of superconducting condensate.

The quantitative conditions for the entry of vortices into type-II superconductors being in the Meissner and/or mixed state, will be described in the framework of two different approaches. Specifically, a detailed theory of geometrical barrier will be constructed, on the basis of which the shape of the vortex fragment will be described, and the vortex-entry criterion will be obtained. The analytical and numerical study of the Ginzburg-Landau equations reveal the existence of a modulation instability of superconducting condensate. The instability threshold is reached when superconducting order parameter turns to zero at the sample surface and the supervelocity (or gauge-invariant vector potential) of the condensate reaches some a threshold value at the superconductor edge or surface. For superconductors with high value of Ginzburg-Landau parameter this criterion practically coincides with criterion based on the depairing current density. Due to this instability, vortices start to nucleate at the edge or surface and then penetrate deeper into the sample.

Amorphous Nb_xG_{e1-x} thin films have recently attracted the attention of researchers interested in basic properties of vortex dynamics. These samples are conventional, isotropic superconductors in the dirty limit, with a high Ginzburg-Landau parameter $\kappa \sim 80-100$, the transition temperature $T_c \sim 2.7-3$ K and the upper critical magnetic field B_{c2} at zero temperature of $4-5$ T. Their amorphous structure and homogeneity lead to a low density of defects and consequently to a very weak pinning. Together with the experimental accessibility of the entire magnetic field vs temperature phase diagram of the mixed state, this makes them suitable for investigations of those aspects of vortex motion which are general rather than sample specific. I shall highlight several experimental results the understanding of which has benefited from this unique combination of the simplicity of vortex matter and marginal role of pinning in the studied amorphous $Nb_{0.7}Ge_{0.3}$ microbridges. The phenomena under discussion are of a fundamental character and are not manifested in such a clear form in most other type II superconductors, where they are usually hindered by other mechanisms - such as strong pinning in classical superconductors or fluctuation phenomena inherent in the depinned vortex state of high-T_c compounds. The experimental methods used in the studies presented in Chapter 4 are measurements of the time-averaged voltage (magnetoresistance, voltage-current characteristics) and the related voltage noise, whereas the analysis relies on the Larkin-Ovchinnikov theory of flux flow in its original or modified form. Another advantageous property of amorphous Nb-Ge films is that they maintain their superconducting properties when deposited onto a substrate through a mask produced by standard electron-beam lithography. Thus the samples can be made in various shapes containing features as small as a few hundred nanometres. Since we already understand well how vortices in amorphous Nb-Ge films move, it is perhaps not too bold to predict an upcoming development in fabricating mesoscopic devices which would involve vortex motion as the principle of operation.

Vortices in type-II superconductors are usually considered as macroscopic entities and treated with classical mechanics because they have large size. However, a number of experimental results show that vortices can undergo quantum effects, which imply that, in some sense, the nature of vortices should be described by quantum mechanics. In Chapter 5, the quantum effects of vortex in type-II superconductors is studied philosophically by reviewing the ideas in elementary quantum mechanics, in particular, the origin of the quantum nature of vortex is discussed and is concluded to be of that vortex has small mass per unit length. Other vortex matter phenomenology in type II superconductors, such as quantum creep, quantum depinning and quantum melting are also reviewed. It is believed that quantum effects will manifest themselves at sufficiently low temperature because in high temperature (large quantum numbers) the energy differences become arbitrarily small and the energies form almost a continuum. The quantum fluctuations of vortices drive various types of phase transitions and define much the superconducting properties of real materials at low temperature. The extension of the quantum description to the fields beyond the micro world, such as flux pinning classification and economics, is also given.

Superconducting thin films and micro- and nanostructured superconducting elements have enormous potential for applications in electronics and as highly sensitive detectors. The authors present, in Chapter 6 recent studies of the superconductivity of nanostructured Nb and NbN thin films as well as their potential use as very fast, sensitive and energy-dispersive photon-counters. Using photo- and electron-beam-lithography *5* to *20*~nm thick films were patterned into single bridges and meander structures with widths of the conduction paths as narrow as *50*~nm. Dependencies of the critical temperature and critical current density on variations in bridge width and thickness can be modelled rather well using existing theories applicable to dirty superconductors with very short coherence lengths.

The meander structures have already been shown to be very fast and sensitive single photon detectors in the visible and near-infrared range. Their ultimate sensitivity is hampered however, by electric noise caused at least partly by fluctuations in the superconducting state. The detection mechanism exploits the creation of a localized hot-spot by the absorption of a photon with sufficient energy. Albeit being a relatively small perturbation of the superconducting state, it can be detected as a voltage transient given the right experimental conditions. These conditions require cross-sectional areas of the conduction paths comparable with the cross-section of the hot-spot and an applied bias current only slightly less than the critical current at the operating temperature. Under these conditions fluctuations, like e.g. phase-slip centers, fluctuations in the number density of Cooper-pairs or fluctuations within the vortex system, may also cause voltage pulses very similar in magnitude and shape to those caused by absorbed photons and may be a significant contribution to the dark count rate. Measured dark count rates have been analyzed with respect to these various noise sources. Besides the noise characteristics or noise equivalent power the performance of the single-photon detectors was investigated concerning the quantum efficiencies as a function of the incident photon energy and material parameters of the superconducting meander line.

Melt-processed Gd-Ba-Cu-O bulk superconductors have drawn worldwide interest due to their superior superconducting performance. In particular, air processing is attractive from the economical viewpoints. In Chapter 7, the authors report on the superconducting performance of $GdBa_2Cu_3O_{7-\delta}$ (Gd123) fabricated in air with focus placed on the effects of the addition of BaO_2, refined Gd_2BaCuO_5 (Gd211), and a new second phase $Gd_2Ba_4CuMoO_y$ (Gd2411). First, they discussed the solubility and the stability of $Gd_{1+x}Ba_{2-x}Cu_3O_{7-\delta}$ solid solutions and the relation with the properties of single domain Gd-Ba-Cu-O bulks prepared in air, in that the doping of BaO_2 was found to be crucial for the suppression of Gd substitution on Ba sites. They further discussed the relation between the sizes of the second phase Gd211 particles and the critical current (J_c) of Gd123 bulks. The addition of Gd2411 powders sintered in air was also performed and its effect on the superconducting properties and microstructure of Gd-Ba-Cu-O was studied. With the proper amount additions of BaO_2, Gd2411 and refined Gd211, the J_c of Gd-Ba-Cu-O bulks grown in air reached 90,000 A/cm^2 at 77 K in self-field. The authors also propose potential applications of air-processed Gd-Ba-Cu-O to rotating machines used for motors and generators.

In: New Frontiers in Superconductivity Research
Editor: Barry P. Martins, pp. 1-46
ISBN: 1-59454-850-1

Chapter 1

VORTICES IN LAYERED SUPERCONDUCTORS

Gunadhor S. Okram
UGC-DAE Consortium for Scientific Research, University Campus,
Khandwa Road, Indore 452001, MP, India

Abstract

The aspects of vortices in high temperature superconductors are interesting. These superconductors (in cuprates, for example) are layered materials due to the alternating CuO plane and charge reservoir layers. They therefore have (highly) anisotropic physical properties along and perpendicular to ab-plane. When the coherence length ξ_c along the c-axis drops below the layer separation d at a particular temperature, the vortex system is considered to go from three-dimensional (3D) to two-dimensional (2D) regime at which the conventional Abrikosov vortices suggestively disappears, forming novel types of vortices for which the vortex supercurrents are no longer necessarily circular or the axes rectilinear. Instead, pancake, tilted, combine and Josephson vortices are formed with the change of orientation (θ) of the magnetic field (**H**) with respect to the ab-plane. The experimental observations of these vortices are discussed using such data as vortex scanning, and flux flow resistivity as a function of temperature T (below the superconducting transition temperature T_c), H and θ collected on different cuprate single crystals, whereby lock-in transition angles and hence the θ-T and H-T phase diagrams are extracted. Further, the correlations among the vortex lock-in transition, irreversibility field, and dimensional crossover have also been discussed. In particular, there is difficulty to understand the experimental θ versus H data. This has been attributed to the understanding of these vortices in terms of the superconducting order parameter having s-wave symmetry instead of this parameter with d-wave symmetry pertinent to these layered structures.

1 Introduction

1.1 General

The trend in superconductivity research has been changing gradually since Dutch physicist Heike Kamerlingh Onnes discovered it in mercury (Onnes 1911). Initially considered to be the domain of *metallic elemental and isotropic* properties, this phenomenon has been proven to exist in varieties of materials such as other elements, alloys and oxides, which are not necessarily isotropic. Not only simple (continuous), several materials *with layered structure* with anisotropic properties due to structural anisotropy have been discovered till date. In the history of about a century of superconductivity, many theories have been formulated to understand these changing paradigms. In spite of its simple approach, London theory has been a time-tested theory for understanding uniform superconductivity or supercurrent density distribution (London 1935). The theory of Bardeen, Cooper and Schrieffer (BCS) is rather successful to explain microscopic superconductivity (Bardeen, Cooper, Schrieffer 1957). Ginzburg and Landau theory that was initially proposed to work close to the critical temperature, T_c-$T \ll T_c$ has been later on proven to be one of the most successful theory in understanding the quantum phenomenological behavior of the phenomenon (Ginzburg, Landau 1950). In the Ginzburg-Landau theory, a complex position-dependent order parameter $\psi(\mathbf{r}) = |\psi(\mathbf{r})|e^{i\varphi(\mathbf{r})}$ describes the macroscopic properties of a superfluid condensate of *the s-wave symmetry* wherein $\varphi(\mathbf{r})$ is the phase and $|\psi(\mathbf{r})|$ is a modulus. The theory is based on Landau theory of second order phase transition as the symmetries above and below the transition from normal to superconducting state are related due to the continuous symmetry breaking across the transition.

Abrikosov in 1957 studied Ginzburg-Landau theory that describes the situation in which Ginzburg-Landau parameter κ is much larger than unity, thereby he identified superconductors as type I and type II superconductors (Abrikosov 1957). The generic phenomenological mean field vortex diagram for a type II superconductor is shown in Figure 1. In their mixed or vortex state between the lower critical field H_{c1} and the upper critical field H_{c2} ($H_{c1}<H<H_{c2}$), magnetic field penetrates in quanta of flux tubes called vortices. These so-called flux tubes are linear for conventional superconductors, surrounded by the swirling supercurrents, and the magnetic field at the center of the tube is high enough to destroy superconductivity, wherein the superconducting order parameter ψ goes to zero (Figure 2(a)). Each flux quantum carries $\Phi_0= h/2e \approx 2.0678 \times 10^{-15}$ tesla m^2 of flux, where h is Planck's constant 6.6262×10^{-34}Js, and e is charge of an electron 1.60219×10^{-19}C. These flux tubes form a regular array instead of laminar domains and the complex order parameter ψ goes to zero for $h=H_c$ and $x=-\infty$ and ψ attains a maximum $\psi\infty$ for $h=0$ and $x=+\infty$ (Figure 2(b)). The partial penetration of flux requires less diamagnetic energy cost to hold out the field, which is fundamental to high field superconducting magnet applications.

Gor'kov generalized the Ginzburg-Landau theory as a limiting case of microscopic BCS theory (Gor'kov 1959). In the vicinity of T_c, Ginzburg-Landau equations are derived from BCS theory. The Ginzburg-Landau order parameter $\psi(\mathbf{r})$, proportional to the energy gap $\Delta(\mathbf{r})$, is denoted with the pair wave function. In furtherance, Josephson gave the physical significance of phase degree of freedom of the superconducting order parameter ψ, and predicted (and verified experimentally) tunneling of pairs between two superconductors even at zero voltage difference giving a supercurrent density $J=J_c\sin(\varphi_1-\varphi_2)$, where J_c is a constant and φ_i is the phase of ψ in the ith superconductor at tunnel junction (Josephson 1962). He further predicted that a voltage difference V_{12} between the electrodes would cause the phase difference to increase with time as $2eV_{12}t/\hbar$ leading to an oscillating current of frequency $\omega=2eV_{12}/\hbar$.

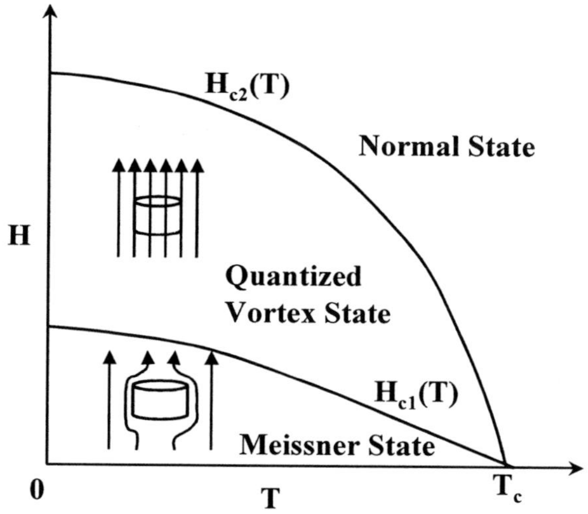

Figure 1 Mean-field theory vortex phase diagram of a type II superconductor. (Meissner state only represents a type I superconductor.) Arrows indicate the flux lines of an external magnetic field interacting with the cylindrical shape superconductor. In vortex state, the sample allows the magnetic field lines to penetrate in flux quanta.

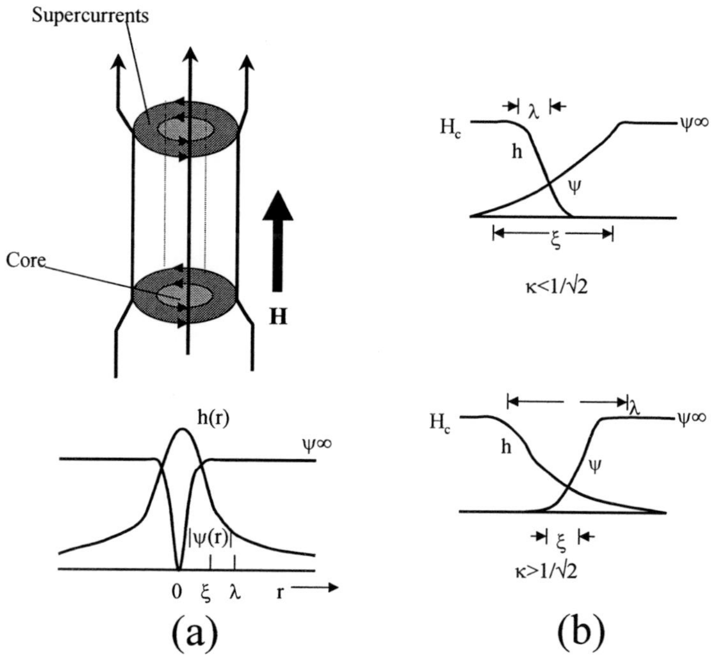

Figure 2. (a) Structure of an isolated (rectilinear) Abrikosov vortex in the presence of an applied magnetic field H (top) and its field and ψ profile (bottom). (b) Variation of h and ψ in a domain wall for a type I, $\kappa<1/\sqrt{2}$ with positive wall energy (top), and a type II, $\kappa >1/\sqrt{2}$ with negative wall energy (b), superconductors.

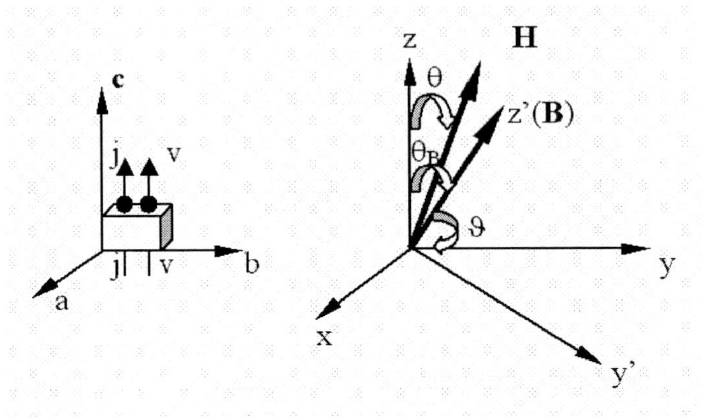

Figure 3 Coordinate system: the x, y and z axes are aligned with crystal symmetry axes a, b and c; c represents axis of approximately uniaxial anisotropy and the ab plane contains the superconducting CuO plane. The magnetic field **H** lies in the yz plane and encloses an angle $\theta=(\pi/2-\vartheta_H)$ with the c axis or an angle ϑ_H with the ab plane. The magnetic induction **B** also lies in the yz plane but not in general aligned with **H** and encloses an angle ϑ with b axis or an angle θ_B with c axis.

The question then came for the case of a stack of superconducting layers. Lawrence-Doniach model, which incorporates the Josephson coupling between the layers in the Ginzburg-Landau theory, explains this (Lawrence, Doniach 1970). The discovery of Bednorz and Muller of high temperature superconductivity at T_c~30K (Bednorz, Muller 1986) then gave birth to a new class of high temperature superconductors wherein the basic structure is usually layered, comprising of metallic CuO plane separated by insulating charge reservoir buffer layers. Therefore, their physical properties are (highly) anisotropic along and perpendicular to the ab plane. In the extreme case, layers are distinctly decoupled depending on the temperature, and the decoupling range differs with the superconductor in hand. They are understood through the Lawrence-Doniach model.

1.2 Dimensionality and Vortices

Question then arises on the vortex structure of these decoupled layered materials. Two important parameters are the Josephson coupling between the layers and the situation in the rectilinear flux tubes. They depend on the angle the applied magnetic field **H** makes with respect to the ab plane (Figure 3). In a layered superconductor, the structure of a single vortex and that of a vortex lattice can be strongly modified. Three dimensional (3D) vortex lines passing through a stack of superconducting layers at a non-vanishing angle with respect to the principal axis have surprisingly complex structure, owing to the tendency of the supercurrents of the vortices to be confined to the easy current directions, parallel to the layers (Kogan 1981; Kogan, Clem 1981). Unlike the continuous rectilinear Abrikosov vortex, an individual vortex in this case is a non-rectilinear object, composed of the so-called pancake vortices directed along the c-axis joined by Josephson vortices whose axes thread through the junctions between the superconducting layers (Figure 4). It has therefore been reasonably well understood that depending on the angle the applied magnetic field **H** subtends with respect to the plane, the degree of layered nature, temperature and strength of **H** used, the energetic of these complicated objects has their preferences to modify themselves to suit particular situations prevailed (Blatter et al. 1994; Bulaevskii et al. 1992; Feinberg, Billard 1990; Ivlev et al. 1990; 1991; Bualevskii 1991).

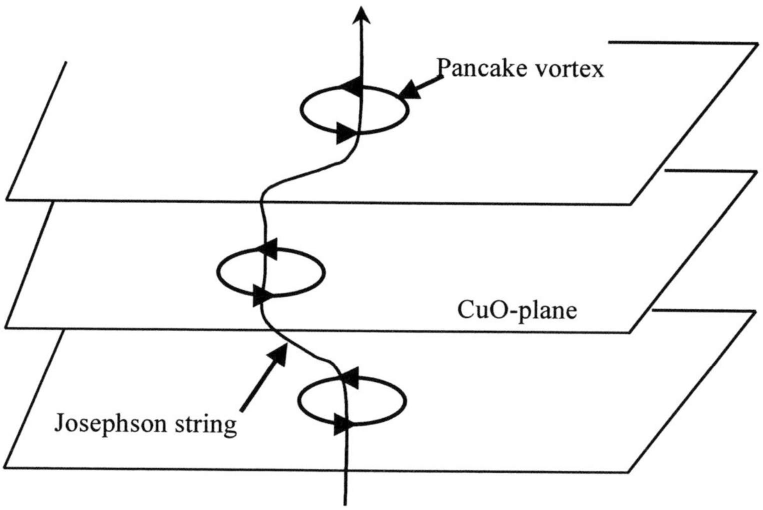

Figure 4 Schematic of an arrangement for a single vortex object in a highly layered superconductor.

Near the ab plane, for **H** not exactly aligned with the plane, the flux lines prefer to pass through the insulating layers with least disturbance to the superconducting plane whereby the internal field is exactly aligned with the ab plane, $\vartheta=0$, and this phenomenon is known as the lock-in transition (Bulaevskii et al. 1992; Feinberg, Billard 1990; Ivlev et al. 1990; 1991; Bualevskii 1991; Maslov, Porovsky 1991), a consequence of the intrinsic pinning of the stacking layers (Tachiki, Takahashi 1989). As a result, four types of vortices, viz., pancake, tilted, combine and Josephson can be considered to develop depending on the angle and material one chooses; for a review see Blatter et al. 1994. Further question arises as to 'If they do exist, then how can one identify them experimentally?' For this, one realizes that these vortices should have different physical parameters such as the line energy, elastic properties, and so on (Blatter et al. 1994; Bulaevskii et al. 1992; Feinberg, Billard 1990; Ivlev et al.1990; 1991; Bualevskii 1991). Such a generic phase diagram for highly layered superconductors when the magnetic field is parallel with c-axis, wherein the vortices are pancake type, is shown in Figure 5. Vortex state can be of solid, liquid, pinned liquid, unpinned liquid and so on. These features are expected to differ with temperature and field chosen. Should there then be a distinct vortex phase diagram for each type or not? If it is, will they possess similar phase diagram as shown in Figure 5 or not?

If they were not, what way, they would differ? In brief, the weakness of the interlayer coupling and the resulting quasi-two dimensional nature may quite strongly influence the thermodynamic properties of layered superconductors (Bulaevskii et al. 1992; Blatter et al. 1994). These types of questions have been reviewed using various experimental data such as scanning tunneling microscopy (STM), scanning superconducting quantum interference devices (SQUID) microscopy (SSM), and flux flow resistivity (ρ) as a function of temperature T below T_c, H and θ collected on several single crystals. In particular, ρ as a function of temperature T below T_c, H and θ on $Tl_2Ba_2CuO_6$ single crystals provided the θ-T and H-T phase diagrams. In the vast literature of vortices, it is difficult to discuss them all. A brief description of the theory of these vortices in general relevant to a distinctly layered superconductor is highlighted first mainly; the symbols used are essentially based on Blatter et al. 1994. This is followed by a brief description of some typical experiments. Then, some of the experimental results are outlined, which are discussed in the context of the present theoretical understanding. Finally, conclusion is given.

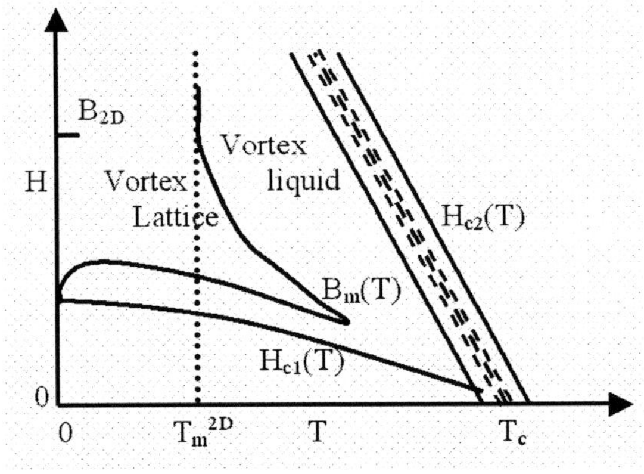

Figure 5 Phenomenological phase diagram of highly layered high temperature superconductors for the field oriented along the c axis.

2 Theory

2.1 The Lawrence-Doniach Model

As in conventional Ginzburg-Landau theory, one defines the Lawrence-Doniach model in terms of a free-energy expression for the stack of superconducting layers coupled by Josephson tunneling. The Gibbs free energy functional describing a discrete set of superconducting layers with order parameter ψ_n with s-wave symmetry separated by a distance d and coupled together by a Josephson term is given by (Lawrence, Doniach 1971)

$$G_{LD} = \int d^2R\left[d\sum_n\left\{\alpha|\psi_n|^2 + \frac{\beta}{2}|\psi_n|^4 + \frac{\hbar^2}{2m}\left|\left(\frac{\nabla^{(2)}}{i} + \frac{2\pi}{\Phi_0}A^{(2)}\right)\psi_n\right|^2\right\}\right]$$
$$+ \int d^2R\left[d\sum_n\left\{\frac{\hbar^2}{2Md^2}\left|\psi_{n+1}\exp\left(\frac{2\pi i}{\Phi_0}\int_{nd}^{(n+1)d} dzA_z\right) - \psi_n\right|^2\right\}\right]$$
$$+ \int d^2R\int dz\left(\frac{B^2}{8\pi} - \frac{B.H}{4\pi}\right) \qquad (1)$$

The ab-planes define the layers and the c-axis is normal to them. We take z-axis along c-axis, with d as the interlayer distance, and x, y as coordinates in the plane. On the assumption that a constant modulus ψ_n within the plane and allowing only the phase degree of freedom of currents the complete functional (1) can be treated in a London-type approximation guarantying applicability at any temperature:

$$G_{LD} = \int d^2R \frac{\varepsilon_0 d}{2\pi}\left[\sum_n \left(\nabla^{(2)}\varphi_n + \frac{2\pi}{\Phi_0}A^{(2)}\right) + \frac{2m}{Md^2}\left\{1-\cos\left(\varphi_{n+1}-\varphi_n+\frac{2\pi}{\Phi_0}\int_{nd}^{(n+1)d} dzA_z\right)\right\}\right]+$$
$$\int d^3r\left(\frac{B^2}{8\pi}-\frac{B.H}{4\pi}\right), \qquad (2)$$

where

$$\varepsilon_0 = \frac{\pi\hbar^2\left|\psi_n{}^2\right|}{m}. \qquad (3)$$

Using (3) we find a description of the layered structure, in which the superconducting properties of the CuO plane are expressed by the period d of the layer structure and planar bulk penetration depth λ. Variation of (2) with respect to the vector potential **A** and the phases φ_n gives (Blatter et al. 1994)

$$\frac{\partial G_{LD}}{\delta A^{(2)}} = 0 \Rightarrow \lambda^2 \Delta A^{(2)} = d\sum_n \delta(z-nd)\left[A^{(2)} + \frac{\Phi_0}{2\pi}\nabla^{(2)}\varphi_n\right], \qquad (4)$$

$$\frac{\partial G_{LD}}{\delta A_z} = 0 \Rightarrow \Delta A_z = \frac{4\pi}{c} j_J \sin\Phi_{n+1,n}, \qquad (5)$$

$$\frac{\partial G_{LD}}{\delta \varphi_n} = 0 \Rightarrow \Lambda^2\left[\Delta^{(2)}\varphi_n + \frac{2\pi}{\Phi_0}\nabla^{(2)}A^{(2)}\right] = \sin\Phi_{n,n-1} - \sin\Phi_{n+1,n}. \qquad (6)$$

Here $\Phi_{n+1,n}$ is the gauge invariant phase difference between n and n+1 layers:

$$\Phi_{n+1,n} = \varphi_{n+1} - \varphi_n + \frac{2\pi}{\Phi_0}\int_{nd}^{(n+1)d} dzA_z. \qquad (7)$$

$$j_J = \frac{c\Phi_0\varepsilon}{8\pi^2\lambda^2\Lambda} \cong j_0\frac{\varepsilon\xi}{\Lambda} \qquad (8)$$

is the Josephson coupling current density between the layers. The Λ is the phase screening length such that

$$\Lambda = d\left[\frac{M}{m}\right]^{1/2} = \frac{d}{\varepsilon}. \tag{9}$$

Here, ε is the anisotropy parameter defined by $(m/M)^{1/2}$ with m and M as the effective mass of the charge carriers along and perpendicular to the ab plane. Also a gauge $\nabla \mathbf{A}=0$ has been used. $\Delta^{(2)}$, $\nabla^{(2)}$ and $\mathbf{A}^{(2)}$ are the planar components of Δ, ∇ and **A**. Eq. (6) is related to Josephson vortices.

2.1.1 Josephson Vortex (at or near ab plane)

The Josephson vortices are aligned along the y-axis when the magnetic field is parallel to the ab plane, pointing along the y-axis (Bulaevskii 1973; Efetov 1979; Clem, Coffey 1990; Carton 1991; Bulaevskii, Clem 1991). Consider a gauge $A_y=0$ and A_x, A_z depending only on x and z. Eq. (6) gives

$$\left(\partial_x^2 + \varepsilon^2 \partial_z^2 - \frac{\varepsilon^2}{\lambda^2}\right)\Phi(r) = 0 \tag{10}$$

Here, we use the fact that outside the core region with dimensions Λ and d, the non-linearities in Eq. (6) and discreteness become irrelevant so that we obtain the screening current density along the z-axis in the continuum anisotropic description

$$j_z = -\frac{2e\hbar}{m}|\psi|^2 \varepsilon \partial_z \varphi \approx \frac{c\Phi_0 \varepsilon}{8\pi\lambda^2 x}. \tag{11}$$

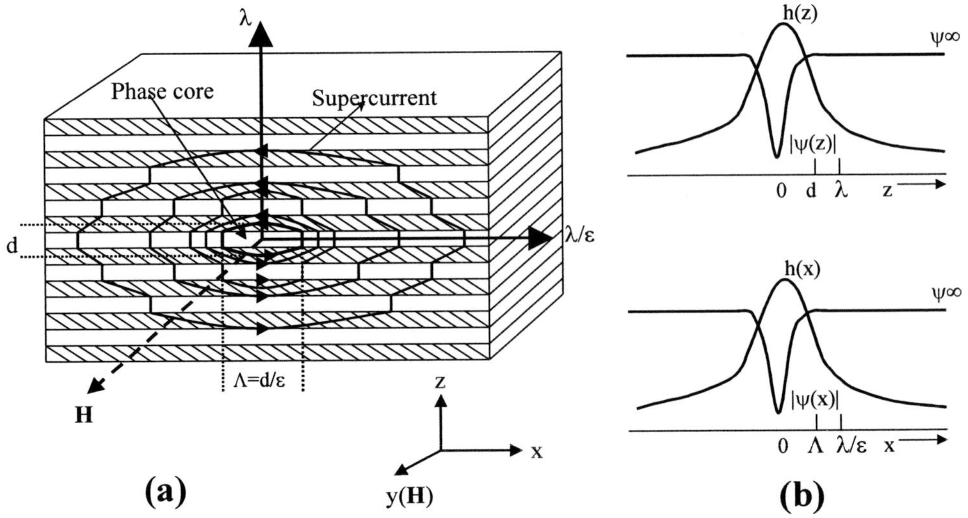

Figure 6 Isolated Josephson vortex in a highly layered superconductor for H along y-axis. (a) The phase core and the supercurrent are indicated. xz (or equivalently ac) plane lies in the plane of the paper and

applied field **H** is perpendicular to it. Different length scales are also indicated as explained in the text. (b) Its field h(r) and ψ profile, in *which ψ does not go to zero.*

Eq. (11) becomes Josephson current density j_J at $x \approx \Lambda$. Therefore, for $x > \Lambda$, the linearized continuum approximation gives a good description. For small distances $x < \Lambda$, a further increase in the current density j_z is limited due to $|j_z| < j_J$, thereby discrete nature prevails. The second critical dimension d along the z-axis is trivially $z = \varepsilon\Lambda = d$.

Eq. (10) tells that the gauge-invariant driving phase differences disappear due to screening at λ/ε for the x-axis and at λ for the z-axis. These length scales thus provide the magnetic extent (Figure 6). Thus, *apart from tiny corrections in the current flow and the magnetic field pattern*, the Josephson vortices in layered superconductors and Abrikosov vortices in anisotropic superconductors are similar (Bulaevskii 1973; Efetov 1979). However, the order parameter ψ at the axis (x=0) of the Josephson vortex is feebly suppressed, $\delta|\psi| \approx \left|\frac{\xi}{\Lambda}\right|^2$ (Clem, Coffey 1990; Bulaevskii, Clem 1991), instead of the complete suppression of ψ at the axis of an Abrikosov vortex (cf. Figures 6 and 7). The core region (Λ, d) is known as *the phase core* of the Josephson vortex, analogous to the vortex core (ξ,εξ) in an Abrikosov vortex (Figure 7). Thus, $d/\Lambda = \varepsilon\xi/\xi = \varepsilon$. Further, Eq. *(6) is extendable for other cases apart from the parallel fields* (Bulaevskii, Ledvij, Kogan 1991).

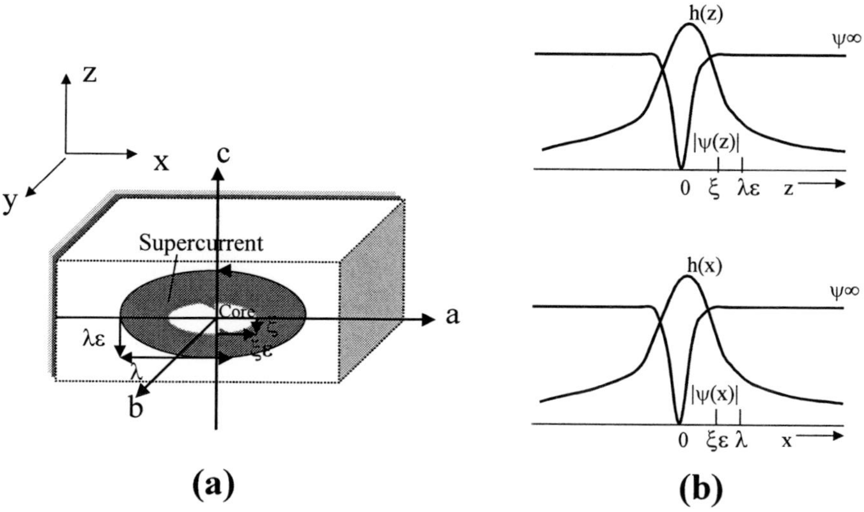

Figure 7 Cross Section of an Abrikosov vortex along the a-axis of an anisotropic superconductor. The core, the supercurrent, different penetration depths (λ) and coherence lengths (ξ) are also indicated. (b) Its field h(r) and ψ profile, which goes to zero at the center.

The line energy of Josephson vortices is

$$e_J \approx \varepsilon\varepsilon_0 \ln(\lambda/d). \tag{12}$$

Here $\varepsilon_0 = (\Phi_0/4\pi\lambda)^2$ is the basic energy scale. More accurate calculations of Clem, Coffey, Hao (1991) on (12), and for an Abrikosov vortex directed along the plane (Hu 1972), respectively, give

$$e_J \approx \varepsilon\varepsilon_0 \ln(\lambda/d+1.12) \tag{13}$$

$$\text{and } e_l \approx \varepsilon\varepsilon_0 \ln(\lambda/\varepsilon\xi+0.50). \tag{14}$$

Due to isotropy within the plane the elastic tension of a Josephson vortex is equal to its line energy. The lower critical field along the plane is simply $H_{c1}=4\pi e_J/\Phi_0$ at $\vartheta=0$. The relation between the viscous drag coefficient and the normal state resistivity (along the c-axis ρ_n^c) for the Josephson (J) and Abrikosov (l) vortices are respectively (Clem, Coffey 1990; Bardeen, Stephen 1965),

$$\rho_n^c(J)=2.226\frac{\Phi_0^2}{2\pi\eta_J c^2\varepsilon\Lambda^2}, \tag{15}$$

$$\text{and} \quad \rho_n^c(l)=\frac{\Phi_0^2}{2\pi\eta_l c^2\varepsilon\xi^2}, \tag{16}$$

which are therefore rather very similar.

2.1.2 Pancake Vortex (at or near c-axis)

These vortices are formed in an isolated thin film (Pearl 1964), an uncoupled stack of parallel films (Effetov 1979; Artemenkno, Krugolov 1990; Buzdin, Feinberg 1990; Feigel'man et al. 1990; Clem 1991; Fischer 1991), and a layered superconductor (Artemenkno, Krugolov 1990; Chakravarty et al. 1990a; Feigel'man et al. 1990; Bulaevaskii et al. 1992). A single vortex core in such a system constitutes a pancake vortex (Pearl 1964; Abrikosov 1988). We consider the case for the latter two types of material systems, wherein the electromagnetic coupling and the Josephson coupling between the superconducting layers and their consequences are relevant. The electromagnetic coupling is deduced basically from the Eq. (4). A layered superconductor is considered as the stack of thin films of thickness d. An individual vortex line threading the stack is defined by the positions $\mathbf{R}_n$ of the vortex cores within each layer such that each core generates a driving phase field (Artemenkno, Krugolov 1990; Buzdin, Feinberg 1990; Feigel'man et al. 1990; Clem 1991; Fischer 1991). As a consequence, a screening current starts to flow in this layer. The screening effect due to the other layers in the stack squeezes the field into a narrow strip of size λ, along the z-axis, and the magnetic field lines escape parallel to the layers (Figure 8).

Two consequences for the modified magnetic field lines are that (i) the current flow never screen the driving phase field in the central layer, and (ii) the logarithmic dependence of the interaction potential between two vortices placed in the same layer continues to infinity:

$$V_{em}(R,z=0)=2d\varepsilon_0\left[\ln\frac{\xi}{R}+O(d/\lambda)\right],\ \xi<R<\infty. \tag{17}$$

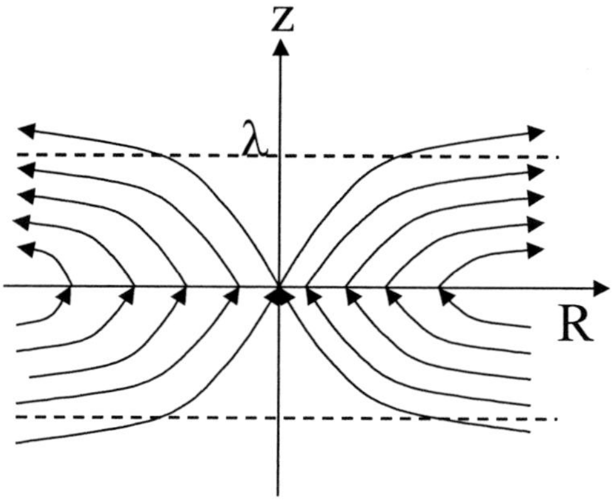

Figure 8 Schematic of a pancake vortex for a layered superconductor with negligible interlayer Josephson coupling.

2.1.3 Kinked and Tilted Vortices

When Josephson coupling is also involved, the total energy consists of the Josephson coupling energy contribution (Chakravarty et al. 1990; Feigel'man et al. 1990; Bulaevskii et a. 1992). Define $\Phi=\varphi_{n+1}-\varphi_n=\varphi_{n-1}-\varphi_n$ for two vortices at n±1 layers. For the case Φ is small, and for $R>\Lambda=d/\varepsilon$, the discreteness of the system does not affect the currents along z direction thereby a continuous anisotropic 3D character is sufficient.

If we consider small Φ, the Josephson interaction energy of the vortex pair is

$$V_J(R)=d\varepsilon_0\left(\frac{R}{\Lambda}\right)^2\ln\frac{\Lambda}{R},\ R\ll\Lambda. \tag{18}$$

When the separation is very large $R\gg\Lambda$, two Josephson strings connecting the pair are created between the layers n and n±1 and the interaction of the strings dominates:

$$V_J(R)=2d\varepsilon_0\left(\alpha\frac{R}{\Lambda}-\frac{1}{4}\frac{\Lambda}{R}\right),\ \Lambda\ll R\ll\lambda/\varepsilon \tag{19}$$

where $\alpha\sim 1$ (Cataudella, Minnhagen 1990; Glazman, Koshelev 1990). Eq. (19) describes ***kinked vortices*** for $|z|>d$ and $R>\Lambda$ away from the core region (Chakravarty et al. 1990a). Further, for $R\gg\lambda/\varepsilon$

$$V_J(R)=2d\varepsilon_0\left(\alpha\frac{R}{\Lambda}\right),\ R\gg\lambda/\varepsilon \tag{20}$$

From a comparison of Eqs. (19) and (20) for the Josephson coupling interaction energy, and Eq. (17) for the electromagnetic interactions, it is clear that the former is dominant for large distances R>Λ and the latter dominates at small distances R<Λ. In other words, at short distances (R<Λ), the Josephson currents have not yet developed wherein the electromagnetic energy dominates and the interaction is logarithmic. At large distances (R>Λ), Josephson currents are established, and the magnetic field between the two vortices gets redirected into the two Josephson strings, thereby the interaction grows linearly.

For a double-kink structure in a vortex line directed parallel to the plane (Figure 9), the interaction energy for two oppositely charge vortices is (Chakravarty et al. 1990a; Blatter et al. 1994)

$$V_{K,-K} \approx 2d\varepsilon_0 \begin{cases} \ln\frac{R}{\xi}, \xi < R < \Lambda, \\ \ln\frac{\Lambda}{\xi} - \frac{\Lambda}{4R}, \Lambda < R < \lambda/\varepsilon, \\ \ln\frac{\Lambda}{\xi}, \lambda/\varepsilon < R. \end{cases} \quad (21)$$

The term ln(R/ξ) for R<Λ is for the currents flowing within the plane comprising the kinks. Additional term corresponding to the Josephson coupling to neighboring plane has been ignored. For R>Λ, the in-plane core contributions of the two kinks and Josephson coupling correction term are there. As the string connecting the two kinks is shifted by a distance d only without any new creation, there is no confinement term in Eq. (21).

When the two vortices are of the same charge, kinks cause a shift of the vortex line by 2d (Figure 9). This arrangement may be considered for the origin of the building block of a ***tilted vortex***. Their interaction energy is (Artemenko, Kruglov 1990)

$$V_{K,K} \approx d\varepsilon_0 \begin{cases} \frac{d}{2\lambda}\frac{R}{\lambda} + \frac{1}{2}\left[\frac{R}{\Lambda}\right]^2 \ln\frac{\Lambda}{R}, \xi < R < \Lambda, \\ \frac{\Lambda}{2R}, \Lambda < R < \lambda/\varepsilon. \end{cases} \quad (22)$$

The two terms for R<Λ arise due to the electromagnetic and Josephson coupling between the two vortices.

The discussions so far sufficiently suggest that we can study the arbitrary single vortex configuration in layered superconductors. We illustrate this here taking the case for elastic tensions for the in-plane and out-of-plane tilt modes. The line energy of a vortex tilted at an angle ϑ with the ab plane is given by the sum of the energies of the individual pancake vortices that include their interactions provided in Eqs. (12) and (22). We concentrate on the major term due to the Josephson coupling between the layers. At large angle ϑ>ε, separation between the pancake vortices is less than Λ. The line energy is (Blatter et al. 1994; Blatter, Geshkenbein 1993; Bulaevskii et al. 1992)

$$\varepsilon_l(\vartheta > \varepsilon) = \varepsilon_0 \sin\vartheta\left[\ln\frac{\alpha\lambda}{\xi\sin\vartheta} + \frac{\varepsilon^2}{2\tan^2\vartheta}\ln\frac{\alpha'\lambda\tan\vartheta}{d}\right], \tag{23}$$

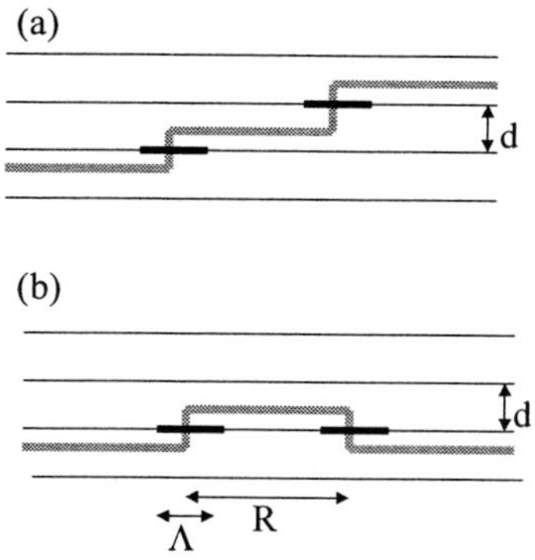

Figure 9 Schematic of the kinked vortex configurations in highly layered superconductor for (a) a double-kink arrangement representing a vortex line tilted at a small angle with respect to the plane, and (b) kink-antikink arrangement for the nucleus of a vortex motion across the superconducting layer.

where α and α' are of order unity. In this angular regime, the Josephson strings are not yet formed, and the interlayer coupling produces only the term Eq. (18). At small angles, $0<\vartheta<\varepsilon$, Josephson contribution to the energy is large. The combination of Eqs. (19) and (22) gives

$$\varepsilon_l(0 < \vartheta < \varepsilon) = \varepsilon_0 \sin\vartheta\left[\ln\frac{\Lambda}{\xi} + \frac{\varepsilon}{\tan\vartheta}\ln\frac{\lambda}{d} + \frac{1}{2\varepsilon}\tan\vartheta\right]. \tag{24}$$

The out-of-plane and the in-plane line tensions (elastic constants) in layered superconductor given by (Klemm, Clem 1980; Balatskii, Burlachkov, Gor'kov 1986; Blatter et al. 1994)

$$\varepsilon_l^{\perp}(\vartheta) \cong \frac{\varepsilon_0}{\varepsilon}\begin{cases}\dfrac{1}{\cos^3\vartheta}, 0<\vartheta<\varepsilon,\\ \left(\dfrac{\varepsilon}{\sin\vartheta}\right)^3, \varepsilon<\vartheta<\dfrac{\pi}{2},\end{cases} \tag{25}$$

and

$$\varepsilon_l^{\parallel}(\vartheta) \cong \varepsilon\varepsilon_0\begin{cases}\cos\vartheta, 0<\vartheta<\varepsilon,\\ \dfrac{\varepsilon}{\sin\vartheta}, \varepsilon<\vartheta<\dfrac{\pi}{2}\end{cases} \tag{26}$$

are distinct from those $\varepsilon_l^{\perp}(\vartheta) \cong \frac{\varepsilon_0 \varepsilon^2}{\varepsilon_\vartheta^3}$ and $\varepsilon_l^{\parallel}(\vartheta) \cong \frac{\varepsilon_0 \varepsilon^2}{\varepsilon_\vartheta}$ [Here, $\varepsilon_\vartheta^2 = \varepsilon \cos^2\vartheta + \sin^2\vartheta$] of an anisotropic superconductor. This is the consequence of dominance of the line tension by the current flow in the vortex, which is equal for layered and continuous cases outside the core region. In the above discussion, the variations in the logarithmic factors with different core sizes have been ignored. It is noted for $\varepsilon_l^{\perp}$ that it diverges as ϑ goes to zero. This is due to the intrinsic pinning effect, which was ignored in its calculation. When the vortex segment is entirely locked to the plane, then this contribution has to be considered. In the presence of kinks, the tilt modulus is finite.

2.1.4 Vortex Lattice and Combine Vortices

What would be the vortex lattice in a layered superconductor? The difference one may expect from the discreteness of the material and the weak Josephson coupling is the introduction of two important modifications into the free energy of the system. These two corrections are the logarithmic corrections due to the different core structure of the kinked vortices from those of the rectilinear Abrikosov vortices and the appearance of a linear term $\propto\vartheta$ due to the trapping of the vortices between the CuO planes. For the Abrikosov vortices, contributions from the supercurrents and cores are involved. For the kinked vortices, there are contributions from the Josephson and pancake segments, wherein Josephson and pancake vortices and their contributions prevail (Figure 10). The line energies of a rectilinear Abrikosov vortex and kinked vortex after the inclusion of the core energy contributions for small angles $\vartheta<\varepsilon$ are

$$e_l^r \approx \varepsilon_0 \left\{\varepsilon + \frac{\vartheta^2}{2\varepsilon}\right\}\left[\ln\frac{\lambda}{\varepsilon\xi} + 0.5\right], \tag{27}$$

$$e_l^k \approx \varepsilon_0 \left[\varepsilon \ln\frac{\lambda}{d} + |\vartheta|\left(\ln\frac{\Lambda}{\xi} + 0.5\right)\right]. \tag{28}$$

Notably, the result of the logarithmic corrections to the free energy of the vortex lattice can be the new physical properties such as the evolution from kinked vortex lattice to *a combine lattice of independent Josephson and Abrikosov vortices running parallel and perpendicular to the plane* (Kes et al. 1990; Theodorakis 1990; Bulaevskii et al. 1992). The kinked structure causes the term linear in $|\vartheta|$. The origin of the term linear in energy is assigned to the lock-in transition of the vortices and hence to trapping of flux lines (Feinberg, Villard 1990; Bulaevskii 1991; Ivlev et al. 1991; Maslov, Pokrovsky 1991). Such situations are absent in the Abrikosov vortices (27) with corrections of ϑ^2-order.

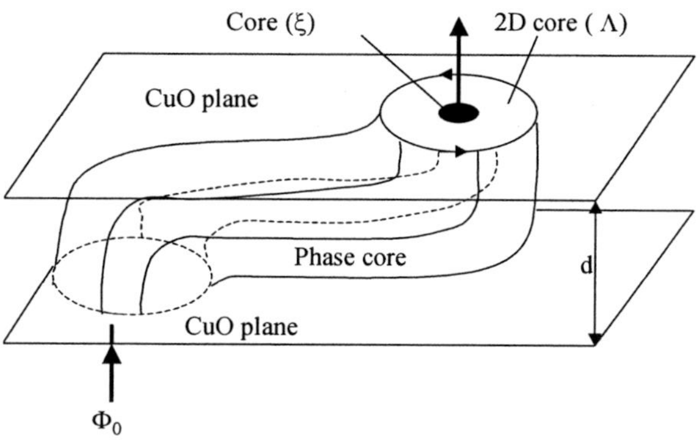

Figure 10 A vortex line tilted at small angle $\vartheta<\varepsilon$ with respect to the plane having fully developed 2D core region of the pancake vortex and the phase core of the Josephson string.

2.2 Types of Vortices, Angular Regimes and Intrinsic Pinning

The above discussion is applicable to a vortex lattice in layered superconductors wherein four regimes of angular ranges can be considered. They are (i) large angle range $d/\xi<\tan\vartheta$, (ii) intermediate angle range $\varepsilon<\tan\vartheta<d/\xi$, (iii) small angle range $\vartheta<\varepsilon$ and (iv) $\vartheta=0$ and nearby angle. In the angular regime (i), the normal cores of the *pancake vortices* still overlap, which can be described by the continuum model. Regime (ii) has the screening currents change from planar to Josephson, i.e. tilted vortex regime. Regime (iii) is associated with the fully developed kinked vortices, i.e. *combine vortices*. Regime (iv) is closely related with the *intrinsic pinning*, i.e. *purely Josephson vortices*.

With the magnetic field H aligned along the plane (y-axis), the vortex lattice tends to accommodate itself to the layer structure in such a way that the vortex cores lie in between the superconducting planes. In order for the vortices to move across the layers, a large expenditure of condensation energy equal to $\varepsilon_0 d\ln(\Lambda/\xi)$ is necessary, which leads to intrinsic pinning barriers and hence the vortices tend to develop sharp kinks as they cross a superconducting layer.

2.3 Lock-in Transition and Vortex Structure

At small angles ϑ_H of applied fields H below the lock-in angle ϑ_L, vortex lattice can gain energy when the H aligns itself with the ab plane. The phenomenon of lock-in transition therefore happens at $\vartheta>0$ instead of $\vartheta=0$. By locking-in the vortices between the layers, superconductor in the layers is least perturbed and pure Josephson vortices are formed. The lock-in angle is given by (Feinberg, Villard 1990; Bulaevskii 1991; Bulaevskii et al.1992; Ivlev et al. 1991; Maslov, Pokrovsky 1991)

$$\vartheta_L \cong 2n\frac{H_{c1}^c}{H}\frac{\ln(\alpha''\Lambda/\xi)}{\ln(\lambda/\xi)}. \qquad (29)$$

Here n ($=n_x=n_y$) is the demagnetization factor along x- or y-axes defined by $L_z/L_y<<1$ such that $n_z=1-2n$, and $\alpha''\sim1$. The expected ϑ_L value is $\sim0.1^o-1^o$ (Kwok et al. 1991).

Away from the lock-in transition, a kinked vortex lattice is expected (Ivlev et al. 1991). In this regime, the field components parallel and perpendicular to the plane may penetrate independent of each other setting up a combine vortex lattice, which is made up of the coexisting but non-interacting Josephson and Abrikosov vortices, respectively. This holds for intermediate fields, $H_{c1}<H<H_J=\Phi_0/d\Lambda$ for small enough coupling, $\varepsilon<d/\lambda$ i.e. $\Lambda>\lambda$. Then, $\vartheta_L \cong 2n\frac{H_{c1}^c}{H}$. At $\vartheta>\vartheta_L$, the parallel lattice evolves into a combine vortex lattice through a second order phase transition. Combine to tilted vortex lattice transition is of first order, which takes place below $\vartheta=\arctan(d/\xi)$. Bulaevskii et al. (1992) argued that for high fields $H>H_J$, and small angles ($\vartheta<\varepsilon$), the Josephson cores in the neighboring layers are expected to start overlapping, thereby combine vortex lattice tend to be favorable irrespective of the ratio Λ/λ. This requires the free energies of the different phases beyond logarithmic accuracy. The combine vortex lattice forms a stable (rescaled) equilateral triangular Abrikosov lattice while other lattices (square or rectangular) are unstable (Ivlev et al. 1990; Campbell, Doria, Kogan 1988).

2.4 The Anisotropic Ginzburg-Landau Limit

From Eq. (1), it is clear that the free energy reduces to a Ginzburg-Landau form with ellipsoidal anisotropy in the long wavelength limit. Then, the variation along z-axis is smooth enough so that $\partial\psi/\partial z$ can replace $(\psi_n-\psi_{n+1})/d$, which leads ∇ and A to 3D quantities, and $1/m_i$ is the reciprocal mass tensor with principal values 1/m, 1/m and 1/M. Under these conditions, variation in $\psi_n{}^*$ gives (Tinkham 1996)

$$\alpha\psi+\beta|\psi|^2\psi-\frac{\hbar^2}{2}\left(\nabla-i\frac{2e}{\hbar c}A\right).\left(\frac{1}{m_i}\right).\left(\nabla-i\frac{2e}{\hbar c}A\right)\psi=0. \tag{30}$$

When the interlayer coupling is weak, M>>m and anisotropy of the coherence length ξ arises due to the anisotropy in the mass so that

$$\xi_i^2(T)=\hbar^2/\{2m_i|\alpha(T)|\}=\hbar^2/\{2m_i|T-T_c|\}. \tag{31}$$

Since $\alpha(T)$ is isotropic and proportional to $(T-T_c)$, ξ_i scales with $m_i^{-1/2}$ and diverges as $|T-T_c|^{-1/2}$. Therefore, adequately close to T_c, ψ will vary sufficiently smoothly to justify Ginzburg-Landau continuum approximation. If we apply the Ginzburg-Landau relation $\xi(T)=\Phi_0/\{2\sqrt{2}\pi H_c(T)\lambda(T)\}$, then

$$2\sqrt{2}\pi H_c(T)\xi_i(T)\lambda_i(T)=\Phi_0. \tag{32}$$

This implies that the anisotropy of $\xi_i(T)$ is inverse of $\lambda_i(T)$ as $H_c(T)$ is isotropic. As a consequence, $\lambda_i(T)$ is the screening due to the supercurrents flowing along the ith axis, not the screening of a magnetic field along the ith axis. To illustrate this, consider an Abrikosov vortex in a sample with the magnetic field H along the a-axis. Had this been an isotropic superconductor, the vortex should have been circular symmetry. However, in an anisotropic superconductor, the core radius along the plane direction will be ξ while that for c-axis will be $\xi_c=\varepsilon\xi<<\xi$. The flux penetration radius along the ab plane will be $\lambda_c=\varepsilon\lambda$ and that for the c-axis will be λ (see Figure 7).

As the upper critical field $H_{c2}=\Phi_0/2\pi\xi^2$ is determined by the vortices of the current flowing in a plane perpendicular to the field, the relevant values of ξ_i are those for the two axes orthogonal to the field direction. Therefore for the two principal axes,

$$H_{c2\perp}=\Phi_0/2\pi\xi^2$$
$$H_{c2\|}=\Phi_0/2\pi\xi\xi_c. \tag{33}$$

As $\xi>>\xi_c$, $H_{c2\perp}<<H_{c2\|}$. By the similar arguments, $H_{c1\perp}<<H_{c1\|}$. We recapitulate,

$$\varepsilon=(m/M)^{1/2}= \lambda/\lambda_c=\xi_c/\xi= H_{c2\perp}/H_{c2\|}= H_{c1\perp}/H_{c1\|}. \tag{34}$$

The typical values of ε are approximately 1/7 for YBCO (Farrell et al. 1988,Welp et al. 1989, Okram et al. 1999), 1/100 for Tl-2212 (Kim et al. 1988) and 1/150 for Bi-2212 (Hazen et al. 1988; Naughton et al. 1988; Farrell et al. 1989; Palstra et al. 1988; Chikumoto et al. 1992; Martinez et al. 1992).

Using an anisotropic generalization of the harmonic oscillator calculation, we obtain an ellipsoidal form for the angular dependence of the upper critical field in three dimensions (3D) that obeys Lawrence-Doniach model as (Lawrence, Doniach 1971)

$$\left(\frac{H_{c2}(\vartheta)\sin\vartheta}{H_{c2\perp}}\right)^2+\left(\frac{H_{c2}(\vartheta)\cos\vartheta}{H_{c2\|}}\right)^2=1. \tag{35}$$

Here ϑ is the angle between the field and the ab-plane. In the case for a thin film, the relation is (Tinkham 1963)

$$\left|\frac{H_{c2}(\vartheta)\sin\vartheta}{H_{c2\perp}}\right|+\left(\frac{H_{c2}(\vartheta)\cos\vartheta}{H_{c2\|}}\right)^2=1. \tag{36}$$

In Eq. (35), the first derivative $dH_{c2}/d\vartheta=0$ for parallel field and the curve is smooth and bell-shaped. From Eq. (36), it follows that $H_{c2}(\vartheta)$ has a cusp at $\vartheta=0$ and this is used to describe the 2D nature.

2.5 Dimensionality

As Eq. (31) indicates, close to T_c, $\xi_c(T)$ will be large enough to justify Ginzburg-Landau approximation. However, as one lowers the temperature, $\xi_c(T)$ contracts toward a limiting value. When that limit is less than the interplanar distance d, smooth variation condition to obtain Eq. (30) has to give way to 2D condition at some lower temperature T* (Klemm et al. 1975). Below T*, the system will be 2D. In this limit, Md^2 tends to infinity and layers are completely decoupled, giving $\xi_c(T^*) = d/\sqrt{2}$.

This simple analysis provides the existence of a regime for 2D and 3D depending on temperature once the sample is chosen as highly layered. These features combine with Eqs. (35) and (36) have been used to identify the dimensionality of a system. However, applicability of these models for the stacks of superconducting layers separated by the insulating layers, wherein different types of vortices are expected to prevail depending on the angle one chooses, was an open question because thin film is basically different from such stacks. Hence, will these models still hold true or not? These situations are also closely related with the intrinsic pinning and the lock-in transition (Okram et al. 2001a). Such complicacies were in fact not taken care of in these models.

Further, the estimation of $H_{c2}(0)$ is also uncertain and hence $\xi_c(0)$ (Okram et al. 1999). Also, it is believed that $H_{c2||}$ does not exist in these type of superconductors (Xiao, Tachiki 1998). The disappearance of $H_{c2||}$ is the most intriguing aspect for both the models, Eqs. (35) and (36). A pertinent question comes in: should the disappearance of $H_{c2||}$ coincide with the lock-in transition discussed in Section 2.3? The lock-in transition would lead to an abrupt change in the $H_{c2}(\vartheta)$ curve leading to the failure of these models. However, determination of $H_{c2}(\vartheta)$ at very close intervals of the angle is also impractical, regardless of the controversy in the proper determination of $H_{c2||}$. This means that the expected abrupt change in the behavior of $H_{c2}(\vartheta)$ with ϑ is likely to miss in all probability and hence inadvertently apply these two models to identify the regimes of dimensionality.

In fact, fluctuation rounding of the superconducting transition in high temperature superconductors has resulted in differing values of H_{c2}, which lie between irreversibility field $H_{irr}(t)$ and H_{c2} (Okram et al. 1999). This in turn led to the supposedly H_{c2} values lying either in the 2D or 3D regime. This is the point where these models have been used to identify the existence of the 2D to 3D crossover (see e.g. Fastumpa et al. 1991; Marcon et al. 1992). Marcon et al. (1991) found that the $H_{irr}(\vartheta)$ is analogous with $H_{c2}(\vartheta)$, and $H_{irr}(\vartheta)$ follows either model depending on the given temperature. It is also well known that the flux flow resistivity ρ is associated with some critical magnetic field because $\rho \sim \rho_n B/H_{c2}$, where ρ_n is the normal state resistivity and $B \sim H$ (Bardeen, Stephen 1965). These two conditions then imply that $\rho(\vartheta) \propto 1/\ H_{c2}(\vartheta) \sim 1/\ H_{irr}(\vartheta)$ at constant H and T. One may, as a consequence assume that $1/\rho(\vartheta)$ at constant H and T can be used to verify the applicability of these two models. This contention is motivated by the fact that $\rho(\vartheta)$ can be recorded at very small angular steps.

2.6 Vortex Lock-in Scenario

It is thus clear that the vortex physics and vortex phase diagram are expected to be rather involved. Their understanding has been growing with time. The cases for the direct observations of the pancake vortices and Josephson vortices have been reported (Hess et al. 1989; Renner et al. 1998; Moler et al. 1998), as they are comparatively easier orientations. However, other details such as other types of vortices are obviously difficult. Indeed, it is noted from equations (15) and (16), in particular, and from the foregoing energetic of the different vortices that have individual characteristic interactions such as elastic properties, line energy, and so on, as described above, in general, that the flux flow resistivity should vary with the orientation of H with respect to the ab plane. In simplified picture, these vortices having different interaction energies have correspondingly differing Lorentz forces and hence dissipations with angle ϑ. These parameters therefore change with angle ϑ. Then, the flux flow resistivity as a function of angle $\rho(\vartheta)$ is one such probe that may perhaps provide maximum details about these vortices. Therefore, indirect mapping of these vortices and hence determination of their phase diagrams such as ϑ-H at constant T, and ϑ-T at constant H for these vortices in highly layered compounds such as $Tl_2Ba_2CuO_6$ single crystals is desirable (Okram et al. 2001a; 2001b; 2001c; 2001d).

Interestingly, the transition from combine to Josephson vortices was considered to associate with lock-in of H at a critical angle ϑ_L (Bulaevskii et al. 1992; Feinberg, Billard 1990; Ivlev, Ovchinnikov, Porovsky 1990; Ivlev, Porovsky, Ovchinnikov 1991; Bualevskii 1991; Maslov, Porovsky 1991). The rather vastly different predicted ϑ_L values (Blatter et al. 1994; Bulaevskii et al. 1992; Feinberg, Billard 1990; Ivlev, Ovchinnikov, Porovsky 1990; Ivlev, Porovsky, Ovchinnikov 1991; Bualevskii 1991) were considered to confirm experimentally on $YBa_2Cu_3O_{7-\delta}$, organic superconductor and $La_{1.9}Sr_{0.1}CuO_4$ (Kwok et al. 1991; Deligiannis et al. 1999; Mansky et al. 1993; Bugoslavsky et al. 1997). Results on $YBa_2Cu_3O_{7-\delta}$ (Kwok et al. 1991) that confirm quite near to the predictions of Blatter et al. (1994) and Maslov and Porovsky (1991) are however seemingly basically different from those on organic superconductors (Mansky et al. 1993) that match those predicted by Feinberg and Billard (1990). A close examination suggests that these results should be compatible or at least both features should be present in both the cases, in line with the prospective phase transitions from one to another type of the four vortex types (Okram et al. 2001a; 2001b; 2001c; 2001d).

3 Vortex Mapping Techniques

3.1 Scanning

Scanning of vortices has been done extensively using various approaches, which include bitter decoration, the scanning tunneling microscopy, magnetic force microscopy, scanning electron microscopy, Lorentz microscopy, scanning Hall probes, Magneto-optical imaging and scanning SQUID microscope (SSM). For example, SSM with pickup coils of diameter several millimeters or fewer, is a scanning system that is used to acquire images of a magnetic fields above sample surfaces; for details, see for example Kirtley, Wikswo 1999. In this method, sample scanning, optimization of spatial resolution, and image deconvolution techniques, is directly applicable to the micron size pickup coils of the smallest systems. It has the advantage of high sensitivity and bandwidth but disadvantages of relatively modest spatial resolution and the requirement of cooled SQUID sensor.

3.2 Flux Flow Resistivity

Single crystals of $Tl_2Ba_2CuO_6$ (Tl2201) of varied T_cs grown by the contamination controlled flux method were used (Moler et al. 1998). The crystals had negligible dislocations and nearest to the nominal $Tl_2Ba_2CuO_6$ composition with no twinning. Good contacts were made for the standard four-probe ac ab-plane $\rho_{ab}(T)$ or c axis $\rho_c(T)$ resistivity measurements; typical contact resistance was ~1Ω. The current density j applied was ~28mA/cm^2 with frequency ~18Hz. The stability of temperature was better than 3mK; for details, see Okram et al. (2001).

A superconducting solenoid generates the magnetic field **H** in the vertical direction for a maximum applied value of 14T; the uniformity of **H** was 10^{-5} per 10mm diameter spherical volume. As the crystal (holder) rotates about the a-axis with respect to the z or **H** direction, the polar angle θ changes. The value of θ was varied from –90° to +90° (Figure 3). The crystal was rotated about the c axis for the angle range –200° to +200°. A personal computer controlled the stepper motor motion of the rotation through the axle and gear system coupled to a stepper motor. The resistivity versus angle was thus recorded. The instrumental resolutions for reading θ and φ were 0.0004° and 0.002° respectively. Crucial region for θ was identified to be near 90°, the angular step maintained was 0.05°. The angular step was increased to 0.05°, 2° and 5° to cover the θ range from 0 to 110° (Figure 3). Thus, the flux flow resistivity $\rho_c(\theta)$ and $\rho_{ab}(\theta)$ was recorded as a function of θ at different temperatures for a fixed value of H and vice-versa.

4 Experimental Observations of Vortices

4.1 Scanning

For scanning the Abrikosov vortices in conventional superconductors, and pancake and Josephson vortices in layered superconductors, there are extensive data available in the literature using bitter decoration (Essmann, Trauble 1967; Gammel etal. 1987; Bolle et al. 1991; Marchevsky et al. 1995; Bezryadin et al. 1996), the scanning tunneling microscopy (Hess et al. 1989; Eskildsen et al. 2002), magnetic force microscopy (Moser et al. 1995), scanning electron microscopy (Straub et al. 2001), Lorentz microscopy (Harada et al. 1992), scanning Hall probes (Oral et al. 1997; Grigorenko et al. 2001), magneto-optical imaging (Goa et al. 2001) and scanning SQUID (superconducting quantum interference devices) microscopy (Kirtley et al. 1996). For example, scanning hall probe images of vortices of $YBa_2Cu_3O_{7-\delta}$ at different temperatures (Oral et al. 1997) are shown in Figure 11(a). As expected, the vortex diameter decreases and contrast improves at lower temperatures owing to the reduction in the superconducting penetration depth; the images were scanned at 0.1mT field. Bolle et al. (1991) have used bitter decoration to scan the vortex lattice of single crystal $Bi_{2.1}Sr_{1.9}Ca_{0.9}Cu_2O_{8+\delta}$ for the magnetic field applied an angle with respect to the c-axis. For the angle range 60° to 85° and fields 23-97G, the equilibrium structure is one with uniformly spaced chains with a higher line density of vortices than the background lattice (Figure 11(b)). These observations were considered to be in qualitative agreement with theories, which suggest that in strongly anisotropic materials, the vortices develop an attractive interaction in tilted magnetic fields (Buzdin, Siminov 1990; Kogan et al. 1990). Essmann, Trauble (1967); Gammel etal. (1987); Bolle et al. (1991); Grigorieva et al. (1995); Marchevsky et al. (1995); and Bezryadin et al. (1996) have also carried out similar works.

Grigorenko et al. (2001) scanned the vortices at a tilted field using scanning Hall probe microscopy to explore the crossing lattice phase diagram in $Bi_2Sr_2CaCu_2O_8$ crystals with T_c=90K under independently applied c-axis (H_c) and in-plane ($H_{||}$) magnetic fields. Magnetic images were generated parallel to ab surface of the freshly cleaved single crystals with a lateral resolution of a few hundred nanometer, and Hall probe typically detects the top 300 pancake vortices in a stack whose flux threads along the c-axis. The location of stacks of Josephson vortices was inferred from the fact that they were decorated by the pancake vortices due to their mutual attraction. This is similar to the Bitter decoration technique with fine ferromagnetic particles described above. Grigorenko et al. (2001) have particularly collected images of how the evolution of the pancake vortex lattice and the chains of these vortices, takes place as the angle between the applied field and the c-axis was gradually increased. In a typical experiment, with H_c=12G field parallel to c-axis and $H_{||}$=0 at 81K, a well-ordered hexagonal lattice of pancake vortices was observed (Figure 12(i)). However, as the in-plane field $H_{||}$ increases from zero, chains of pancake vortex stacks was formed. This separates the domains of approximately hexagonal pancake lattice (Figure 12(ii) and (iii)). The presence of incommensurate chains frustrates hexagonal vortex lattice, giving rise, for example, to the seven-fold and five-fold polygons (A, B respectively) in Figure 12(iii). In

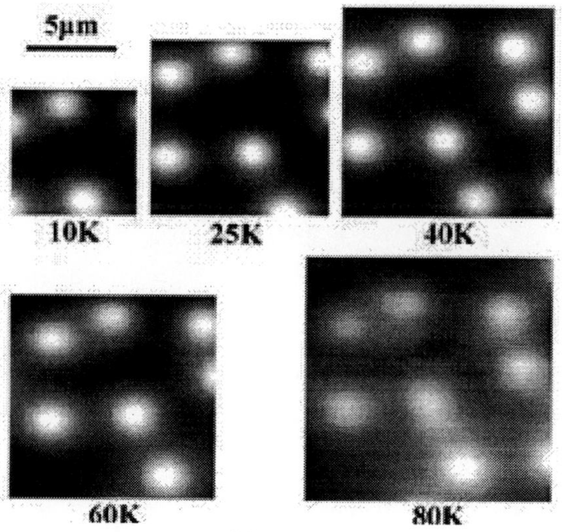

(a)

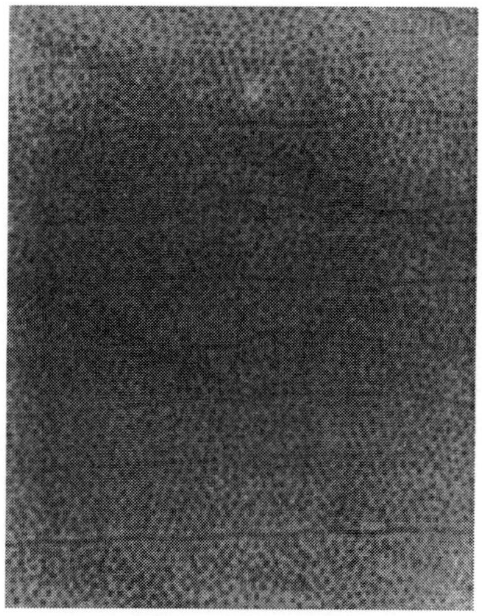

(b)

Figure 11 (a) Scanning Hall probe microscope image of vortices of $YBa_2Cu_3O_{7-\delta}$ at different temperatures. Reprinted from Oral et al.(1997) with permission © 1997 IOP Publishing Ltd. (b) Bitter decoration scan for the vortex lattice of single crystal $Bi_{2.1}Sr_{1.9}Ca_{0.9}Cu_2O_{8+\delta}$ for the magnetic field of 35G applied an angle 70° with respect to the c-axis. The dark spots are the vortices, which have an average separation of 1.4μm. Chain-like (horizontal line-like) features of the pancake vortices run roughly perpendicular to the axis of rotation: view area 75μmx60μm. Reprinted from Bolle et al. (1991) with permission. © 1991 American Physical Society.

confirmation with the earlier reports (Bolle et al. 1991; Grigorieva et al. 1995), the chains were aligned along the in-plane field direction, and the vortex separation within the chains was smaller than that of Abrikosov lattice.

Below a field called the ordering field, the field at which intervortex interactions become comparable to the pinning forces, quenched disorder in the crystal destroys the six-fold symmetry of the Abrikosov vortex domains (Figure 12(iv)). On decreasing the field further, a phase transition to an ordered 1D vortex chain state occurs, wherein Abrikosov vortex lattice domains disappeared completely (Figure 12(v) and (vi)).

This picture was considered to be the confirmation of the theories on flux crossing lattice picture of Bulaevski et al. (1992), Huse (1992) and Koshelev (1999). Based on these theories, chains are developed due to the attraction of the pancake vortices to the underlying Josephson vortices in the rhombic Josephson vortex lattice with lateral separation $c=[\sqrt{3}\Phi_0/(2\varepsilon H_{\|})]^{1/2}$. On increasing the tilt angle further, Grigorenko et al. have further shown that the composite vortex lattice of chains in a hexagonal matrix at low tilt angles gives way to an unstable state composed of 1D chains associated with sublimed pancake vortex stacks when H_c is varied and $H_{\|}$ is fixed at 38G.

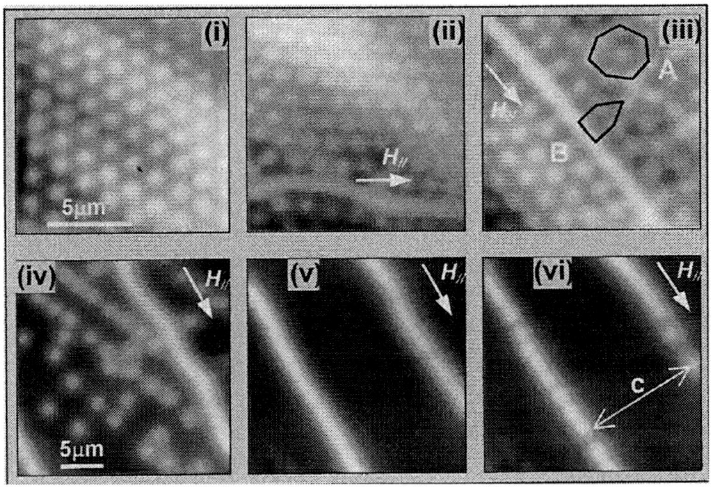

Figure 12 Scanning Hall probe microscopy images of how the evolution of the pancake vortex lattice and the chains of these vortices takes place as the angle between the applied field and the c-axis was gradually increased in a $Bi_2Sr_2CaCu_2O_8$ crystal under independently applied c-axis (H_c) and in-plane ($H_{\|}$) magnetic fields. (i) Well-ordered hexagonal lattice of pancake vortices for H_c=12G and $H_{\|}$=0 at 81K. (ii) Composite vortex lattice for H_c=14G and $H_{\|}$=32G at 81K. (iii) For a different orientation of $H_{\|}$ at H_c=10G and $H_{\|}$=32G (77K). (iv) The composite lattice at H_c=2G and $H_{\|}$=27G (81K). (v) The 1D vortex chain state at H_c=1.2G and $H_{\|}$=40G (81K). (vi) Lower density of pancake vortices at H_c=0.5G and $H_{\|}$=35G (81K). See text for more details. Reprinted from Grigorenko et al. (2001) with permission © 2001 Nature Publishing group.

Moler et al. (1998) have scanned the Josephson vortices of $Tl_2Ba_2CuO_{6+\delta}$ (Tl2201) using scanning SQUID microscope. Some typical images are shown in Figure 13. This shows interlayer vortices emerging parallel to the plane from Tl2201single crystals. The image is oriented so that the CuO plane is 9° away from the vertical and is pointed out of the plane of the figure. Orientation of the 8.2μm x 8.2μm pick up loop is indicated in Figure 13A. Figure

13B shows the resolution limited two vortex images of $YBa_2Cu_3O_7$ emerging perpendicular to the plane. Figure 13C shows three vortices (I, II, II) and two vortices (III, IV), all ac plane, obtained from two different Tl2201 single crystals, respectively. The images (Figure 13C) are not resolution limited parallel to the plane. The crystal axes were rotated by 9° from the scan axes (Figure 13). Fitting such images to the expected magnetic fields for a vortex emerging from an anisotropic vortex gives the penetration depth parallel to the plane, which for Tl2201 is ~20μm. The consistency in the fits led Moler et al. to believe, within their instrument resolution, the intrinsic shape for the ac plane vortices in Tl2201.

The forgoing results on the scanning of the vortices, say comprising Figures 11 through 13 and gallery of such pictures cited above are the representative images available in the literature. These images provide surface morphology of these vortices because the scanning could be done in two dimensions or on the surface of the crystal only but do not furnish the 3D nature of the vortices; that is, the complexity and the different types of vortices it may develop with angle as understood from the theories described in Section 2. Therefore, such situations suggest necessity for other methods, which give the bulk scenario of the vortices wherein the theories described above have more close compatibility of the picture of various vortices expected. One of these is the flux flow resistivity as a function of angle, temperature and magnetic field, which can be obtained in very close angular intervals.

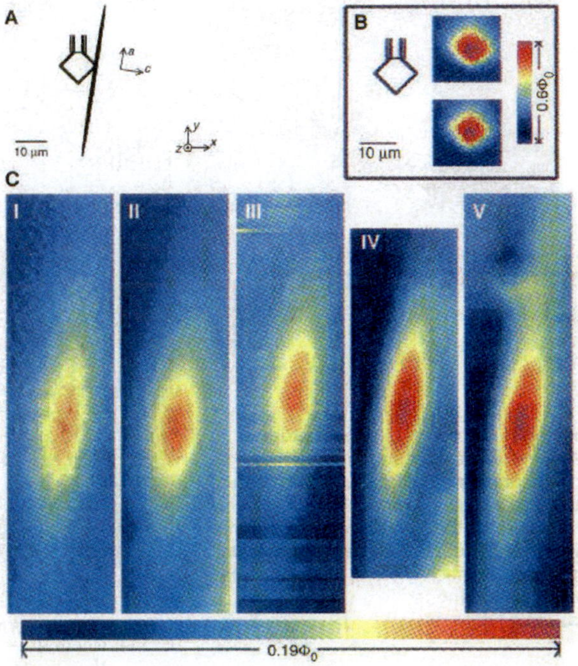

Figure 13 (A) Arrangement of the 8.2μm x 8.2μm pick up loop along with magnetic fields of a highly anisotropic vortex. The flux through the pickup loop is the integrated magnetic field at a height z_o=3μm over the pickup loop area; a and c indicate the Tl2201 single axes. (B) The resolution limited two vortex images of $YBa_2Cu_3O_7$; their apparent shape takes the shape of pickup loop. (C) Three vortices (I, II, II) obtained from one Tl2201 single crystal, and two vortices (III, IV) obtained from another Tl2201 single crystal, all ac plane. The images are not resolution limited parallel to the plane. Reprinted from Moler et al. (1998) with permission. © 1998 AAAS.

4.2 Flux Flow Resistivity

4.2.1 Vortex Types

Figure 14a is flux flow resistivity $\rho_c(\theta)$ of $Tl_2Ba_2CuO_6$ single crystal as a function of angle θ at 10T and 30K. Figure 14a inset shows the resistivity $\rho_c(T)$ data without magnetic field; T_c~37K and superconducting transition width is about 4K. A close look in the $\rho_c(\theta)$ indicates that as θ increases, $\rho_c(\theta)$ increases rather slowly until it attains the maximum at ~83° at which $\rho_c(\theta)$ suddenly drops very fast as θ increases. $\rho_c(\theta)$ then falls to the minimum at ~89.65°. $\rho_c(\theta)$ then increases suddenly until it reaches the next peak at ~89.85° and then it abruptly falls to the next minimum at 90° (Figure 14 a and b). The origin of these features were attributed to the different types of vortices that may be formed as the angle changes because their associated energy, pinning and other vortex parameters are different (Section 2; see also Okram et al. 2001b; Blatter et al. 1994; Bulaevski et al. 1992). The angles at which these turning points take place are therefore assigned to the vortex transitions from one type to another. The marginal change of slope of $\rho_c(\theta)$ between two θ values at which these minima and maxima occur are correlated to the change in the vortex dynamics of the one type of vortex to another. These are expected from the theories presented in Section 2. For example, the pancake vortex flow must vary with angle due to the variation in the line energy.
Therefore, these minima and maxima were identified in the following manner. The first maximum at ~83° was defined by θ_{LT1}, the first minimum at ~89.65° by θ_{LT2}, and the second maximum at ~89.85° by θ_{LT3} (Figure 14). The $\rho_c(\theta)$ is larger in the angle range 0 to ~83° compared to any other regime. The slope also remains almost the same in this angle range. This range is assigned to the pancake vortex dynamics wherein the vortices wander at random and the dissipation increases due to perhaps increasing instability of the pancake vortices [cf. Eq. (17)]. It is noted here that this dissipation is in mixed state, not in normal state as can be easily understood from a comparison of the resistivity ρ_c without magnetic field (Figure 14a). In other words, these features are absent in the normal state as also seen in $\rho_c(\theta)$ for 45K at 10T (Figure 15(a)).The step like feature at 87° is due to an experimental artifact, related to the change in angular step. The decreasing trend in $\rho_c(\theta)$ in the angle range ~83 to 89.65° is assigned to the titled vortices, for which the screening currents change from predominantly planar to Josephson type (Blatter, Gheshkenbein 1993; Bulaevski et al. 1992; Ivlev et al. 1991). This situation leads to reduction in the dissipation gradually with angle.

The marginal increase in $\rho_c(\theta)$ in the angle range ~89.65 to 89.85° is assigned to the combine vortices, which can be expected for 10T (Bulaevski et al. 1992). The combine vortices having independent perpendicular (pancake) and parallel (Josephson) vortex components are expected to offer negligible dissipation from the Josephson vortices compared to the pancake vortices. However, the pancake vortex part is seemingly still dominant. The marginal rise in $\rho_c(\theta)$ can be understood in terms of the instability of pancakes in the combine vortices that are arranged along the c-axis. The in-plane currents due to the Josephson vortices lead to the Lorentz force-induced instability to the pancake vortex state (Bulaevski et al. 1992).

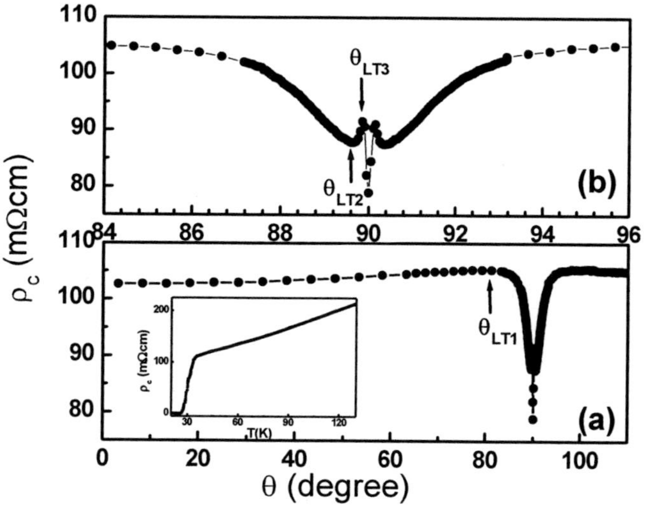

Figure 14 (a) The vortex state c-axis resistivity $\rho_c(\theta)$ of $Tl_2Ba_2CuO_6$ single crystal as a function of angle θ at 10T and 30K. The arrow is the angle θ_{LT1} at which $\rho_c(\theta)$ changes slope. Inset: resistivity $\rho_c(T)$ at H=0T. (b) High angle region of (a). Arrows have been used to identify the change in slope of $\rho_c(\theta)$: θ_{LT2} and θ_{LT3}. Reprinted from Okram et al. (2001b) with permission. © 2001 Elsevier Science.

The abrupt drop in $\rho_c(\theta)$ from 89.85 to 90° is attributed to the onset of the complete Josephson vortices. The finite width is related to the intrinsic pinning, in which the insulating layers allows the magnetic field to enter the crystal plane without significant disturbance to the superconducting plane before the magnetic field directly comes parallel to the plane (see Section 2.3). In fact, these features of $\rho_c(\theta)$ at different angles were thought to be related to the lock-in transition phenomenon (Okram et al. 2001b). Based on this, Okram et al. (2001b) choose to use θ_{LTY}, where Y=1,2,3 for the systematic steps to scan the lock-in transition, which takes place with the steps: θ_{LT1}, θ_{LT2} and θ_{LT3}; the actual lock-in transition being taken place at θ_{LT3}.

Based on these, Okram et al. (2001b) have carried out investigations on $\rho_c(\theta)$ at T= 20, 22, 24, 26, 28, 30, 32, 34, 38 and 45K for a fixed magnetic field 10T as shown in Figure 15, which shows the dramatic change in the features of $\rho_c(\theta)$ as T varies from 20-45K at 10T, and determined θ_{LTY}, where Y=1,2,3 for all T for which such transition features were observed. A systematic analysis of these trends led Okram et al. (2001b) to study for a particular temperature at different fixed fields H=0.005 – 14T, which was chosen to be 30K (Figure 16). These features are the consequences of the transitions of various types of vortices under these constraints for the Tl2201 superconductor, and are considered to be above the irreversibility line for the field parallel to c-axis, which are true both for the temperature variation for a fixed field and vice-versa, which is discussed later. The irreversibility line for H || ab-plane lies higher compared to that for H || c-axis in the H-T phase diagram (e.g. Okram et al. 1999). The characteristic difference between these two lines can be understood to be a consequence of the prevailing vortex type and hence related to the associated flux flow mechanism (Blatter et al. 1994).

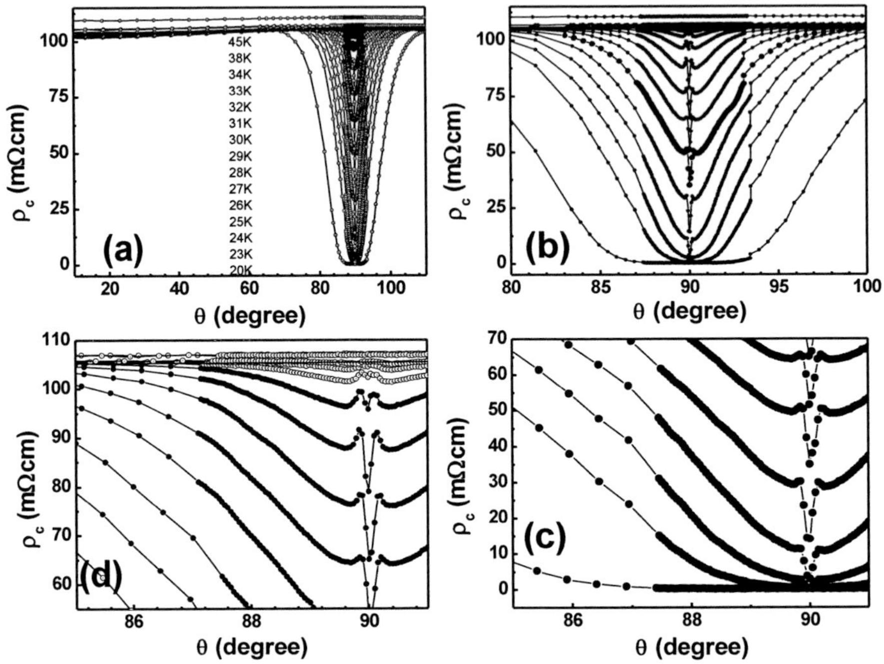

Figure 15 (a) The vortex state c-axis resistivity $\rho_c(\theta)$ of $Tl_2Ba_2CuO_6$ as a function of angle θ for various temperatures as indicated in order as curve at 10T. (b), (c), (d) Various regions of $\rho_c(\theta)$ and θ in (a). Reprinted from Okram et al.(2001b) with permission. © 2001 Elsevier Science.

The plots of θ_{LTY} as a function of t, and of θ_{LHY} as a function of H, where Y=1,2,3 and $t=T/T_c$, reduced temperature are shown in Figure 17. It is clear from these Figures that in principle, one obtains θ-t and θ-H phase diagrams for the pancake, tilted, combine and Josephson vortices, denoted respectively by P, T, C and J (Figure 17). These features are expected to be the consequence of the varying degree of the thermodynamics of each vortex type, which needs critical discussion. Before starting this, one is recalled to note the overall feature of the resistivity $\rho_c(\theta)$ due to dissipation.

Dissipation sets in once the Lorentz force overcomes the flux pinning and flux motions starts. Lorentz force is defined as jHsinθ, where θ is the angle between the transport current j and the applied magnetic field H. This means that value of jHsinθ increases with θ for fixed j and H. This situation is contrary to the results presented in Figures 16 and 17 on $\rho_c(\theta)$, which is larger at θ=0 than that at θ=90°. These two limits correspond to the pancake and Josephson vortices, respectively. This means that the normal core-associated pancake vortices dissipate more than that of the phase core-associated Josephson vortices, as discussed in Section 2. Similarly, the dissipation in the interval between these limits can be understood, albeit with caution. These results suggest that in highly layered superconductors such as Tl2201, the angle between the direction of H and ab plane decides the flux flow resistivity. It is not simply decided by the angle between H and j. These data and many others such as those of Iye et al. (1989, 1990, 1991) prove this situation clearly (see also Chakravarty et al. 1990b). The independence of the flux flow resistivity on the Lorentz force is related to the layered nature of the superconductor.

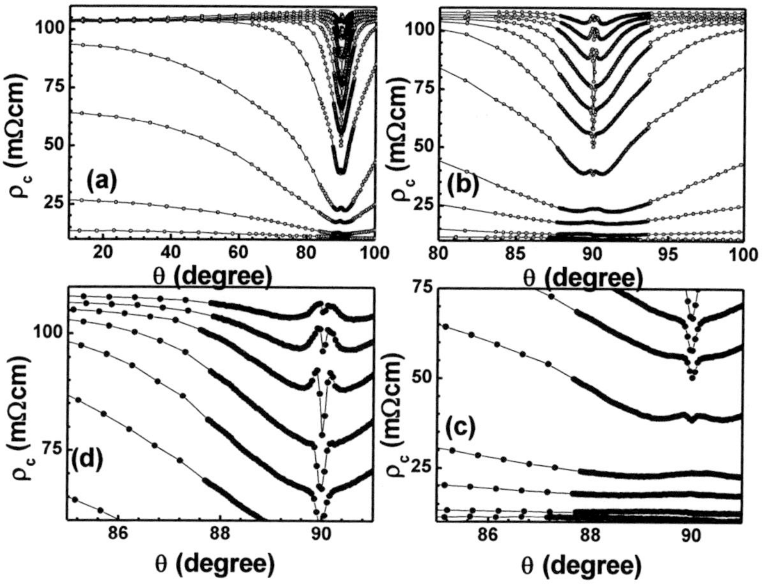

Figure 16 (a) The vortex state c-axis resistivity $\rho_c(\theta)$ of $Tl_2Ba_2CuO_6$ single crystal as a function of angle θ at 30K for 0.005, 0.5, 0.2, 0.5, 2, 4, 6, 8, 10, 12 and 14T (bottom to top curves). (b), (c), (d) Various regions of $\rho_c(\theta)$ and θ in (a). Reprinted from Okram et al. (2001b) with permission. © 2001 Elsevier Science.

4.2.1.1 Pancake Vortex Regime

In Figures 14-16, the region from θ=0 to $\theta_{LX1}(X)$, where X=H or T of $\rho_c(\theta)$ has been assigned due to the presence of pancake vortices. For 20K at 10T, the increase in $\rho_c(\theta)$ in this region is marginal and is almost linear albeit with slight change in slope as θ increases (Figure 15b). The marginal increase in $\rho_c(\theta)$ in the same region progressively decreases as the temperature is increased. Consequently, the $\rho_c(\theta)$ curves at lower temperatures have higher slopes than those at higher temperature as θ increases. This trend is suggestive of increasing flux flow and reduced pinning as θ increases. This increasing trend can be correlated with the changing electromagnetic interaction energy while Josephson coupling energy is still negligibly small. In other words, there is varying disturbance in the degree of pancake vortices with increase in θ and T for 10T.

For different fixed fields H=0.005 – 14T at 30K (Figure 16b), $\rho_c(\theta)$ in the region from θ=0 to θ_{LX1}, where X=H or T, changes significantly compared to those of different fixed temperatures T= 20, 22, 24, 26, 28, 30, 32, 34, 38 and 45K at 10T (Figure 15b). The slope of $\rho_c(\theta)$ monotonically diminishes as H decreases. This slope becomes negative at ≤2T. Also, the value of $\rho_c(\theta)$ reduces significantly. For ≤0.5T, even near θ=0, $\rho_c(\theta)$ decreases significantly indicative of better pinning. For 0.005T, no distinct features on the maxima and the minima are seen. Thus the trends in $\rho_c(\theta)$ (Figures 15b and 16a, b) are attributed to the influence of increasing angle, temperature and magnetic fields (Okram et al. 2001b).

4.2.1.2 Tilted Vortex Regime

From θ_{LX1} to θ_{LX2}, where X=H or T, $\rho_c(\theta)$ shows a monotonic decrease indicative of enhanced pinning of the vortices, which Okram et al. (2001b) attributed to the tilted vortices in line with Blatter et al. (1994). In this angle range, the trend in $\rho_c(\theta)$ for different fixed T at 10T is comparable to that of different fixed fields at 30K (Figures 15c, 16c). However, taking the same cue, at 30K for ≤0.5T, there seems to have signs of prevailing tilted vortices even at θ=0. In overall, the resistivity trend is such that tilted vortex pinning increases with reduction in temperature and magnetic field but the pinning increases with increasing angle.

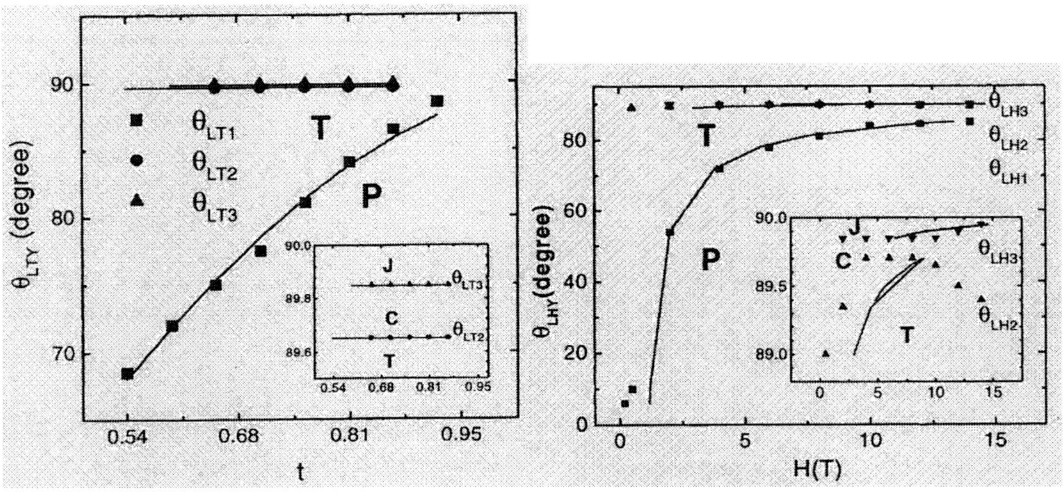

(a) (b)

Figure 17 (a) Plots of θ_{LTY} of $Tl_2Ba_2CuO_6$ single crystal as a function of t, where Y=1,2,3 and t=T/T_c, reduced temperature as represented by filled rectangle, circle and triangle respectively. The solid curve is the best fit of Eq.(40) for Y =1, and the horizontal solid lines are the best fits of the same equation for Y=2, 3. Inset: the latter two curves in expanded forms with corresponding fits for clarity. (b) Plots of θ_{LHY} as a function of H, where Y =1,2,3 as represented by filled rectangle, up-triangle and down-triangle respectively. The solid curve/ line is the best fit of Eq.(38) for Y=1, 2 and 3. Inset: the latter two curves in expanded forms with corresponding fits for clarity. The letters, P, T, C and J have been assigned as the regimes for pancake, tilted, combine and Josephson vortices respectively. More details are discussed in text. Reprinted from Okram et al. (2001b) with permission. © 2001 Elsevier Science.

4.2.1.3 Combine and Josephson Vortex Regimes

In the angle range θ_{LX2} to θ_{LX3}, where X=H or T, $\rho_c(\theta)$ shows a monotonic increase (Figures 15c,d, and 16c, d). This indicates drop in pinning of the vortices, and is considered to be due to combine vortices (Okram et al. 2001b). Okram et al. considered that such an increasing trend of $\rho_c(\theta)$ to be related to the instability of the component of the c-axis oriented pancake vortices associated with the combine vortices. The in-plane currents of parallel vortex component lead to Lorentz-force induced instability to the pancake vortex (Bulaevski et al. 1992) and hence justify the rise in $\rho_c(\theta)$. This trend tends to hold for T≥28K at 10T (Figure 15c, d) and for H≥0.5 at 30K (Figure 16c, d).

On the other hand, below 28K at 10T, there is change in rising trend of $\rho_c(\theta)$ (Figure 15c, d). In this low temperature regime, the tilted vortices seem to merge with combine vortices. If it is not, either both these types of vortices are indistinguishable or combine vortices cannot be developed (for T≤28K and 10T). Further, the situation $\rho_c(\theta)\to 0$ for $\theta\to 90^\circ$ suggests the formation of solidified vortices. Moreover, flux flow regime or irreversibility line for field aligned along the c-axis, the lock-in transition behavior to be discussed below should not be there.

The abrupt fall of $\rho_c(\theta)$ for $\theta=\theta_{LX3}$ to 90° (Figures 15c, d and 16c, d) was attributed to the prevailing Josephson vortices and hence to the lock-in transition. This assignment is consistent with the absence of normal cores in the Josephson vortices, wherein the dissipation should in principle be small [see Eq. (15)].

4.2.1.4 Four-Types of Vortices and their Phase Diagrams

Thus, the resistivity features are attributed to the four types of vortices that can be formed with angle (Okram et al. 2001a, b, c, d). Their vortex dynamics has been considered to be directly reflected in the $\rho_c(\theta)$ behavior. Each vortex type except the pancake vortices is suggestively contributory to the lock-in transition, which however takes place in entirety as the Josephson vortices sets in. These details shall now be discussed.

4.2.1.4.1 The θ-t Phase Diagram

In Figure 17a are the plots of $\theta_{LTY}(t)$ for Y=1, 2 and 3 at H=10T as a function of the reduced temperature t. The curve $\theta_{LT1}(t)$ increases almost linearly with t suggesting almost linear enhancement of dominance of pancake vortices (denoted by P) for large $\theta_{LT1}(t)$ at increasing t. In contrast, $\theta_{LT2}(t)$ and $\theta_{LT3}(t)$ are unchanged with increase in t (Figure 17a inset). These features imply rigidity of the vortex transition from the tilted vortices to combine vortices and from the combine vortices to Josephson vortices, respectively. Tilted, combine and Josephson vortices are denoted by T, C and J, respectively in the figure. Hence, Figure 17a shows the $\theta_{LTY}(t)$-t vortex phase diagram indicating the regimes of pancake, tilted, combine and Josephson vortices.

4.2.1.4.2 The θ-H Phase Diagram

In Figure 17b are the plots of $\theta_{LHY}(H)$ for Y=1, 2 and 3 at T=30K as a function of magnetic field strength H. Initially, the curve $\theta_{LH1}(H)$ increases dramatically fast until about 2.5°, beyond which $\theta_{LH1}(H)$ increases rather gradually albeit with a tendency for a saturation as H increases. This trend suggests that pancake vortices (denoted by P) give way to stronger magnetic fields and hence the regime of pancake vortices is very large (Figure 17b). That is, the pancake vortices in liquid state prevail at stronger fields (Okram et al. 2001a). As a consequence, the regimes of tilted, combine and Josephson vortices (denoted respectively by T, C and J) are relatively smaller compared to those in t dependence (Figure 17a). Therefore, the curves $\theta_{LH2}(H)$ and $\theta_{LH3}(H)$ are shown expanded (Figure 17b inset). The $\theta_{LH2}(H)$ rises rather very fast initially reaching to a maximum and falls slowly and non-linearly, which makes it to fall rather quickly beyond 10T. This curve therefore conveys the meaning that the combine vortices maintain a very small region at low magnetic fields below 4T and their regime increase marginally for H≥8T (Bulaevski et al. 1992). In contrast, the $\theta_{LH3}(H)$ curve is constant for H≤10T, beyond which $\theta_{LH3}(H)$ increases marginally. These features are indicative of the influence of stronger fields H≥10T on the lock-in transition closure to ab plane.

In summary, the curves $\theta_{LT1}(t)$ and $\theta_{LH1}(H)$ demarcating the pancake vortices from the tilted vortices indicate that the pancake vortex regimes increase to larger θ for either t or H increase. Their magnitudes are comparable to those found in other compounds (Mansky et al. 1993). Contrastingly, the magnitudes of $\theta_{LTY}(t)$ and $\theta_{LHY}(H)$, where Y=2, 3 are nearly the same as those of Kwok et al. (1991); Iye et al. (1991); Kufer et al. (1996). The tilted (T) vortices are found in the regimes between the curves $\theta_{LX1}(X)$ and $\theta_{LX2}(X)$, where X=t(T), H. Similarly, the combine (C) vortices exist in the regime $\theta_{LX2}(X)$ and $\theta_{LX3}(X)$. Finally, the Josephson (J) vortices prevail in the regime $\theta_{LX3}(X)$ and $\theta=90^o$. The θ-H phase diagram is qualitatively similar to that predicted (Bulaevski et al. 1992).

4.2.1.5 Lock-in Transition Theories

From the theoretical understanding (Section 2) and experimental situations (Section 4) on the angular regime of these vortices, we realize that they ought to form at different angle ranges, as they are highly dependent on the material anisotropy ε, temperature, and magnetic field. For d=11.5Å, ε=1/26 and $\xi_{||} \sim 37$ (Hussey et al. 1996; Uemura et al. 1993) in $Tl_2Ba_2CuO_6$, the three angular regimes denoted by (i) $d/\xi<\tan\vartheta$, (ii) $\varepsilon<\tan\vartheta<d/\xi$, (iii) $\vartheta<\varepsilon$ become $\vartheta<72.7^o$, $72.7^o<\vartheta<87.8^o$ and $\vartheta>87.8^o$ respectively. These angles are clearly of order $\theta_{LX1}(X)$, where X=t or H but are much larger than $\theta_{LX3}(X)$. The values of $\theta_{LX2}(X)$ lie roughly in the third regime ($\vartheta>87.8^o$). This simple analysis indicates the widely varying predictions of these angles and the lock-in transition (Blatter et al. 1994; Bulaevski et al. 1992; Ivlev et al. 1990e 1991; Blatter et al. 1993; Feinberg, Villard 1990; Bulaevski 1991; Maslov et al. 1991). This scenario compelled Okram et al. (2001b) to assume all these three angle ranges $\theta_{LXY}(X)$, where X=t or H, Y=1, 2, 3, are related to the lock-in transition (angle). They therefore assumed that $\theta_{LHY}(H)$ should be described either by Eq. (29) written as

$$\theta'_L(H)=p1\ln(1/\varepsilon)/H \tag{37}$$

or

$$\theta_L(H)=\cos^{-1}\left\{2\alpha_1 H^*\left(1+\frac{H^*}{\varepsilon^2 H}\right)/H\right\}. \tag{38}$$

Here p1 is a constant (related to the c-axis lower critical field H^c_{c1}, penetration depth $\lambda_\perp$, coherence length $\xi_\perp$, d and the demagnetization factor n), α_1 is the barrier height, $H^*\sim H_{c1}$, the lower critical field, and $\theta'_L=(90^o-\theta_L)$.

The temperature dependent relation for the lock-in transition was not available. Okram et al. (2001) therefore considered the t-dependent factors like $\lambda_\perp$ and $\xi_\perp$ in (37):

$$\theta'_L(t)=p2(1-t)^{p3}\ln\{(1-t)^{1/2}/\varepsilon\}. \tag{39}$$

Here p2 and p3 are constants. Similarly, Eq. (38) can be written as (Okram et al. 2001)

$$\theta_L(t)=\cos^{-1}\left\{2H^*\exp\left\{\frac{-0.631c}{(1-t)^{1/2}}\right\}\left(1+\frac{H^*}{\varepsilon^2 H}\right)/H\right\}. \tag{40}$$

Here c is a constant. Eqs. (39) and (40) were chosen for the t-dependent lock-in angle and Eqs. (37) and (38) as the H-dependent case (Okram et al. 2001b).

4.2.1.5.1 Theoretical Fits to the θ-t and θ-H Data

The solid curves in Figure 17a are the fits of Eq. (40). The theoretical fit (solid curve) is marginally deviated from the apparently linear experimental $\theta_{LT1}(t)$ data (Okram et al. 2001a, b). In contrast, the theoretical fits (solid curves) of the same equation to the $\theta_{LT2}(t)$ and $\theta_{LT3}(t)$ data are rather good. Quite identical fits of Eq. (39) to these data are also clearly seen (Figure 18A). That is, the corresponding fits viz, fits to the $\theta_{LT1}(t)$ and $\theta'_{LT1}(t)$; $\theta_{LT2}(t)$ and $\theta'_{LT2}(t)$; $\theta_{LT3}(t)$ and $\theta'_{LT3}(t)$ data are identical. On the contrary, the fit of Eq. (38) to $\theta_{LH1}(H)$ comes out to be reasonably good (Figure 17b). The fits of this equation to $\theta_{LH2}(H)$ and $\theta_{LH3}(H)$ come out to be not so good (Figure 17b; inset for clarity). Indeed, the fit of Eq. (37) to the $\theta'_{LH1}(H)$ also comes out to be reasonably good (Figure 18B). The fits of the same equation to the $\theta'_{LH2}(H)$ and $\theta'_{LH3}(H)$ data come out to be not so good (Figure 18b; inset for clarity) again comparing with that of Eq. (38). These fits suggest deviation in adequate explanation of these data theoretically, especially for the fully lock-in angle, $\theta_{LH3}(H)$. This calls for some remarks.

4.2.1.5.2 Remarks

The angles at which the deviations in the flux flow resistivity data of $Tl_2Ba_2CuO_6$ are near to the theoretical predictions (Okram et al. 2001a, b; Blatter et al. 1994; Bulaevskii et al. 1992; Feinberg, Billard 1990; Ivlev, Ovchinnikov, Porovsky 1990; Ivlev, Porovsky, Ovchinnikov 1991; Bualevskii 1991). A little concern is related to the full range variations of these angles $\theta_{LXY}(X)$ with X=t or H, Y = 1, 2, or 3 with respect to temperature for a fixed field or vice-versa. The three angle regimes can also considered to be in line with theories (Blatter et al. 1994; Bulaevskii et al. 1992). The reasons for not able to fit the data well for the lock-in angles, i.e. the $\theta_{LXY}(X)$ for X=t or H, Y = 2, or 3 is presently not so clear. It may also be related to the presently unexplored intricacies of these vortices in these layered superconductors, wherein s-waves are no longer valid but the d-waves (Tsuei, Kirtley 2005).

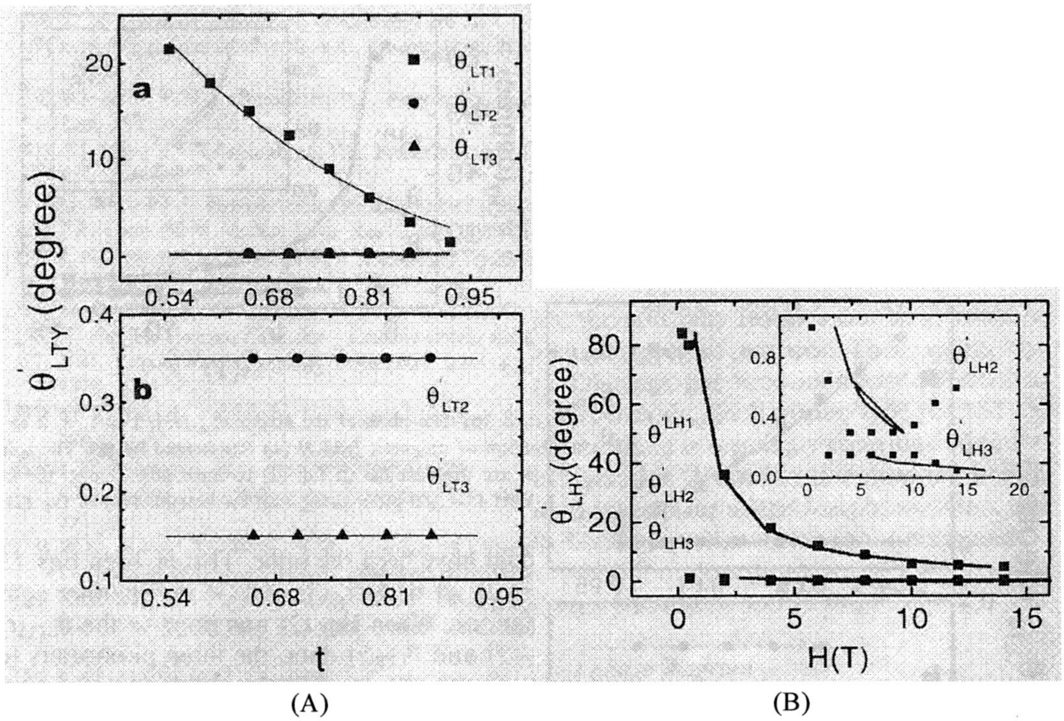

Figure 18 (A) The plots of $\theta'_{LTY}(t)$, Y=1,2,3 of $Tl_2Ba_2CuO_6$ (a). The expanded view of $\theta_{LTY}(t)$, Y=2,3 (b). (B) The plots of $\theta_{LHY}(H)$, Y=1,2,3 of $Tl_2Ba_2CuO_6$. Inset expanded view of $\theta_{LHY}(H)$, Y=2,3. Solid curves/ lines are the theoretical fits to experimental data. Reprinted from Okram et al. (2001b) with permission. © 2001 Elsevier Science.

4.3 Irreversibility Line versus Lock-in Transition

4.3.1 Irreversibility Field Lines

Okram et al. (2001d) investigated the correlation between the irreversibility line and the lock-in transition in $Tl_2Ba_2CuO_6$ single crystals. The irreversibility lines were obtained using the temperature dependence resistivity $\rho_c(T)$ (and $\rho_{ab}(T)$) data for the field H || c-axis (and ab plane) and for various H values (Figure 19). When H || c-axis, $H_{irr\perp}(t)$ obtained independently from $\rho_c(T)$ and $\rho_{ab}(T)$ almost coincide each other. These were considered to be due to the identical flux-flow behaviors for these configurations. This happens so in spite of the fact that J and H are parallel for $\rho_c(T)$, and J and H are perpendicular for $\rho_{ab}(T)$. These features therefore imply that the Lorentz-force is independent of the angle between H and J. This looks to be an intuitive example of such a situation consistent with other reports (e.g. Iye et al. 1989, 1990, 1991) and above wherein the classical angular dependence of the Lorentz force does not follow.

The vortices involved in these layered superconductors are noticeably different from the classic superconductors. In the former, the vortex line, unlike the continuous Abrikosov vortex, is not a rectilinear object. The flux line composes of the so-called pancake vortices directed along the c-axis joined by Josephson vortices whose axis threads through the junctions between the superconducting layers (Doniach 1989). In the Josephson vortex, the vortex line is rectilinear as Abrikosov vortex (Abrikosov 1957) but the cross section is not circular. Thus, one has to consider the corresponding vortex dynamics taking into account of such differences. As seen from Figure 20, when H || ab plane, $H_{irr||}(t)$ obtained from $\rho_c(T)$ data is quite higher than that of either of the two $H_{irr\perp}(t)$ data. These results clearly show that the irreversible lines and their behaviors depend on the orientation of H. These irreversible lines must be related to the respective flux dynamics of pancake vortex, and of Josephson vortex. Therefore, these features of irreversibility lines are suggestive of different vortex dynamics involved.

The vortex dynamics discussed above pertains to the particular cases when the magnetic field is parallel and perpendicular to the ab-plane. The irreversibility fields at the same reduced temperature t for these two cases are remarkably different (Figure 20). It would therefore be interesting to see "what happens to the flux flow resistivity above the flux flow onset as a function of angle θ and how the lock-in transition features are related to the irreversibility lines?" From the definition of the irreversibility fields (Figure 20) and observation of lock-in transition in the flux flow regime (Okram et al. 2001b, c), it seems apparent that the lock-in transition phenomenon can be observed only above the irreversibility line $H_{irr\perp}(t)$ for H || c-axis. In addition, one has to check the H and t values appropriately. To verify this, the resistivity $\rho_c(\theta)$ measurements were thus carried out at several H values at fixed T and vice-versa.

4.3.2 Lock-in Transition

Figure 21 shows the vortex state $\rho_c(\theta)$ of $Tl_2Ba_2CuO_6$ at 10T for several values of T. From these curves, $\theta_{LT1}(T)$, $\theta_{LT2}(T)$ and $\theta_{LT3}(T)$ were determine for all T as described above to identify the vortex phase diagram. Also, for 30K recorded at several magnetic fields, $\theta_{LHY}(H)$, where Y=1,2,3 were determined. The $\theta_{LTY}(t)$ and $\theta_{LHY}(H)$ graphs are plotted in Figure 22a and b, where t is the reduced temperature T/T_c. They are identical to those shown in Figure 17 and hence their explanation also is the same. Thus, these lock-in transition features are displayed for specific values of H and T.

4.3.3 The Correlation

We show here the correlation among the loci of $\theta_{LXY}(X)$ and $H_{irr\perp}(t)$. We consider the data point (0.75,82°) at 10T in Fig.22a. This point corresponds to (0.75,10T) in Figure 20, which lies above the $H_{irr\perp}(t)$ line. Another data point (0.54,68°) at 10T in Figure 22a corresponds to (0.54,10T) that also lies above the $H_{irr\perp}(t)$ line (Figure 20). Similarly, all other $(\theta_{LTY}(t),t)$ points in Fig.4a lie above $H_{irr\perp}(t)$ in Figure 20. Similar argument can be applied to prove that the points (θ_{LHY},H) on Figure 22b lie above $H_{irr\perp}(t)$ in Figure 20. Hence, in magneto-resistive measurements, the lock-in transition can be observed above the irreversibility line $H_{irr\perp}(t)$ for H || c-axis. Thus, the irreversibility lines (H_{irr}) and lock-in transition are correlated for over-doped $Tl_2Ba_2CuO_6$ single crystals.

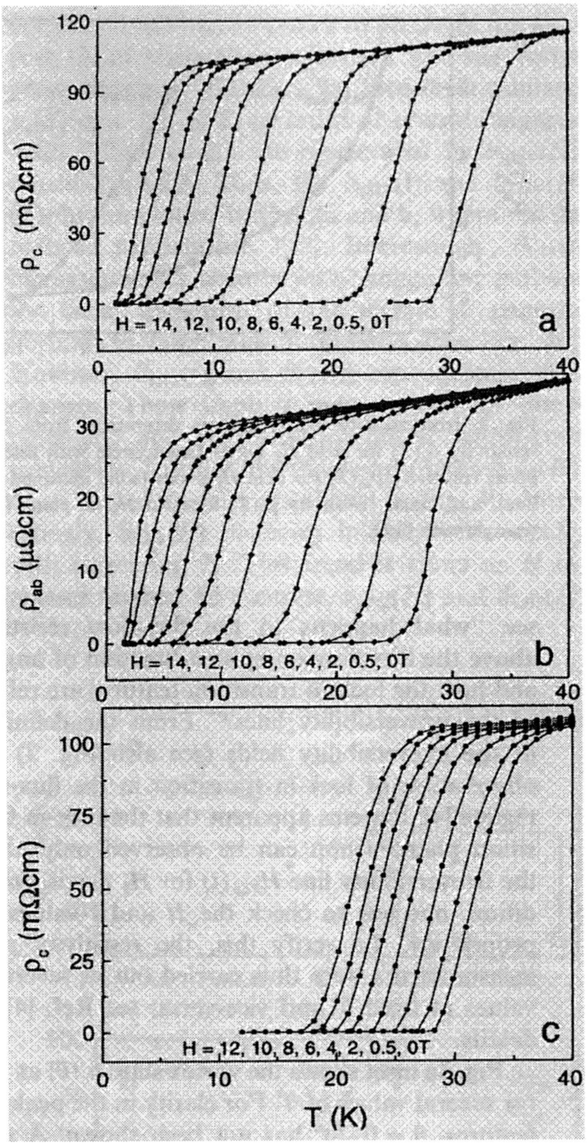

Figure 19 Resistivity ρ of $Tl_2Ba_2CuO_6$ at various magnetic fields H for a fixed orientation. (a) c-axis $\rho_c(T)$ for H || c-axis. (b) ab-plane $\rho_{ab}(T)$ for H || c-axis. (c) c-axis $\rho_c(T)$ for H || ab-plane. Reprinted from Okram et al. (2001d) with permission. © 2001 Elsevier Science.

4.4 Dimensionality and Lock-in Transition

Now, the question arises "What would happen to the two 3D and 2D models (Eqs. (35) and (36) in Section 2.4) when one realizes the upper critical field variation $H_{c2}(\theta)$ with angle (θ) in the context of the complicated (θ) features as seen in Figures 14, 15, 16 and 21?" We address this situation in this section. The main message in Figures 14, 15, 16 and 21 is the

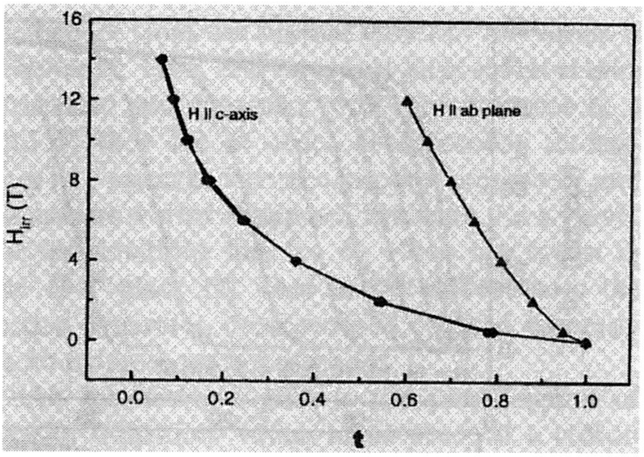

Figure 20 Irreversibility field lines H_{irr} (t) determined from Figure 19. For H|| c-axis, the two curves for the resistivity along and perpendicular to ab-plane (solid diamond and circle) are identical, and H_{irr} (t) determined from $\rho_c(\theta)$ along c-axis and H || ab-plane (solid triangle). Reprinted from Okram et al. (2001d) with permission. © 2001 IOP Publishing Ltd.

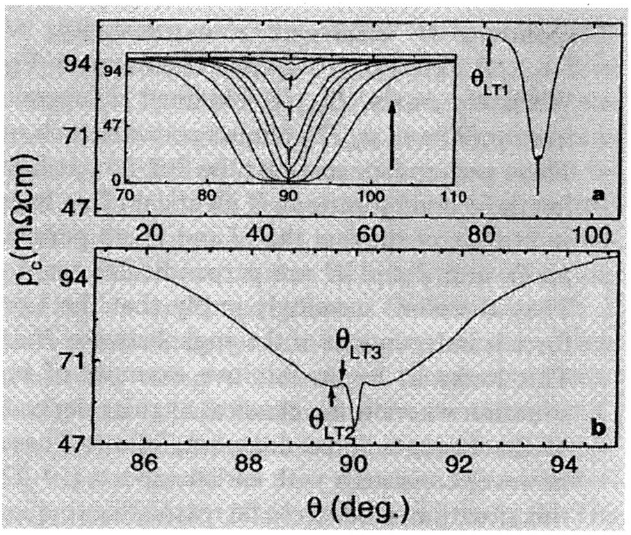

Figure 21 (a) Angle dependence $\rho_c(\theta)$ for fixed 10T at 28K. Arrow indicates the angle θ_{LT1}(28K). Inset: Angle dependence $\rho_c(\theta)$ for fixed 10T at 20, 22, 24, 25, 26, 28, 30, 32, 34 and 38K (bottom to top). (b) Angle dependence $\rho_c(\theta)$ nearby 90°. Reprinted from Okram et al. (2001c) with permission. © 2001 IOP Publishing Ltd.

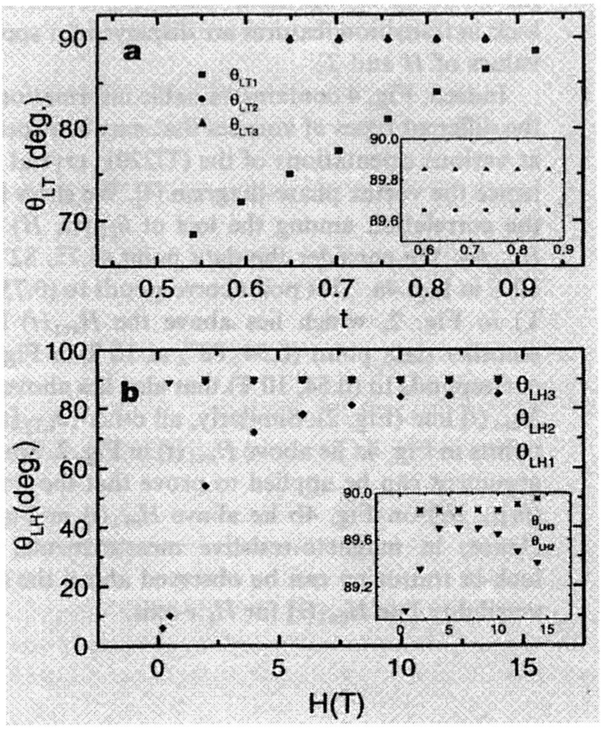

Figure 22 (a) The $\theta_{LTY}(t)$, Y =1, 2 or 3 plots (solid square, circle, triangle respectively) of $Tl_2Ba_2CuO_6$. Inset: The expanded $\theta_{LTY}(t)$, Y =2 or 3 plots (solid circle, triangle respectively). (b) The $\theta_{LHY}(H)$, Y =1, 2 or 3 plots (solid diamond, inverted triangle and star respectively). Inset: The expanded $\theta_{LHY}(H)$, Y =2 or 3 plots (solid inverted triangle and star respectively). Reprinted from Okram et al. (2001c) with permission. © 2001 IOP Publishing Ltd.

existence of four types of vortices that may prevail, as the magnetic field and temperature are chosen properly (Okram et al. 2001c). This is clear especially from the ones associated with the dip and the peak in the $\rho_c(\theta)$, which are not exhibited in all the curves. These suggest that for 10T, there is certain range of temperatures within which the four distinctive features in $\rho_c(\theta)$ can be observed or vice-versa. The question of development of different types of vortices depending on the angle, the field strength and the temperature is very intriguing in particular reference to the dimensional behavior of the superconductor because these vortices are based on the 2D nature of these materials. In this context, these superconductors are 2D materials. However, 2D and 3D models described above were based on the rectilinear vortex i.e. the so-called Abrikosov vortex. Further, the idea of lock-in transition was not apparent by then. These two contrasting consequences need to be investigated.

Here is an attempt to correlate the trend shown in Figures 14, 15, 16 and 21 in terms of the dimensionality. Now, as discussed earlier, for this, we show here as typical examples for 24 and 30K cases at 10T (Figure 23), which in fact was found to be true for all other curves for fitting the 2D and 3D models to the angular dependence of $1/\rho_c(\theta)$. It should be stressed here that for the whole angle range 0-90°, the two models were unable to fit the experimental data. That is why the region of fitting was considered near 90° only. Even with this restriction, it is interesting to note from Figure 23a that both the models are indicative of inadequacy to fit the experimental data. This is not the exception in the fits in Figure 23b. The only difference is in the qualitative agreement in the shape of experimental feature with those of the models. However, one should also realize the fact that the angular region attempted for these fittings is very narrow (Figure 23), which is contrary to that theoretically predicted for 0-90° (Tinkham 1963; Lawrence, Doniach 1970).

The fitted parameters for 2D model at 30K and 10T were $H_{c2\parallel} = 5.95T$ and $H_{c2\perp} = 0.97T$. For 3D model, such parameters were $H_{c2\parallel} = 11.0T$ and $\gamma = 6.5$. The parameters corresponding to 24K at 10T were 1506T, 0.97T for 2D model and 2351T and 642 for 3D models respectively (Okram et al. 2001c).

Interestingly, the lock-in transition, by definition, should take place in the 2D regime. This means that the 2D model should have been able to explain the observed feature(s). This however is not so because the vortex involved is clearly not the usual Abrikosov type. The possible inadequacy to explain the dimensionality directly is expected for all the related physical parameters. Although much more clearer in the present data for the reasons already described above, such situation is reported in the $J_c(\theta)$ or $H_{irr}(\theta)$ data (e.g. Marcon et al. 1992; Fastampa et al. 1991). There is a distinct jump, for example, at ~0.5° of Fig.4b of Fastampa et al. (1991). The better picture such as the present ones is expected to have missed in part due to the very limited data points obtainable in that kind of measurements (Fastampa et al. 1991). This can be easily overcome if one measures $\rho(\theta)$ accurately at close intervals of θ as presented in Figures 14, 15, 16, 21 and 23.

The origin of the failure, to explain the behavior of $1/\rho(\theta)$ or $H_{irr}(\theta)$ in particular for highly layered superconductor is expected to be related to the earlier lack of the idea of possible development of various types of vortices with angle (Okram et al. 2001b; Bulaevski et al. 1992; Blatter et al. 1994) and lock-in transition (Villard, Feinberg 1991; Ivlev et al. 1990 1991). The two models were however based on rectilinear Abrikosov vortices of the superconductor. In addition, the 3D model was constructed based on the thin film (Tinkham 1963) and the 2D model was based on the layered nature of the superconductor (Lawrence, Doniach 1970). However, in reality, the type of vortices concerned is non-rectilinear as distinct from the usual rectilinear Abrikosov vortex (Okram et al. 2001b; Bulaeveski et al. 1992; Blatter et al. 1994; Villard, Feinberg 1991; Ivlev et al. 1990 1991). Therefore, theory of the vortex dynamics and anisotropy of upper critical fields of highly layered superconductors may have to depend on the deeper understanding of the associated vortices. Interestingly, however, upper critical field is considered to be disappearing in these types of layered superconductors (e.g. ref. Hu, Tachiki 1998). Naively, this seems to imply inadequacy of application of the two models to these new superconductors (Tinkham 1963; Lawrence, Doniach 1970). This suggests further theoretical investigations.

Therefore, the lock-in transition can be used as a pointer to the discontinuity in the irreversibility fields and as a consequence the upper critical fields. This experimental fact clearly demonstrates the inadequacy of the 2D and 3D models to explain the angular dependence of the upper critical fields in high T_c cuprates.

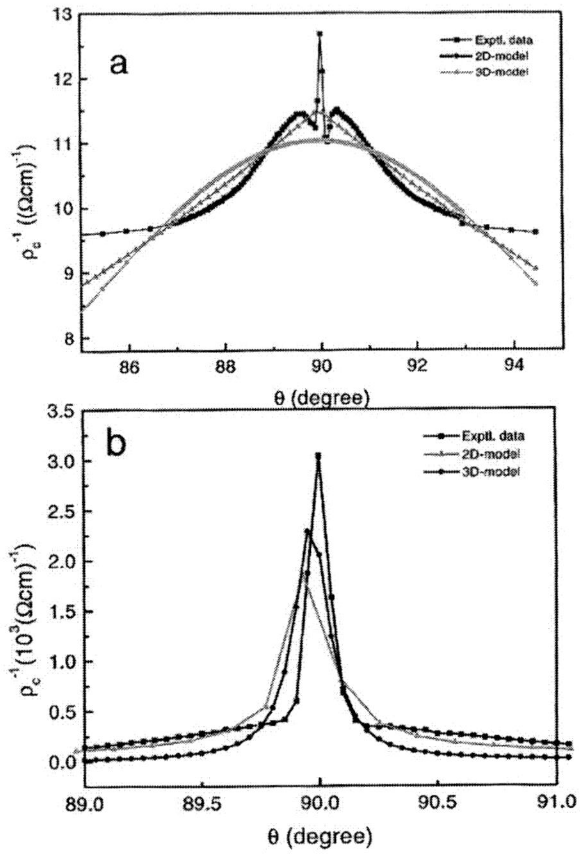

Figure 23 The angular dependence of $1/\rho_c(\theta)$ of $Tl_2Ba_2CuO_6$ for 24 and 30K at 10T along with the 2D and 3D model fits. (a) $1/\rho_c(\theta)$ for 30K at 10T. Solid square, circle and triangle curves are the experimental data, 2D model, and 3D model fits, respectively. (b) $1/\rho_c(\theta)$ for 24K at 10T. Solid square, circle and triangle curves are the experimental data, 2D model, and 3D model fits, respectively. Reprinted from Okram et al. (2001c) with permission. © 2001 IOP Publishing Ltd.

5 Pertinent Issues

5.1 Pairing Symmetry

Thus, in overall the existing s-wave based theories on the vortices (Section 2; references therein) can explain almost all the features of physical parameters of these layered superconductors. We believe that those features not able to describe is related to the theory based on the s-wave symmetry of the superconducting order parameter ψ. The basis for the mean-field theory of superconductivity is the s-wave complex superconducting order parameter ψ, which is true of the classic superconductors. The Ginzburg-Landau order parameter $\psi(\mathbf{r})$, proportional to the energy gap $\Delta(\mathbf{r})$, denotes the pair wave function and the energy gap $\Delta(\mathbf{r})$ as the microscopic identity of the Ginzburg-Landau order parameter is valid in the temperature range $T<T_c$. The essential ingredient for macroscopic quantum phenomena

such as superconductivity and superfluidity is the presence of off-diagonal long-range order that has no classical analog (Yang 1992; Anderson 1966). The use of the pair amplitude $\langle c_{k\uparrow}c_{-k\downarrow}\rangle \cong \Delta_k$ as the order parameter maintains the fundamental nature of the macroscopic phase-coherent superconducting state. The *symmetry of gap parameter* can be determined experimentally, which provides a solid guide for the symmetry of the pair condensate (Tsuei, Kirtley 2000). Indeed, the crystal structure of the cuprate superconductors has inversion symmetry, which can be specified using the Pauli principle.

In the k-space representation, high ratios of c-axis to a-axis lattice constants c/a (3-10) causes flattened Brillouin zones representing the basic symmetry properties of the unit cell of the square/ rectangular lattice. This is evident from the band structure calculations, which show energy bands predominantly derived from the CuO plane with no significant dispersion in the c-axis (Krakauer et al. 1988; Krakauer, Pickett 1988; Hamann, Matheiss 1988; Singh 1993a, b; Singh, Pickett 1994). This and many other studies provide enough evidence that superconductivity in cuprates basically originates from the CuO plane (Kleiner, Muller 1994; Mros et al. 1998). This implies the pairing symmetry a reflection from the underlying symmetry of the crystal CuO square/ rectangular lattices. The observation of the *half-integer flux-quantization effect* in the tetragonal crystal geometry say in $Tl_2Ba_2CuO_6$ represents the model-independent evidence for pure $d_{x^2-y^2}$ pairing symmetry; the phase sensitive test probes the sign changes in the pair critical currents (Wollman et al. 1993; Tsuei et al. 1994; Brawner, Ott 1994; Mathai et al. 1995; Wollman et al. 1995; Miller et al. 1995); for reviews see Tsuei, Kirtley (2000); Annett et al. (1990). In this context, there seems to have fundamental problem in tackling the phenomenology of the high temperature superconductors, wherein square/ tetragonal layer structure is the basic unit and hence the d-wave symmetry prevails.

5.2 Consequences of d-wave Pairing Symmetry

The vortex structure calculations indicate a slight *fourfold* distortion from cylindrical symmetry in the vortex core and in the magnetic field away from the core (Morita et al. 1998; Maki et al. 1999). While decoration experiments at low magnetic fields ~50G indicated that the vortices in $YBa_2Cu_3O_{7-\delta}$ are conventional hexagonal lattice (Dollan et al. 1989), neutron scattering (Keimer et al. 1994) and scanning tunneling microscope (Maggio-Aprile et al. 1995) experiments at 1T revealed an oblique lattice with nearly equal lattice vectors spaced by about 77°. Scanning tunneling microscopy further revealed only one bound state in $YBa_2Cu_3O_{7-\delta}$ (Karai et al. 1992) or perhaps none in $Bi_2Sr_2CaCu_2O_{8+\delta}$ (Renner et al. 1998) contrary to thousands of bound states in the core in conventional superconductors (Caroli et al. 1964; Hess et a. 1989). Moreover, for a periodic vortex lattice the natural low-energy quasiparticle states are the Bloch waves instead of the usual Landau levels (Franz, Tesanovic 2000).

For the d-wave pairing for example, the **k**-dependent BCS gap potential takes the form $\Delta(\mathbf{k}) = \Delta_0(\cos k_x - \cos k_y) \sim \cos 2\theta$, where θ is measured from the x-axis in the ab plane. Consequently, the sign change with angle can introduce an intrinsic phase shift by π into the Josephson effect formalism (Sigrist, Rice 1995). An s-wave superconductor has no such angular dependence. Therefore, the physics of these cuprates is expected to modify. In particular, three aspects can be considered for the vortices relating to (i) their structure, (ii) their interactions and (iii) the electronic bound states within the vortex (Xu et al. 1996; Maki et al. 1999; Franz, Tesanovic 2000). The d-wave vortex structure is different from that of s-wave (Caroli et al. 1964; Gygi, Schluter 1991) or p-wave (Ambegaokar et al. 1974; Scharnberg, Klemm 1980). The core of the d-wave vortex should contain all the possible gap functions that are consistent with the maximal symmetry group of the vortex line and should contain the conventional s-wave pairing with the opposite winding phase (Volovic 1993). Far from the axis, a pure d-wave exists while near the axis is a normal 'inner core' in which both s- and d- waves vanish, and in the middle region, s- and d- waves coexist (Soninen et al. 1994). Thus, Ginzberg-Landau equations for the vortex structure of d-wave superconductors were deduced (Ren et al. 1995), which too noted the opposite winding s-wave component induced near the core, but far away from the core, s-wave component becomes more complicated associated with strong anisotropy. Howsoever, to our knowledge, there is so far no d-wave theory of the angular dependence of the vortices with respect to the ab plane similar to the s-wave theory of vortices as in Section 2.

5.3 Coherence Length

The question of dimensionality based on the behavior of the coherence length $\xi(t) \propto (1\text{-}t)^{-1/2}$ (Section 2.5) is expected to be different from what it actually is in layered superconductors. In spite of its difficulty in experimental determination, the coherence length ξ can be estimated from relation $\xi(t)\lambda(t)H_c(t)$=constant, where λ is the experimentally measurable penetration depth, $t=T/T_c$ and $H_c(t)=1.73H_c(0)(1\text{-}t)$ is the BCS result. $\lambda(t)$ can be determine from $1/\lambda^2 \propto n_s$, the superfluid density. At low temperatures, in BCS case, $[\lambda(t)\text{-}\lambda(0)] \rightarrow e^{-\Delta/kT}$ as the quasi-particles above the gap freeze out. In layered high temperature superconductors wherein the d-wave symmetry holds true, the exponential freeze out is replaced by a linear temperature dependence $n_s \propto (1\text{-}aT)$ at low temperatures (Hardy et al. 1993; Sonier et al. 1994); a~0.0072/K.

Therefore, the inadequacy of the models such as the 2D and 3D models for the angular dependence of the upper critical fields, critical current density, etc., and the θ versus H related to the lock-in transition phenomenon to explain the experimental data is attributed to the expected modified physics for the d-wave superconducting order parameter in the Ginzburg-Landau theory or the Lawrence-Doniach model used to explain the vortices of these layered superconductors.

6 Conclusion

The existence of broadly four types of vortices theoretically has been considered based on the s-wave symmetry of the superconducting order parameter and possible experimental confirmation is described. The existence of these vortices leads to lock-in transition. This in particular causes discontinuous change in the physical parameters such as the flux flow resistivity, the irreversibility field and the upper critical field as a function of angle of the applied magnetic field with respect to the ab-plane of the superconductor. This discontinuous change for example in the irreversibility field or the flux flow resistivity as a function of angle raises a question on the validity of the two dimensional (2D) and three dimensional (3D) models for confirming the transition from the 3D to 2D as a layered superconductor is cooled down. Even though the 3D to 2D transition is basically based on the decrease of the coherence length as the temperature drops below a certain value less than the separation between the superconducting planes, this simple theory seems to be inadequate. Consequently, even though the predicted angular regimes approximately match the experimental observations, their behavior in particular for the lock-in transition angle as a function of the applied magnetic field indicates possible inadequacy perhaps in line with the inappropriate assumption of the s-wave function while it should have been the d-wave.

Acknowledgement

The author is grateful to Prof. H. Aoki, Ms. C. Terakura and Dr. S. Uji for invaluable support in the course of high field measurements; Dr. M. Xu, Prof. D. G. Hinks and Dr. T. W. Li for providing the single crystals; Dr. P. Chaddah and Prof. A. Gupta for their encouragement during the course of this writing; Elsevier Science, Netherlands, Institute of Physics Publishing, UK, American Physical Society, USA, the American Association for the Advancement of Science, USA, and Nature Publishing Group, UK as well as the authors of the Figures on the experimental results used for granting permission.

References

[1] Abrikosov, A.A. Sov. *Phys.-JETP* 1957, 5, 1174.

[2] Abrikosov, A.A. *Fundamentals of the Theory of Metals* 1988 (North Holland, Amsterdam).

[3] Ambegaokar, V.; de Genne, P. G.; Rainer, D. *Phys. Rev. A* 1974, 9, 2676.

[4] Anderson, P.W. *Rev. Mod. Phys.*1966, 38, 298.

[5] Annett, J.F.; Goldenfeld, N.; Renn, S. R. "The pairing state of $YBa_2Cu_3O_{7-\delta}$" in Ginsberg, D.M. (ed.), *Physical Properties of High Temperature Superconductors II*, World Scientific, Singapore (1990), p. 571.

[6] Artemenko, S.N.; Kruglov, A.N. *Phys. Lett.* A 1990, 143, 485.

[7] Balatskii, A.B.; Burlachkov, L.I.; Gor'kov, L.P. Sov. *Phys.-JETP* 1986, 63, 866.

[8] Bardeen, J.; Cooper, L.N.; Schrieffer, J.R. *Phys. Rev.* 1957, 108, 1175.

[9] Bardeen, J.; Stephen M.J. *Phys. Rev. A* 1965, 140, 1197.

[10] Bednorz, G.; Muller, K.A. Z. *Phys. B* 1986, 64, 189.

[11] Berezinskii, V.L. Sov. Phys. JETP 1970, 32, 493; Sov. *Phys. JETP* 1971, 34, 610.

[12] Bezryadin, A.; Ovchinnikov, Y.N.; Pannetier, B. *Phys.Rev. B* 1996, 53, 8553-60.

[13] Blatter, G., Feigel'man, M.V.; Geshkenbein, V.B.; Larkin, A.I.; Vinokur, V. M. *Rev. Mod. Phys.* 1994, 66, 1125-1388.

[14] Blatter, G. Geshkenbein V.B. *Phys.Rev. B* 1993, 47, 2725

[15] Bolle, C. A.; Gammel, P.L.; Grier, D.G.; Murray, C.A.; Bishop, D.J. *Phys. Rev. Lett.* 1991, 66, 112.

[16] Brawner, D.A.; Ott, H. R. *Phys. Rev. B* 1994, 50, 6530.

[17] Bugoslavsky, Y.V.; Zhukov, A. A.; Perkins, G. K.; Caplin, A. D.; Kojima, H.; Tanaka, I. *Phys. Rev. B* 1997, 56, 5610.

[18] Bulaevskii, L.N. Sov. *Phys. JETP* 1973, 37, 1133.

[19] Bulaevskii, L.N. *Phys. Rev. B* 1991, 44, 910.

[20] Bulaevskii, L.N., Clem J.R. *Phys. Rev.* B 1991, 44, 10234.

[21] Bulaevskii, L.N., Ledvij, M.; Kogan, V.G. *Phys. Rev. B* 1992, 46, 366.

[22] Bullut, N.; Scalpino, D.J. *Phys. Rev. B* 1996, 54, 14971.

[23] Buzdin, A.; Feinberg, D. *J. Phys.* (Paris) 1990, 51 1971.

[24] Buzdin, A.I.; Siminov, A.Y. JETP Lett. 1990, 51 191.

[25] Caroli, C.; de Genne, P.G.; Marticon, *J. Phys. Lett.* 1964, 9, 307.

[26] Cataudella,V.; Minnhagen, P. *Physica C* 1990, 166, 442.

[27] Carton, J.P. *J. Phys. I* (Paris) 1991, 1, 113

[28] Chakravarty, S.; Ivlev, B.I.; Ovchinikov, Y. N. *Phys. Rev. B* 1990a, 42, 2143.

[29] Chakravarty, S.; Ivlev, B.I.; Ovchinikov, Y. N. *Phys. Rev. Lett.* 1990b, 64, 3187.

[30] Chikumoto, N.; Monezykowski, M.; Motohira, N.; Kishio, K. *Physica C* 1992, 199, 32.

[31] Clem, J.R. *Phys. Rev. B* 1991, 43, 7837.

[32] Clem, J.R.; Coffey, M.W. *Phys. Rev.* B 1990, 42, 6209.

[33] Clem, J.R.; Coffey, M.W.; Hao, Z. *Phys. Rev.* B 1991, 44, 2732

[34] Deligiannis, K.; Kokkaliaris, S, Oussena, M.; de Groot, P.A.J.; Fruchter, L.; Gagnon, R.; Taillefer, L. *Phys. Rev. B* 1999, 59, 14772-14780.

[35] Dollan,G.J.; Holtzberg, F.; Field, C.; Dinger, T.R. Phys. Rev. Lett. 1989, 62, 2184.

[36] Doniach, S. in The Phenomenology of Flux Motion in High Temperature Superconductors, Proc. Los Alamos Symposium - 1989, ed. K.S. Bedell et al. (Addison-Wesley, Reading, MA), p.406.

[37] Efetov, K.B. *Sov. Phys. JETP* 1979, 49, 905.

[38] Eskildsen, M. R.; Kugler, M; Tanaka, S; Jun, J.; Kazakov, S. M.; Karpinski, J.; Ø. Fischer, O. *Phys. Rev. Lett.* 2002, 89, 187003.

[39] Essmann, U.; Trauble, H. *Phys. Lett.* A 1967, 24, 526.

[40] Farrell, D.E.; Bonham, S.; Foster, J.; Chang, Y.C.; Jiang, P.J.; Vandervoort, K.G., Lam, D.J.; Kogan, V.G. *Phys. Rev. Lett.* 1988, 61, 2805.

[41] Farrell, D.E.; Bonham, S.; Foster, J.; Chang, Y.C.; Jiang, P.J.; Vandervoort, K.G., Lam, D.J.; Kogan, V.G. *Phys. Rev. Lett.* 1989, 63, 782.

[42] Fastumpa, R.; Giura, M.; Marcon, R.; Silva, E. *Phys. Rev. Lett.* 1991, 67, 1795.

[43] Feigel'man, M.V.; Geshkenbein, V.B.; Vinokur, V.M. Sov. Phys. JETP 1990, 52, 546.

[44] Feigel'man, M.V.; Geshkenbein, V.B.; Larkin, A.I. *Physica C* 1990, 167, 177.

[45] Feinberg, D; Billard, C. *Phys. Rev. Lett.* 1990, 65, 919.

[46] Fischer, K.H. *Physica C* 1991, 178, 161.

[47] Franz, M.; Tesanovic, Z. *Phys. Rev. Lett.* 2000, 84, 554.

[48] Gammel, P.L.; Bishop, D. J.; Dolan, G. J.; Kwo, J. R.; Murray, C. A.; Schneemeyer, L. F.; Waszczak, J. V. *Phys. Rev. Lett.* 1987, 59, 2592-95.

[49] Ginzburg, V.L.; Landau, L.D. Zh. Experim. i Teor. Fiz. 1950, 20, 1064.

[50] Glazman, L.I.; Koshelev, A.E. Sov. Phys.-JETP 1990, 70, 774.

[51] Goa, P.E.; Hauglin, H.; Baziljevich, M.; Il'yashenko, E.; Gammel, P.L.; Johansen, T.H. Supercond. *Sci. Technol.* 2001, 14, 729-31.

[52] Gor'kov, L.P. Sov. Phys.-JETP 1959, 9, 1364.

[53] Grigorenko, A.; Bending, S.; Tamegai, T.; Ooi, S.; Henini, M.; *Nature* (London) 2001, 414, 728.

[54] Grigorieva, I. V.; Steeds, J.W.; Balakrishnan, G.; Paul, D.M. *Phys. Rev.* B 1995, 51, 3765-71.

[55] Gygi, F.; Schluter, M. *Phys. Rev.* B 1991, 43, 7609.

[56] Hamann,D.R.; Matheiss, L.F. *Phys. Rev.* B 1988, 38, 5138.

[57] Harada, K.; Matsuda, T.; Bonevich, J.; Igarashi, M.; Kondo, S.; Pozzi, G.; Kawabe, U.; Tonomura, *A. Nature* (London) 1992, 360, 51.

[58] Hardy, W. N.; Bonn, D. A.; Morgan, D. C.; Liang, R.; Zhang, K. *Phys. Rev. Lett.* 1993, 70, 3999.

[59] Hazen, C.W.; Griessen, R. *Phys. Rev. Lett.* 1989, 62, 2857.

[60] Hess, H.F.; Robinson, R.B.; Dynes, R.C.; Valles, Jr., J.M.; Wasziczak, J.V. *Phys. Rev. Lett.* 1989, 62, 214.

[61] Hu, C.R. *Phys. Rev. B* 1972, 6, 1756.

[62] Hussey, N.E.; Cooper, J. R.; Doyle, R.A.; Lin, C.T.; Ling, W. Y.; Sinclair, D.C.; Balakrishnan, G.; Paul, D.; Revcoevschi, A. *Phys. Rev. B* 1996, 53, 6752.

[63] Ivlev, B.I.; Kopnin, N.B. *Phys. Rev. Lett.* 1990a, 64, 1828.

[64] Ivlev, B.I.; Kopnin, N.B. *J. Low Temp.* Phys. 1990b, 80, 161.

[65] Ivlev, B.I.; Kopnin, N.B. *Phys. Rev.* B 1990c, 42, 10052.

[66] Ivlev, B.I.; Kopnin, N.B.; Pokrovsky, V.L. J. *Low Temp. Phys.* 1990d, 80, 187.

[67] Ivlev, B.I.; Ovchinnikov, Y.N.; Pokrovsky, V.L. *Europhys. Lett.* 1990e, 13, 187.

[68] Ivlev, B.I.; Pokrovsky, V.L.; Ovchinnikov, Y.N. *Mod. Phys. Lett.* 1991, 5, 73.

[69] Iye, Y.; Nakamura, S.; Tamegai, T. *Physica* C 1989, 159, 433.

[70] Iye, Y.; Nakamura, S.; Tamegai, T.; Terashima, T.; Yamamoto, K.; Bando, Y. *Physica C* 1990, 166 62.

[71] Iye, Y.; Tamegai, T.; Nakamura, S. *Physica C* 1991, 174, 227.

[72] Josephson, B.D. *Phys. Lett.* 1962, 1, 251.

[73] Keimer, B.; Shih, W. Y.; Erwin, R. W.; Lynn, J. W.; Dogan, F.; Aksay, A. I. *Phys. Rev. Lett.* 1994, 73, 3459.

[74] Kes, P.H.; Aarts, J.; Vnokur, V.M.; van der Veek, C.J. *Phys. Rev. Lett.* 1990, 64, 1063.

[75] Kim, D.H.; Goldman, A.M.; Kang, J.H.; Gray, K.E.; Kampwirth, R.T. *Phys. Rev. B* 1988, 39, 12275.

[76] Kirtley, J.R.; Wikswo, Jr., J.P. Annu. *Rev. Mater. Sci.* 1999, 29, 117-48.

[77] Kirtley, J.R.; Tsuei, C.C.; Rupp, M.; Sun, J. Z.; Yu-Jahnes, L.S.; Gupta, A.; Ketchen, M. B. *Phys. Rev. Lett.* 1996, 76, 1336-39.

[78] Kleiner, R.; Muller, P. *Phys. Rev. B* 1994, 49, 1327.

[79] Klemm, R.A.; Luther, A.; Beasley, M.R. *Phys. Rev. B* 1975, 12, 877.

[80] Klemm, R.A.; Clem, J.R. *Phys. Rev. B* 1980, 21, 1868.

[81] Kogan, V.G. *Phys. Rev. B* 1981, 24, 1572.

[82] Kogan, V.G.; Campbell, L.J. *Phys. Rev. Lett.* 1989, 62, 1552.

[83] Kogan, V.G.; Clem, J.R. Phys. *Rev. B* 1981, 24, 2497.

[84] Kogan, V.G.; Nakagawa, K.; Thiemann, L.S. *Phys. Rev.* B 1990, 42, 2631.

[85] Kosterlitz, J.M. *J. Phys. C* 1974, 7, 1046.

[86] Kosterlitz, J.M.; Thouless, D.J. *J. Phys.* C 1973, 6, 1181.

[87] Krakauer, H.; Pickett, W.E. *Phys. Rev. Lett.* 1988, 60, 1665.

[88] Krakauer, H.; Pickett, W.E.; Cohen, R.E. J. *Supercond.* 1988, 1, 111.

[89] Kufer, H.; Zhukov, A.A.; Will, A.; Jahn, W.; Meier-Hirmer, R.; Wolf, T.; Voronkova, V. I.; Klaser, M.; Saito, K. *Phys. Rev.* B1996, 54 (1996).

[90] Kwok, W.K.; Welp, U. V.; Vinokur, M.; Flesher, S.; Downey, J.; Crabtree, G.W. *Phys. Rev. Lett.* 1991, 67, 390.

[91] Kwok, W.K., Fendrich, J. A.; Vinokur, V. M.; Koshelev, A. E.; Crabtree, G. W. *Phys. Rev. Lett.* 1996, 76, 4596.

[92] Lawrence,W.E.; Doniach, S. in *Proceedings of the Twelfth International Conference on Low Temperature Physics* 1971, Kyoto 1970, Edited by E. Kanda (Academic, New York), p.361.

[93] London, F., and H. Proc. Roy. Soc. (London) 1935, A149, 71.

[94] Maggio-Aprile, I.; Renner, C.; Erb, A.; Walker. E.; Fischer, O.; *Phys. Rev. Lett.* 1995, 75, 2754.

[95] Maki, K.; Won, H.; Kohmoto, M.; Shiraishi, J.; Morita, Y.; Wang, G.-F. *Physica C* 1999, 317-318, 353.

[96] Mansky, P.A.; Chaikin, P.M.; Haddon, R.C. *Phys. Rev. Lett.* 1993, 70, 1323-1326.

[97] Marchevsky, M.; Gurevich, L. A.; Kes, P. H.; Aarts, *J. Phys. Rev. Lett.* 1995, 75, 2400-03.

[98] Marcon, R., Silva, E.; Fastumpa, R.; Giura, M. 1992, *Phys. Rev. B* 46, 3612.

[99] Martinez, J.C.; Brongersma, S.H.; Koshelev, A.; Ivlev, B.; Kes, P.H.; Griessen, R.P.; de Groot, D.G.; Tarnavski, Z.; Menovsky, A.A. *Phys. Rev. Lett.* 1992, 69, 2276.

[100] Maslov, S.S.; Pokrovsky, V.L. *Europhys. Lett.* 1991, 14, 591.

[101] Mathai, A.; Gim, Y.; Black, R. C.; Amar, A.; Wellstood, F. C. *Phys. Rev. Lett.* 1995, 74, 4523.

[102] Miller, Jr, J.H.; Ying, Q. Y.; Zou, Z. G.; Fan, N. Q.; Xu, J. H.; Davis, M. F.; Wolfe, J. C. *Phys. Rev. Lett.* 1995, 74, 2347

[103] Moler, K.A.; Kirtley, J.R.; Hinks, D.G.; Li, T.W.; Xu, M. Science 1998, 279, 1193.

[104] Morita, Y.; Kohmoto, M.; Maki, K. Int. *J. Mod. Phys.* B 1998, 12, 989.

[105] Moser, A.; Hug, H. J.; Parashikov, I.; Stiefel, B.; Fritz, O.; Thomas, H.; Baratoff, A.; Güntherodt, H.-J.; Chaudhari, P. *Phys. Rev. Lett.* 1995, 69, 2276.

[106] Mros, N.; Krasnov, V. M.; Yurgens, A.; Winkler, D.; Claeson, T. *Phys. Rev. B* 1998, 57, R8135.

[107] Naughton, M.J.; Yu, R.C.; Davies, P.K.; Fischer, J.E.; Chamberlin, R.V.; Wang, Z.Z.; Jing, T.W. *Phys. Rev. B* 1988, 38, 9280.

[108] Okram, G.S.; Aoki, H.; Nakamura, K. Solid State Commun. 1999, 110, 327.

[109] Okram. G.S.; Aoki, H.; Xu, M.; Hinks, D.G.; Li, T.W. Solid State Commun. 2001a, 118, 47.

[110] Okram. G.S.; Aoki, H.; Xu, M.; Hinks, D.G. *Physica* C 2001b, 355, 65.

[111] Okram. G.S.; Terakura, C.; Uji, S.; Aoki, H.; Xu, M.; Hinks, D.G. Supercond. *Sci. Technol.* 2001c, 14, 452.

[112] Okram. G.S.; Terakura, C.; Uji, S.; Aoki, H.; Xu, M.; Hinks, D.G. *Physica* C 2001d, 356, 115.

[113] Onnes, H. K. *Leiden Comm.* 1911, 120b, 122b, 124c.

[114] Oral, A.; Bending, S. J.; Humphreys, R. G.; Henini, M. Supercond. Sci. *Technol.* 1997, 10, 17-20.

[115] Palstra, T.T.; Batlogg, B.; Schneemeyer, L.F.; van Dover, R.B.; Waszczak*, J.V. Phys. Rev B* 1988, 38, 5102.

[116] Pearl, *J. Appl. Phys. Lett.* 1964, 5, 65.

[117] Ren, Y.; Xu, J.; Ting, C. S. *Phys. Rev. Lett.*1995, 74, 3680.

[118] Renner, C.; Revaz, B, Kadowaki, K.; Maggio-Aprile, I.; Fischer, O. *Phys. Rev. Lett.* 1998, 80, 3606-3609.

[119] Scharnberg, K.; Klemm, R. A. *Phys. Rev. B* 1980, 22, 5233.

[120] Sigrist, M.; Rice, T.M. J. Phys. Soc. Jpn. 1992, 61, 4283; also *Rev. Mod. Phys.* 1995, 67, 503.

[121] Singh, D. J. *Physica C* 1993a, 212, 228.

[122] Singh, D. J. *Phys. Rev. B* 1993b, 48, 3571.

[123] Singh, D. J.; Pickett, *W.E. Phys. Rev. Lett.* 1994, 73, 476.

[124] Sonier, J. E.; Kiefl, R. F.; Brewer, J. H.; Bonn, D. A.; Carolan, J. F.; Chow, K. H.; Dosanjh, P.; Hardy, W. N.; Liang, R.; MacFarlane, W. A.; Mendels, P.; Morris, G. D.; Riseman, T. M.; Schneider, *J. W. Phys. Rev. Lett.* 1994, 72, 744.

[125] Soninen, P.I.; Kallin, C.; Berlinsky, *A. J. Phys. Rev. B* 1994, 50, 13883.

[126] Straub, R.; Keil, S.; Kleiner, R.; Koelle, *D. Appl. Phys. Lett.* 2001, 78, 3645.

[127] Tachiki, M.; Takahashi, *S. Solid State Commun.* 1989, 70, 291-295.

[128] Theodorakis, *S. Phys. Rev.* B 1990, 42, 10172.

[129] Tinkham, *M. Phys. Rev.* 1963, 129, 2413.

[130] Tinkham, M. in *Introduction to Superconductivity,* 2nd Edition 1996, (McGraw-Hill, Inc. New York, USA) p. 322.

[131] Tsuei, C.C.; Kirtley, J. R., Chi, C.C.; Yu-Jahnes, L.S.; Gupta, A.; Shaw, T.; Sun, J. Z.; Ketchen, *M. B. Phys. Rev. Lett.* 1994, 73, 593.

[132] Tsuei, C.C.; Kirtley, *J. R. Rev. Mod. Phys.* 2000a, 72, 969-1016.

[133] Tsuei, C.C.; Kirtley, J. R. 2000b, http://www.arXiv.org/abs/cond-mat/0004/85.

[134] Uemura, Y. J.; Keren, A.; Lee, L.P., Luke, G.M.; Wu, W.D.; Kubo, Y.; Manako, T.; Shimakawa, Y. Subramanium, M.; Cobb, J.L.; Markert, *J. T. Nature* 1993, 364, 605.

[135] Volovic, *G. E. JETP Lett.* 1993, 58, 469.

[136] Welp, U.; Kwok, W.K.; Crabtree, G.W.; Vandervoort, K.G.; Liu, *J.Z. Phys. Rev. Lett.* 1989, 62 1908.

[137] Wollman, D. A.; Harlingen, D. J. V.; Lee, W. C.; Ginsberg, D. M.; Leggett, *A. J. Phys. Rev. Lett.* 1993, 71, 2134

[138] Wollman, D. A.; Harlingen, D. J. V.; Giapintzaskis, J.; Ginsberg, *D. M. Phys. Rev. Lett.* 1995, 74, 596.

[139] Xiao, H.; Tachiki, *M. Phys. Rev. Lett.* 1998, 80, 4044.

[140] Xu, J.H.; Ren, Y.; Ting, *C.S. Phys. Rev.* B 1996, 53, R2991.

[141] Yang, C. N. *Rev. Mod. Phys.*1962, 34, 694.

In: New Frontiers in Superconductivity Research
Editor: Barry P. Martins, pp. 47-80

ISBN: 1-59454-850-1

Chapter 2

CRITICAL CURRENTS AND VORTEX DYNAMICS IN PERCOLATIVE SUPERCONDUCTORS CONTAINING FRACTAL CLUSTERS OF A NORMAL PHASE

Yuriy I. Kuzmin *

Ioffe Physical Technical Institute of the Russian Academy of Sciences,
Polytechnicheskaya 26 St., Saint Petersburg 194021 Russia

Abstract

The effect of fractal clusters on magnetic and transport properties of percolative superconductors is considered. The superconductor contains percolative superconducting cluster, carrying a transport current, and clusters of a normal phase, which act as pinning centers. A prototype of such a structure is a high-temperature superconducting (HTS) wire. The superconducting core of first generation HTS wires is the conglomeration of superconducting micro-crystallites containing normal-phase clusters inside. There are the clusters of columnar defects in superconducting layer, which is the current-carrying element of second generation HTS wire. It is found that normal-phase clusters can have essential fractal features that affect the vortex dynamics. The fractal dimension of the boundary of normal-phase clusters in YBCO films is estimated and the cluster statistics is analyzed. Depinning and transport of vortices in fractal superconducting structures are investigated. Transition of the superconductor into a resistive state corresponds to the percolation transition from a pinned vortex state to a resistive state when the vortices are free to move. It is revealed that a mixed state of the vortex glass type is realized in the superconducting system involved. The fractal distribution of critical currents is derived and its peculiarities are studied. It is found that there is the range of fractal dimension where this distribution has anomalous statistical properties, specifically, its dispersion becomes infinite. The current-voltage characteristics of superconductors containing fractal clusters are obtained. Dependencies of the free vortex density on the fractal dimension as well as the resistance on the transport current are studied. It is found that the fractality of the cluster boundary intensifies pinning and thereby raises the critical current. This feature enables the current-carrying capability of a superconductor to be enhanced without changing of its chemical composition.

PACS: 74.81.-g; 74.25.Fy; 74.81.Bd

* E-mail address: yurk@mail.ioffe.ru, iourk@yandex.ru, Tel.: +7 812 2479902; Fax: +7 812 2471017

1 Introduction

An essential feature of the clusters of defects in superconductor lies in their capability to trap a magnetic flux [1 - 3]. Holding in place the vortices driven by the Lorentz force, such clusters can act as effective pinning centers [4, 5]. This feature can be used in making new composite superconducting materials of enhanced current-carrying capability [6 - 8]. The cluster structure affects the vortex dynamics in superconductors, especially when clusters have fractal boundary [9 - 13]. In the present paper the magnetic and transport properties of composite superconductors with fractal clusters of a normal phase will be considered as well as the phenomena limiting the current-carrying capability of such superconductors will be discussed.

A prototype of superconductor containing inclusion of a normal phase is a superconducting wire. The first generation high-temperature superconducting (HTS) wires are fabricated following the powder-in-tube technique (PIT). The metal tube is being filled with HTS powder, then the thermal and deformation treatment is being carried out. The resulting product is the wire consisting of one or more superconducting cores sheathed by a normal metal (Fig.1).

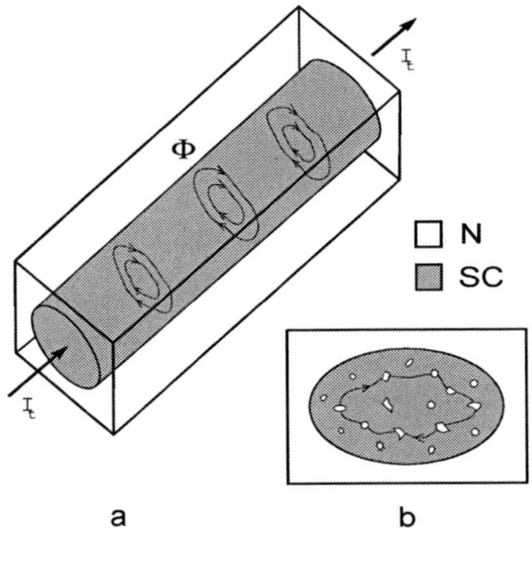

Fig.1. Schematic representation for the first generation superconducting wire. The superconducting core is sheathed by the normal metal. The transport current creates the magnetic flux concentrated along irregular-shaped rings passing through the normal-phase inclusions. The inset shows the cross-section of the wire with normal-phase inclusions as well as one cluster defined as a set of pinning centers, which are united by the common trapped flux and are surrounded by the superconducting phase.

The sheath endows the wire with the necessary mechanical (flexibility, folding strength) and electrical (the possibility to release an excessive power when the superconductivity will be suddenly lost) properties. At present, the best results are obtained for the silver-sheathed bismuth-based composites, which are of practical interest for energy transport and storage. In view of the PIT peculiarity the first generation HTS wire has a highly inhomogeneous structure [14]. Superconducting core represents a dense conglomeration of BSCCO micro-crystallites containing normal-phase inclusions inside (see inset in Fig.1). These inclusions primarily consist of a normal metal (silver) as well as the fragments of different chemical composition, grain boundaries, micro-cracks, and the domains of the reduced superconducting order parameter [15 - 17]. The volume content of a normal phase in the core is far below the percolation threshold, so there is a percolative superconducting cluster that carries the transport current.

The second generation HTS wires (coated conductors) have multi-layered film structure consisting of the metal substrate (nickel-tungsten alloy), the buffer oxide sub-layer, HTS-layer (YBCO), and the protective cladding made from a noble metal (silver). Superconducting layer, which carries the transport current, has the texture preset by the structure of an oxide sub-layer. In the superconducting layer there are clusters of columnar defects (see Fig.2) that can be created during the film growth process as well as by the heavy ion bombardment [5, 18]. Such defects are similar in topology to the vortices, therefore they suppress effectively the flux creep so it is possible to get the critical current up to the depairing value [1, 19].

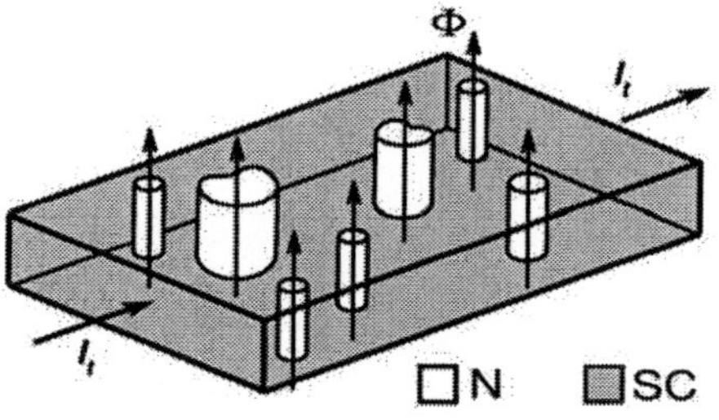

Fig.2. Superconducting film with columnar clusters of a normal phase. Such a film represents the current-carrying element of the second generation superconducting wire (coated conductors).

2 Magnetic Flux Trapping in Percolative Superconductors

Let us consider a superconductor containing inclusions of a normal phase, which are out of contact with one another. These inclusions may be formed by the fragments of different chemical composition, as well as by the domains of reduced superconducting order parameter. The columnar defects of such a kind can readily be created in superconducting film at the sites of defects on the boundary with the substrate (see Fig.2). A passage of electric current through a superconductor is linked with the vortex dynamics because the vortices are subjected to the Lorentz force created by the current. In its turn, the motion of the magnetic flux transferred by vortices induces an electric field that leads to the energy losses. In HTS's the vortex motion is of special importance because of large thermal fluctuations and small pinning energies [20]. Here we will consider the simplified model of one-dimensional line pinning when a vortex filament is trapped by the set of pinning centers [21]. A superconductor containing isolated clusters of a normal phase allows for effective pinning, because the vortices cannot leave them without crossing the surrounding superconducting space. The clusters present the sets of normal-phase inclusions, which are united by the common trapped flux and surrounded by the superconducting phase. The flux can be created both by an external source (e.g., during magnetization in field-cooling regime) and by the transport current (in self-field regime). The magnetic flux remains to be trapped in the normal-phase clusters till the Lorentz force created by the transport current exceeds the pinning force.

If the relative portion of superconducting phase exceeds the percolation threshold, there is a superconducting percolation cluster carrying a transport current. In such a structure magnetic flux is locked in finite clusters of a normal phase (see Fig.3,a).

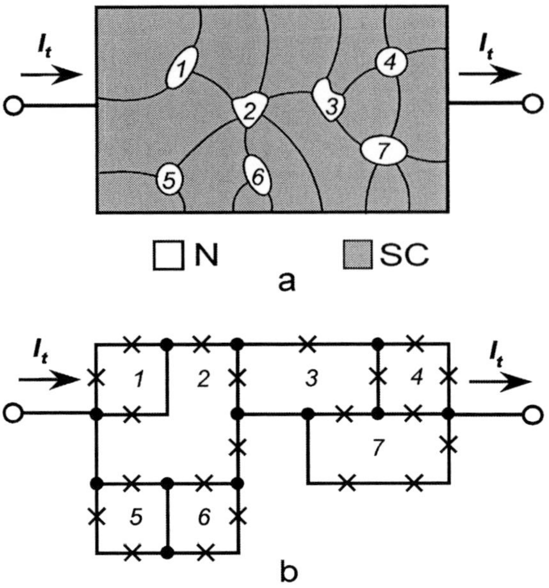

Fig.3. Equivalent weak-link-network circuit representation of superconductor with normal-phase clusters. (a) - the percolation chart (superconducting space is shaded, weak links are shown by the curves joining together the normal-phase clusters); (b) - the electric circuit (the superconducting loops are shown by lines; the crosses on them denote weak links).

If the transport current is passed through the sample, the trapped flux remains unchanged as long as the vortices are still held in the normal-phase clusters. The larger clusters have to allow for a weaker pinning than the smaller ones, because the larger the cluster size, the more entry points into weak links, through which the vortices can pass, are located along its boundary. When the current is increased, the magnetic flux will break away first from the clusters of smaller pinning force, and therefore, of larger size. Magnetic flux trapped into a single cluster is proportional to its area A. Therefore, the decrease in the total trapped flux $\Delta\Phi$ is proportional to the number of the clusters of area larger than a preset value A. So this value can be expressed with the cumulative probability function $W(A) = \Pr\{\forall A_j < A\}$, which is equal to the probability to find the cluster of area A_j smaller than an upper bound of A:

$$\frac{\Delta\Phi}{\Phi} = 1 - W(A) \tag{1}$$

The left hand side of this formula is equal to the relative decrease in the total trapped flux caused by the transport current of the same amplitude as the depinning current of the cluster of area A, and the right hand side gives the provability to find the cluster of area greater than A in the whole population.

When the current is increased, the magnetic flux starts to break away from the clusters of the pinning force weaker than the Lorentz force created by the transport current. The vortices, driven by this force, must cross the surrounding superconducting space, and they will first do that through the weak links, connecting the normal-phase clusters (see Fig.3,a where weak links are shown by the black curves connecting the white normal-phase clusters). In such a system depinning has a percolative nature [22 - 24], because unpinned vortices move through the randomly generated channels created by weak links. The equivalent weak-link-network circuit representation of such a system is shown in Fig.3,b. The normal-phase clusters are presented by the cells of the net interlaced by the superconducting loops. Each of the loops contains weak links (shown by the crosses), which join the adjacent cells and so enable the vortices to pass from one cluster to another.

Weak links form readily in HTS characterized by an extremely short coherence length [20, 25, 26]. Various structural defects, which would simply cause some additional scattering at long coherence length, give rise to the weak links in HTS. On a mesoscopic scale twin boundaries are mainly responsible for weak link existence [27 - 30]. Twins form especially readily in YBCO superconductors inasmuch as their unit cell is only close to the orthorhombic one. The twins can be spaced up to several nanometers apart, so even single crystal may have the fine substructure caused by twins. Magnetic and transport properties of HTS depend strongly on the orientation of twin planes with respect to the applied magnetic field [31, 32]. The flux can easily move along the weak links formed by twins [33 - 36]. At last, on a macroscopic scale there are manifold structural defects which can form weak links: that may be grain or crystallite boundaries as well as barriers arising from the secondary degrading the non-stoichiometric crystal into the domains with a high and low content of oxygen [26, 37 - 40]. Moreover, a magnetic field further reduces a coherence length, thus resulting in more easy weak links formation [41]. In conventional low-temperature superconductors weak links can be formed due to the proximity effect at the sites of minimum distance between the nearest normal-phase clusters.

As soon as the transport current is turned on, this one is added to all the persistent currents, maintaining the constant trapped magnetic flux. The situation is sketched out in Fig.4. Each of these currents (like I_Φ in Fig.4) is circulating through the superconducting loop around the normal-phase cluster wherein the corresponding portion of the magnetic flux is locked (Fig.4,a). The loop contains weak links that join the adjacent normal-phase clusters. As the transport current is increased, there will come a point when the overall current flowing through the weak link will exceed the critical value, so this link will switch into a resistive state. During this process the space distribution of the currents throughout the superconducting cluster is changed in such a way that the resistive subcircuit will be shunted by the superconducting paths where weak links are not damaged yet. Magnetic field created by this re-distributed transport current acts via the Lorentz force on the current circulating around the normal-phase cluster. As a result, the magnetic flux trapped therein will be forced out through the resistive weak link, which has become permeable to the vortices (Fig.4,b). So the cluster will not contain the magnetic flux any more.

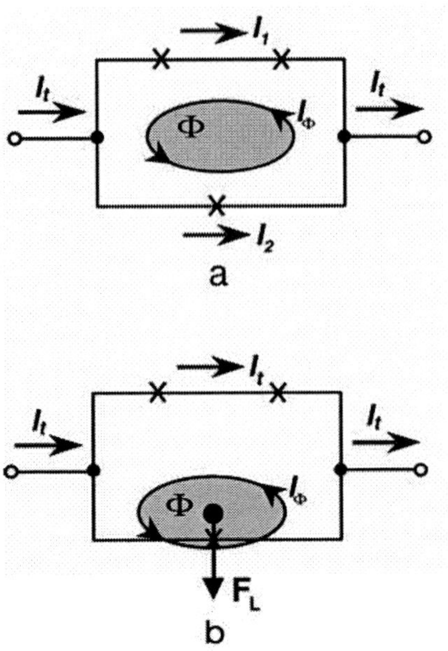

Fig.4. A simple illustration how the magnetic flux leaves a normal-phase cluster through the weak link. (a) - magnetic flux remains to be locked in the cluster as long as $I_2+I_\Phi<I_c$, where I_c is the critical current of the weak link; (b) - as soon as $I_2+I_\Phi\geq I_c$ the Lorentz force expels the flux through the right weak link which has become resistive. $I_t=I_1+I_2$ is the transport current; I_Φ is the current related to the trapped magnetic flux Φ; F_L is the Lorentz force.

Thus, whatever the microscopic nature of weak links may be, they form the channels for vortex transport. It appears that according to their configuration each normal-phase cluster has its own value of the critical current, which contributes to the overall statistical distribution. By the critical current of the cluster we mean the current of depinning, namely, such a current at which the magnetic flux ceases to be held inside the cluster of a normal phase. When a transport current I is gradually increased, the vortices will break away first from clusters of weaker pinning force, and therefore, of smaller critical current. Thus the decrease in the trapped magnetic flux is proportional to the number of all the normal-phase clusters of critical currents less than a preset value. Therefore, the relative change in the trapped flux can be expressed with the cumulative probability function $F(I)=\Pr\{\forall I_j<I\}$ for the distribution of the critical currents of clusters:

$$\frac{\Delta\Phi}{\Phi}=F(I) \tag{2}$$

The critical current distribution $F=F(I)$ is related to the cluster area distribution $W=W(A)$, because the cluster of a larger size has more weak links over its boundary with the surrounding superconducting space, and, consequently, the smaller depinning current.

3 Geometric Probability Analysis of the Distribution of Weak Links over the Cluster Boundary

In order to find out the relationship between the distribution of the critical currents of the clusters and the distribution of their areas, the distribution of entry points into weak links over the boundary of a normal-phase cluster should be analyzed. We will examine the geometric properties of the normal-phase clusters in the cross-section by the plane where the transport current flows. The vortices are moving transversally to this plane so in order to leave the normal-phase cluster they have to get over the barrier at the boundary of that cross-section. We will also suppose that the boundaries of the cluster cross-section are statistically self-similar.

The problem of exit of a vortex from a normal cluster represents the two-dimensional analogue of a problem of a random walk particle reaching a border [42, 43]. At the same time, unlike the classic problem on the distribution of the exit points, here the boundary of the area is not absorbing all over. In order to leave the cluster the vortex has to enter into one of the weak links, which are randomly arranged along the cluster perimeter. There are only discrete absorption points, which are located just at the sites where weak links are going out on the cluster boundary. In other words, these points are the points of the entry of vortices into weak links, or simply, entry points. For simplicity it will have been supposed that after the vortex reaches the entry point, it passes all the way between two adjacent normal-phase clusters without being trapped inside the weak link itself. Here the magnetic flux is transferred by Josephson vortices. The Josephson penetration depth is large enough in the considered materials, so the size of the region, where the vortex is localized, much exceeds the characteristic length of all possible structural defects that can occur along the transport channel. Thus the probability that such a vortex, driven by the Lorentz force, will be trapped in passing through a weak link is very small. This assumption agrees well with the results of research on the magnetic flux motion along weak links [36, 44, 45] including twins [33 - 35, 46, 47]. At the same time, it allows us to highlight the role played by the cluster boundary in the magnetic flux dynamics.

Let us consider the distribution of entry points over perimeter of a normal-phase cluster. Generally, this distribution varies from one cluster to another, so that each normal-phase cluster has the entry point distribution function $\psi(l)$ of its own, which belongs to some function class Ω. Here l is the co-ordinate measured along the cluster perimeter, so $l \in (0,P)$. In this context the functions of class Ω are random elements of the statistical distribution. The probability distribution of functions $\psi(l)$ over all the clusters can be characterized by the functional $\Pr\{\psi(l)\}$, which is equal to the probability of finding a given function $\psi(l)$.

In the most general way the geometric probability analysis of the entry point into weak link distribution can be carried out by means of path integral technique [48]. The probability that there is the function $\psi(l)$ in class Ω may be expressed by the path integral

$$\Pr\{\psi(l)|\Omega\} = \int_{(\Omega)} D\psi(l) \Pr\{\psi(l)\}$$

Therefore, the most probable function of entry point distribution is the mean over all functions of class Ω

$$\Psi(l) \equiv \overline{\psi(l)} = \int\limits_{(\Omega)} D\psi(l)\psi(l)\Pr\{\psi(l)\} \tag{3}$$

The path integral Fourier transform on the probability functional $\Pr\{\psi(l)\}$ represents the characteristic functional

$$H[k(l)] = \frac{\int\limits_{(\Omega)} D\psi(l)\exp[i\oint dlk(l)\psi(l)]\Pr\{\psi(l)\}}{\int\limits_{(\Omega)} D\psi(l)\Pr\{\psi(l)\}} \tag{4}$$

where k=$k(l)$ are the functions of a reciprocal function set, and integration in the kernel $\exp[i\oint dlk(l)\psi(l)]$ is carried out over the cluster perimeter.

The characteristic functional is the path integration analog for the usual moment-generating function. The probability functional $\Pr\{\psi(l)\}$ can be written as an inverse path integral Fourier transform on the characteristic functional

$$\Pr\{\psi(l)\} = \int Dk(l)\exp[-i\oint dlk(l)\psi(l)]H[k(l)]$$

where the path integration is carried out on the reciprocal function space.

In the simplest case, when all the clusters are of equal entry point distribution, which coincides with the most probable one of Eq.(3), the probability functional $\Pr\{\psi(l)\}$ is zero for all $\psi(l)$ that differ from $\Psi(l)$, whereas $\Pr\{\Psi(l)\}=1$. Therefore, the characteristic functional of Eq.(4) becomes

$$H[k(l)] = \exp[i\oint dlk(l)\Psi(l)] \tag{5}$$

If all entry points had fixed co-ordinates l_j instead of the random ones, their distribution would be $\psi(l) = \beta\sum_{j=1}^{N}\delta(l-l_j)$, where N is the number of entry points along the cluster perimeter, $\delta(l)$ is Dirac delta function. The constant β is being chosen to normalize the distribution function $\psi(l)$ to unity: $\oint dl\,\psi(l) = 1$, so that βN=1.

Now suppose that all the points of entries into weak links are randomly distributed with uniform probability over the cluster perimeter, so the probability to find any j-th point within some interval dl_j is proportional to its length. In that case the characteristic functional of Eq.(4) takes the form

$$H[k(l)] = \frac{\oint ... \oint \prod_{j=1}^{N} dl_j \exp[i\beta \sum_{j=1}^{N} \oint dl k(l) \delta(l - l_j)]}{\oint ... \oint \prod_{j=1}^{N} dl_j} =$$

$$= \frac{1}{P^N} \oint ... \oint \prod_{j=1}^{N} dl_j \exp[i\beta \sum_{j=1}^{N} k(l_j)] = \frac{1}{P^N} \prod_{j=1}^{N} \oint dl_j e^{i\beta k(l_j)} \tag{6}$$

Expanding the function $e^{i\beta k(l)}$ in a power series at N>>1, and taking into account the condition βN=1, we may write

$$\frac{1}{P} \oint dl e^{i\beta k(l)} = \exp[i \frac{\beta}{P} \oint dl k(l)]$$

that, after substitution into Eq.(6), gives

$$H[k(l)] = \exp[i \frac{\beta N}{P} \oint dl k(l)] \tag{7}$$

The found characteristic functional of Eq.(7) has the form of Eq.(5) for the function of uniform distribution of entry points

$$\Psi(l) = \frac{1}{P} \tag{8}$$

This means that all the clusters have the same uniform distribution of the entry points of Eq.(8), for which the probability of finding a weak link at any point of the perimeter is independent of its position.

Let us suppose that concentration of entry points into weak links per unit perimeter length $n = \overline{N} / P$ is constant for all clusters, and all the clusters are statistically self-similar. In this case the mean number of entry points $\overline{N}$ along the cluster perimeter is proportional to its length:

$$\overline{N} = \oint n(l) dl = nP \tag{9}$$

Next step will consist in finding the relationship between the size of a cluster and its critical current. The pinning force corresponds to such a current at which the vortices start to break away from the cluster. As the transport current is increasing, the Lorentz force, which expels the magnetic flux, increases as well. The vortices start to leave the normal-phase cluster when the Lorentz force becomes greater than the pinning force. At the same time, growing in current will result in re-distribution of the magnetic flux, which will penetrate deeper and deeper into a transition layer on that side of the surrounding superconducting space where the Lorentz force is directed (see Fig.2,b). In order to leave the normal-phase cluster, vortices have to reach the entry points into weak links. The exit of the magnetic flux can be considered as the result of random walks of vortices driven by the Lorentz force, which is pushing them into weak links. The similar approach has been successfully applied in analyzing the magnetic flux penetration into SQUID arrays [44]. The mean number of the entry points $\overline{N}$ available on the cluster perimeter gives the probability measure of the random walk outcomes, which are favorable for vortex exit. In the case of uniform entry point distribution, from Eq.(9) it follows that $\overline{N} \propto P$, so the perimeter length also represents the probability measure of the amount of favorable outcomes for vortex to leave the cluster. The more entry points into weak links are accessible for random walk vortices, the smaller is the Lorentz force required to push the flux out. Hence, we may write the following relation between the critical current of the cluster and its geometric size:

$$I \propto \frac{1}{\overline{N}} \propto \frac{1}{P} \tag{10}$$

This expression is true for the simplest case of uniform distribution of entry points, which is assumed to be the same for all clusters. Such a simplification allows us to emphasize that in the case being considered the magnetic flux is held in the normal-phase cluster by its boundary.

Thus, to deal with the distribution function of Eq.(1), the relation between perimeter and area of clusters should be studied. It might be natural to suppose that the perimeter-area relation obeys the well known geometric formula: $P \propto \sqrt{A}$. However, it would be a very rough approximation, because this relation holds for Euclidean geometric objects only. As was first found in Ref.[12], the boundaries of normal-phase clusters can be fractal, i.e. can have non-Euclidean features. The fractal nature of such clusters exerts an appreciable effect on the magnetic flux dynamics in superconductors [13, 49, 50].

4 A Brief Introduction to Fractal Geometry

The notion of a fractal as an object of fractional dimension was first introduced by Mandelbrot [51] in 1967 and since then it has received a lot of applications in various domains of sciences [52 - 55]. This concept is closely connected to ideas of scaling and self-similarity. Self-similarity is invariance with respect to scaling; in other words, an invariance relative to multiplicative changes of scale. Whereas a usual periodicity is invariance with respect to additive translations, self-similarity is a periodicity in a logarithmic scale.

The simplest examples of self-similar objects are Cantor sets [52]. Such a set has measure zero and, at the same time, has so many elements that it is uncountable. The Cantor set may be constructed by such a way (see Fig.5). Let us draw a line segment, and erase its middle third. Then eliminate middle third of each remaining part, and so on. The first six steps of this procedure are shown in Fig.5. The resulting set obtained by endless erasing of the middle thirds of remaining intervals is called "Cantor dust" [52]. The Cantor dust forms a self-similar set: if we will take any line in Fig.5, leave out the left or right half and magnify the remainder three times, this will result in the line segment immediately above it. So the Cantor dust is invariant to scaling by a factor $s=3$.

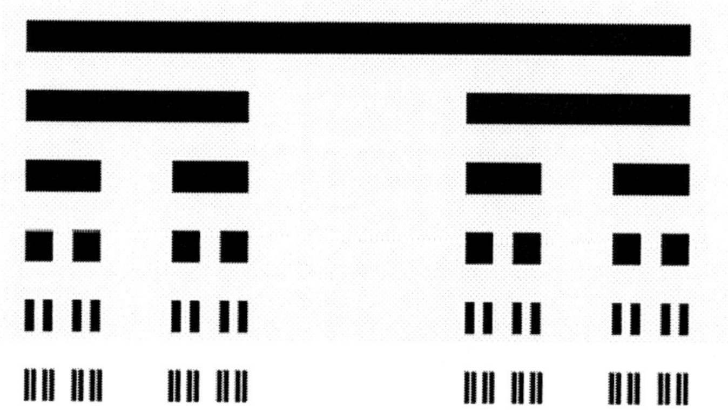

Fig.5. The first six steps of the Cantor dust construction.

This set is not only a self-similar one, but it is a fractal as well. Let us look at that part of the set that falls within one s^{th} part of the original set, where s is a scaling factor, and ask what fraction of the set falls into that portion? In other words, how many subsets, each is similar to the original set, are there if the length is subdivided into s parts? This number of such subsets is equal to $N=2$, so one-half of the original set falls into one third of the initial length. The fractal dimension is defined as the logarithm of the subset number divided by the logarithm of the scaling factor: $D=\ln N/\ln s=\ln 2/\ln 3\approx 0.631$. This formula represents the relation between the subset number and the scaling factor: $N=s^D$. Thus the fractal dimension of the Cantor dust is less than unity. This fact reflects its dust-like consistency compared to a usual line. A Euclidean line is such a set that if we change the length scale, we recover the same set of points. Hence the fractal dimension of a line coincides with the topological dimension, and both of them are equal to unity. Generally, a fractal object has a fractional dimension. A fractal set is such a set for which the fractal dimension strictly exceeds its topological dimension [53].

The Cantor dust gives an example of determinate fractal that is self-similar by deterministic construction. But the same approach can be applied to the stochastic fractals, which are statistically self-similar [54]. The fundamental feature of any fractals, both determinate and the stochastic ones, is that their characteristic measures obey the scaling law that includes an exponent called the fractal dimension.

Fractal dimension of statistically self-similar curve can be estimated by covering the curve by square grid. The size of the grid cell $L\times L$ should be small enough for measuring all the bends of the curve. If the whole curve (the boundary of the normal-phase cluster cross-section) goes in square of side L_{max}, then each of $N(L)$ self-similar subsets of that curve will go in square of side $L=L_{max}/s$, where s is a scaling factor. Let us suppose that the curve is self-similar at any scaling factor. The minimum number of such squares of side L needed to cover the explored curve is equal to

$$N(L) = s^D = \left(\frac{L_{\max}}{L}\right)^D \propto L^{-D}$$

where D is the fractal dimension of the curve (so-called coastline dimension [52]). So we have the formula that determines the fractal dimension

$$D = \frac{\ln N(L)}{\ln(L_{\max}/L)} \tag{11}$$

It is obvious that the fractal dimension of a curve is always below the topological dimension of a plane, which is equal to two.

The ratio between the size of measuring cell L and the big square side L_{max} sets the unit of measurement of length. At $L_{max}=1$ the expression for the fractal dimension of Eq.(11) takes the form $D = -\ln(N(L))/\ln(L)$.

Thus, in order to find the fractal dimension it is necessary to count the total number $N(L)$ of non-empty cells of area L^2, which cover the curve. To attain the highest precision of estimate we have to get wide range of L and to average obtained values of D over all the configurations.

The length of topologically one-dimensional fractal curve of size L_{max}, measured with an accuracy of L, is equal to a product of the number of cells, which cover that curve, and the linear size of the measuring cell

$$P = NL = \left(\frac{L_{\max}}{L}\right)^D L = \frac{L_{\max}^D}{L^{D-1}} \tag{12}$$

In other words, if the whole curve goes in the scale of length L_{max}, then in the smaller scale of $L=L_{max}/s$ (i. e. for the yardstick of a smaller size) the same curve will consist of $N=s^D$ pieces of length L. As is seen from the formula of Eq.(12), the length of a fractal curve is not well defined, because its value depends on the measurement accuracy, and even diverges as the yardstick size L is reduced infinitely. At the same time, the area restricted by any closed (including fractal) curve is well-defined finite quantity, which is proportional to its squared linear size: $A \propto L^2_{max}$. According to expression of Eq.(12), the length of the perimeter of the figure, formed by the closed fractal curve, is proportional to its linear size raised to a power of fractal dimension: $P \propto L^D_{max}$. Hence it follows that the perimeter of fractal figure and enclosed area has to obey the scaling law

$$P^{\frac{1}{D}} \propto A^{\frac{1}{2}} \tag{13}$$

The perimeter-area scaling law of Eq.(13) expresses the generalized Euclid theorem about measures of similar figures, stating that the ratios of corresponding measures are equal when reduced to the same dimension [53]. This assertion is valid both for the Euclidean figures and for the fractal ones.

The fractal approach has been found to be most useful in an investigation of inhomogeneous materials. There are many possibilities both the determinate fractals and the stochastic ones to be formed in composite superconductors. As an example of the first kind the multi-layered structures prepared by electron-beam deposition of superconductor (Nb) and normal metal (Cu) layers with fractal stacking sequence on sapphire substrates can be mentioned [56]. In order to obtain stochastic fractal clusters, it is essential that the process like the diffusion-limited aggregation would take place in the course of the synthesis of material [55]. The similar process can be realized, for instance, when thin films are evaporated. So, in Ref.[57] the films of fractal structure have been grown by vapor deposition of gold on silicon substrate with silicon nitride buffer sub-layer. It is worthy of note that porous, random, or highly ramified clusters do not necessarily all are fractals. A fractal cluster has such a property that its characteristic measures (in what follows - the perimeter and the enclosed area) have to obey the certain scaling law that includes an exponent named fractal dimension [52, 55].

The fractal clusters can be also formed in such highly inhomogeneous materials as high-temperature superconductors. So, the fractal dissipative regime has been observed in high-resolution measurements of dynamical resistance of BSCCO and BPSCCO composites containing normal-phase inclusions of Ag [11]. The fractal properties of the normal-phase clusters contained in YBCO films, which were prepared by magnetron sputtering on sapphire substrate with a cerium oxide buffer sub-layer, have been found in Ref.[12]. Percolation clusters are another example of fractals in superconductors [58, 59]. Although, mathematically, the percolation cluster is a fractal at the threshold point only, the fractal approach works well for any clusters which have a scaling feature [60 - 62]. In that case the normal-phase clusters may be formed by the inclusions of different chemical composition as well as by the domains of reduced superconducting order parameter. The existence of the fractal inclusions of this kind can be demonstrated by the fractal dissipation, which has been observed in non-textured polycrystalline YBCO and GdBCO bulk samples [11]. The fractal structure of clusters near the percolation threshold in epitaxial YBCO films has been fully considered in Ref.[63]. Of special interest are the works of Refs.[10, 64,] where the fractal penetration of magnetic flux in thin HTS films has been investigated by magneto-optical technique. Epitaxial TlBCO films were grown by magnetron sputtering on $SrTiO_3$ substrates, and YBCO films were prepared by pulsed laser deposition on $NdGaO_3$ substrates. The cluster structure of such films is clearly visible on the atomic force microscopy picture published in Ref.[10], whereas the magnetic flux, penetrating into the sample from the outside, has well-definite fractal front.

5 Fractal Geometry of Normal-Phase Clusters

The scaling law of Eq.(13) furnishes a clue to find relation between the critical current of the cluster and its geometric size. Using the formula of Eq.(10), which links the cluster size and its current of depinning, we can get the following expression: $I=\alpha A^{-D/2}$, where α is the form factor, and D is the fractal dimension of the cluster boundary. In the general way the cluster area distribution can be described by gamma distribution, which is appropriate for the most part of practically realizable cases [50, 65 - 67]. This distribution is characterized by the following cumulative probability function:

$$W(A) = \left[\Gamma(g+1)\right]^{-1} \gamma\left(g+1, \frac{A}{A_0}\right) \tag{14}$$

where $\Gamma(\nu)$ is Euler gamma function, $\gamma(\nu)$ is the incomplete gamma function, A_0 and g are the parameters of gamma distribution that control the mean area of the cluster $\overline{A} = (g+1)A_0$ and its variance $\sigma_A^2 = (g+1)A_0^2$.

In accordance with starting formulas of Eq.(1) and Eq.(2), gamma distribution of cluster areas of Eq.(14) gives rise to the critical current distribution of the form:

$$F(i) = \left[\Gamma(g+1)\right]^{-1} \Gamma\left(g+1, Gi^{-\frac{2}{D}}\right) \tag{15}$$

where

$$G \equiv \left[\frac{\theta^{\theta}}{\theta^{g+1} - \frac{D}{2}\exp(\theta)\Gamma(g+1,\theta)}\right]^{\frac{2}{D}}, \quad \theta \equiv g + 1 + \frac{D}{2},$$

$\Gamma(\nu, z)$ is the complementary incomplete gamma function, $i \equiv I/I_c$ is the dimensionless electric current, $I_c = \alpha(A_0 G)^{-D/2}$ is the critical current of the transition into a resistive state. The found cumulative probability function of Eq.(15) gives the probability density $f(i) \equiv dF/di$ for the critical current distribution:

$$f(i) = \frac{2G^{g+1}}{D\Gamma(g+1)} i^{-\frac{2}{D}(g+1)-1} \exp\left(-Gi^{-\frac{2}{D}}\right) \tag{16}$$

This distribution allows us to fully describe the effect of the transport current on the trapped magnetic flux taking into account the fractal properties of the normal-phase clusters. Let us note that the probability density for the critical current distribution of Eq.(16) is equal to zero at i=0, which implies the absence of any contribution from negative and zero currents. This will allow us to avoid any artificial assumption about the existence of a vortex liquid, having finite resistance in the absence of transport currents due to the presence of free vortices: $r(i \to 0) \neq 0$. Such an assumption has to be made, for example, in the case of normal distribution of critical currents [68].

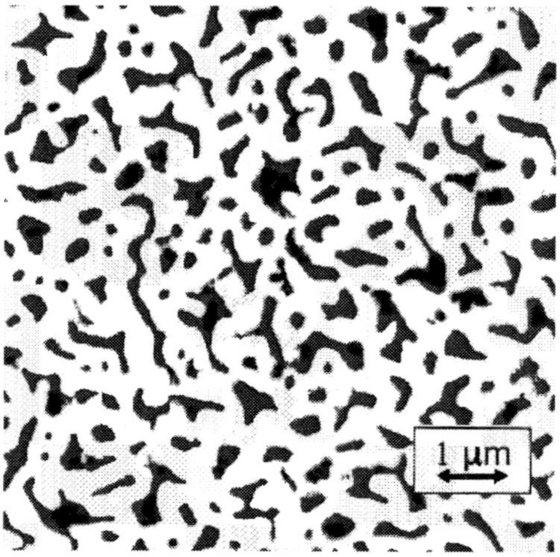

Fig.6. The electron photomicrograph of YBCO superconducting film with clusters of a normal phase (shown by black).

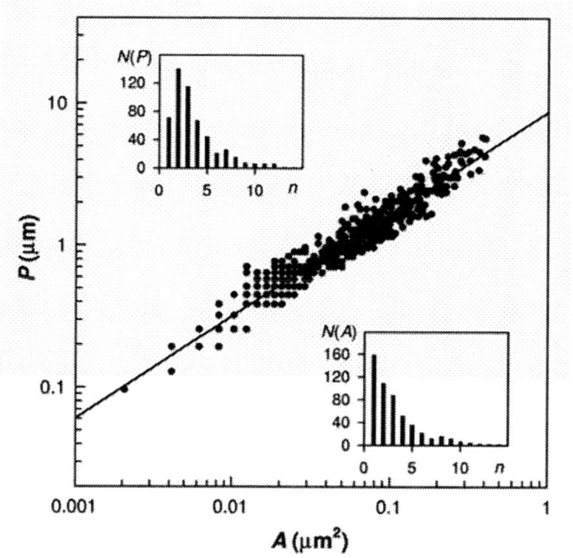

Fig.7. Perimeter-area relationship for the normal-phase clusters with fractal boundary. The chart is built for the primary sampling containing 528 normal-phase clusters. The insets represent histograms of sampling for corresponding distributions.

In order to clear up how the developed approach can be used in practice, the geometric probability analysis of electron photomicrographs of superconducting films was carried out. For this purpose electron photomicrographs of YBCO film prepared by magnetron sputtering on sapphire substrate with a cerium oxide buffer sub-layer have been scanned. One of such photomicrographs is presented in Fig.6. The normal-phase clusters are shown there by black. The normal phase has occupied 20% of the total surface only, so the transport current can flow through the sufficiently dense percolation superconducting cluster. The perimeters and areas of clusters have been measured by covering their digitized pictures with a square grid of spacing 60×60 nm^2. The results of the statistical treatment of these data are presented in Fig.7.

The primary sampling has contained 528 normal-phase clusters located on the scanned region of a total area of 200 μm^2. The distribution of the cluster areas is fitted well to exponential cumulative probability function, as is shown on the histogram in the lower inset of Fig.7. The number of clusters that fall within the assigned rank is plotted on the ordinate of this graph; the rank number is plotted on the abscissa. A high skewness (1.765) as well as the statistically insignificant (5%) difference between the sample mean area of the cluster and the standard deviation also attests that there an exponential distribution of the cluster areas holds true. This distribution is a special case of gamma distribution for which g=0, so the cumulative probability function of Eq.(14) can be simplified to the following form:

$$W(A) = 1 - \exp\left(-\frac{A}{\overline{A}}\right) \tag{17}$$

The exponential distribution has only one characteristic parameter - the mean cluster area $\overline{A}$.

The obtained data allow us to find the perimeter-area relation for the normal-phase clusters. All the points in Fig.7 fall on a straight line in double logarithmic scale with correlation coefficient of 0.929. Accordingly to the scaling relation of Eq.(13), the slope of the regression line gives the estimate of fractal dimension of the cluster perimeter, which is equal to D = 1.44±0.02. The graph in Fig.7 shows that the scaling law for perimeter and area, which is inherent to fractals, is valid in the range of almost three orders of magnitude in cluster area. The perimeter-area scaling behavior means that there is no characteristic length scale in the range of two orders of linear size of the normal-phase cluster. Whatever the shape and size of the clusters may be, all the points fall closely on the same straight line in logarithmic scale; so that there are no apparent kinks or bends on the graph. This point that the found value of the fractal dimension differs appreciably from unity engages a great attention. What this means is that the fractal properties of the cluster boundary are of prime importance here.

The geometric probability properties of the normal-phase clusters are responsible for main features of the critical current statistical distribution. Now, knowing the fractal dimension of the cluster boundaries, the change in the trapped magnetic flux caused by the transport current can be found with aid of Eq.(2). The exponential distribution of the cluster areas of Eq.(17) gives rise to the exponential-hyperbolic distribution of critical currents

$$F(i) = \exp\left[-\left(\frac{2+D}{2}\right)^{\frac{2}{D}+1} i^{-\frac{2}{D}}\right] \tag{18}$$

which follows from Eq.(15) at g=0. The effect of a transport current on the trapped magnetic flux is illustrated in Fig.8 for the case of Euclidean clusters (dotted curve), and for the clusters of found fractal dimension D=1.44 (solid curve).

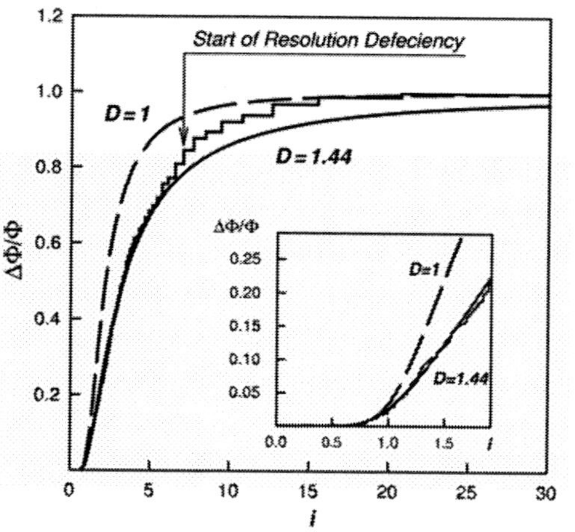

Fig.8. Crossover between the fractal and Euclidean regimes on dependence of the magnetic flux trapped in fractal clusters of a normal phase on a transport current. The solid line shows the decrease in trapped flux for the fractal clusters of boundary dimension $D = 1.44$; the dotted line corresponds to the case of Euclidean clusters ($D = 1$); step line is the sample empirical function of critical current distribution.

In order to get the relationship between the dynamics of the trapped magnetic flux and the geometric properties of the superconducting structure the empirical function of the distribution of the critical currents $F^* = F^*(i)$ has been found. First, the empirical distribution function $W^* = W^*(A)$ for the sampling of the areas of the normal-phase clusters has been obtained. The value of $W^*(A)$ was calculated for each order statistic as the relative number of clusters of area smaller than a given value A. Next, the empirical distribution of the critical currents was computed for the same order statistics using the formulas:

$$\left.\begin{aligned} F^* &= 1 - W^* \\ i &= \left(\frac{2+D}{2}\right)^{\frac{2+D}{2}} \left(\frac{\overline{A}}{A}\right)^{\frac{D}{2}} \end{aligned}\right\}$$

This function, shown in Fig.8 by the step line, gives a statistical estimate of the cumulative probability function of Eq.(18). As is seen from this figure, both distributions coincide well in the range of currents $i<6$. Starting with the value of the current $i=6$ the crossover from the fractal regime to the Euclidean one is observed. This transition into the Euclidean regime is over at large transport currents, when the magnetic flux changes mainly for the breaking of the vortices away from the small clusters (as the smaller clusters have the larger pinning force). The observed crossover has its origin in the finite resolution capability of measuring the cluster sizes. When estimating the fractal dimension, we have to take into account that the resolution of the measurement of any geometric sizes is finite. The peculiarity of the topologically one-dimensional fractal curve is that its measured length depends on the measurement accuracy [52]. In our case such a fractal curve is represented by the boundary of the normal-phase cluster. That is why just the statistical distribution of the cluster areas, rather than their perimeters, is fundamental for finding the critical current distribution. The topological dimension of perimeter is equal to unity and does not coincide with its fractal dimension, which strictly exceeds the unity. Therefore the perimeter length of a fractal cluster is not well defined, because its value depends on the yardstick size. On the other hand, the topological dimension of the cluster area is the same as the fractal one (both are equal to two). Thus, the area restricted by the fractal curve is a well-defined quantity.

Taking into account the finite resolution effect, the perimeter-area relationship of Eq.(13) can be re-written in the following form

$$P(\delta) \propto \delta^{1-D}\left[A(\delta)\right]^{\frac{D}{2}} \tag{19}$$

where δ is the yardstick size used to measure this length, (1-D) is the Hausdorff codimension for the Euclidean one-dimensional space. This relation holds true when the yardstick length is small enough to measure accurately the boundary of all smallest clusters in the sampling. When the resolution is deficient, the Euclidean part of the perimeter length will dominate the fractal one, so there is no way to find the fractal dimension using the scaling relation of Eq.(19). It means that if the length of a fractal curve was measured too roughly with the very large yardstick, its fractal properties could not be detected, and therefore such a geometrical object would be manifested itself as Euclidean one. It is just the resolution deficiency occurs at the crossover point in Fig.8. Starting with the cluster area less than some value (which corresponds to the currents of i>6) it is impossible to measure all "skerries" and "fjords" on the cluster coastlines, whereas all the clusters of area less than the size of the measuring cell (3600 nm^2 that relates to the currents of i>23), exhibit themselves as objects of Euclidean boundaries. This resolution deficiency can be also observed in Fig.7: some points at its lower left corner are arranged discretely with the spacing equal to the limit of resolution (60 nm), because some marks for smallest clusters coincide for the finite resolution of the picture digitization procedure.

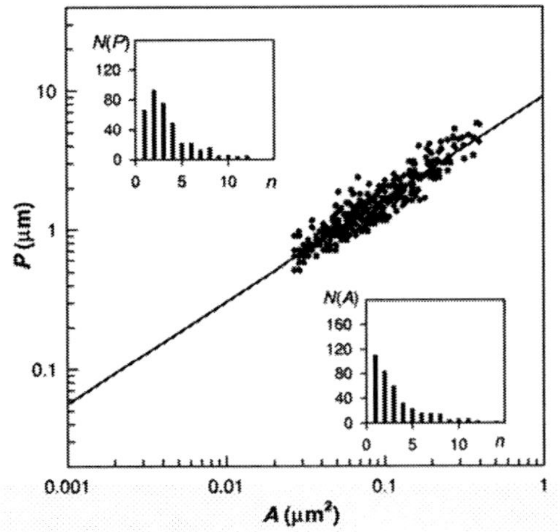

Fig.9. Perimeter-area relationship for the normal-phase clusters with fractal boundary. The chart is built for the truncated sampling containing 380 normal-phase clusters. The insets represent histograms of sampling for corresponding distributions.

The fractal dimension was found above by means of regression analysis of the whole primary sampling, where the very small clusters of sizes lying at the breaking point of the resolution limit were also included. In order to evaluate how the finite resolution affects the accuracy of the estimation of the fractal dimension, all the points, for which the resolution deficiency was observed, were eliminated from the primary sampling. So the truncated sampling has been formed in such a way that only 380 clusters, for which the resolution deficiency is not appeared yet, have been selected from the primary sampling. The regression analysis of the truncated sampling is presented in Fig.9. The least squares treatment of perimeter-area data gives the adjusted magnitude of the fractal dimension: $D=1.47\pm0.03$. The found value virtually does not differ from the previous one within the accuracy of the statistical estimation, whereas the correlation coefficient (which becomes equal to 0.869) falls by six hundredth only. Therefore, we can conclude that the found estimate of the fractal dimension is robust.

6 The Pinning Gain for the Magnetic Flux Trapped in Fractal Clusters of a Normal Phase

It has been revealed that the fractality of the cluster boundaries intensifies the pinning. As can be seen from Fig.10, the decrease in the trapped magnetic flux at the same value of the transport current is less for larger fractal dimension. The relative change in the trapped flux $\Delta\Phi/\Phi$, which, according to formula of Eq.(2), can be calculated from Eq.(18), defines the density of vortices n broken away from the pinning centers by the current i:

$$n(i) = \frac{B}{\Phi_0}\int_0^i f(i')di' = \frac{B}{\Phi_0}\frac{\Delta\Phi}{\Phi} \tag{20}$$

where B is the magnetic field, $\Phi_0 \equiv hc/(2e)$ is the magnetic flux quantum (h is Planck's constant, c is the velocity of light, e is the electron charge). The graph of the change in the trapped flux as a function of transport current, shown in Fig.10, coincides qualitatively with the curves obtained in experiments on the magnetization of YBCO films subjected to current pulses [69, 70]. Figure 10 also displays such an important property of superconducting structure containing fractal clusters of a normal phase: the fractality intensifies the magnetic flux trapping, hindering its breaking away from pinning centers, and thereby enhances the critical current which sample is capable to withstand, remaining in a superconducting state. The pinning amplification can be characterized by the pinning gain factor

$$k_\Phi \equiv 20\log_{10}\frac{\Delta\Phi(D=1)}{\Delta\Phi(D)}\text{, dB} \tag{21}$$

which is equal to relative decrease in the fraction of magnetic flux broken away from fractal clusters of fractal dimension D compared to the case of Euclidean ones. Dependencies of the pinning gain on the transport current for different fractal dimension at $g=0$ are shown in the inset of Fig.10. The highest amplification (about 10 dB) is reached when the cluster boundaries have the greatest possible fractality. Figure 10 demonstrates that with increase in fractal dimension the trapped magnetic flux is changed less and less by the action of the transport current. The pinning gain of Eq.(21) characterizes the properties of a superconductor in the range of the transport currents corresponding to a resistive state ($i > 1$). At smaller current the total trapped flux remains nearly unchanged (see Fig.10) for lack of pinning centers of such small critical currents, so the breaking of the vortices away has not started yet. When the vortices start to leave the normal-phase clusters and move through the weak links, their motion induces an electric field, which, in turn, creates the voltage drop across the sample. Therefore, the passage of electric current is accompanied by the energy dissipation. As for any hard superconductor (that is to say, type-II, with pinning centers) this dissipation does not mean the destruction of phase coherence yet. Some dissipation always accompanies any motion of a magnetic flux that can happen in a hard superconductor even at low transport current. Therefore the critical current in such materials cannot be specified as the greatest non-dissipative current. Superconducting state collapses only when a growth of dissipation becomes avalanche-like as a result of thermo-magnetic instability.

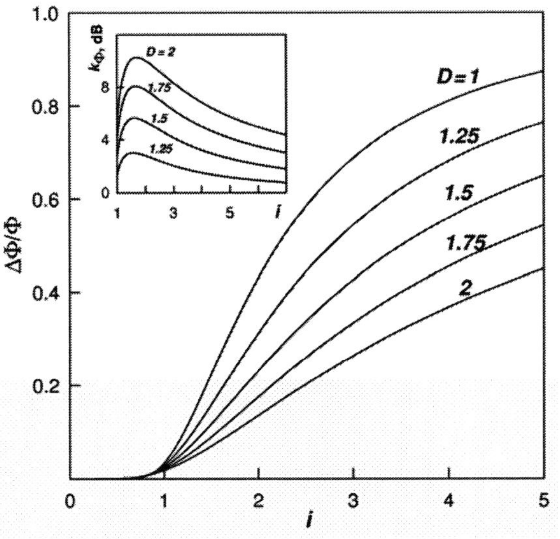

Fig.10. Effect of a transport current on the magnetic flux trapped in fractal clusters of a normal phase. The inset shows the pinning gain for the different fractal dimension of the cluster boundary.

The principal reason of pinning enhancement due to the fractality of the normal-phase clusters lies in the fundamental properties of the critical current distribution. Figure 11 demonstrates the peculiarities of the fractal probability density specified by Eq.(18). As may be seen from this graph, the bell-shaped curve of the distribution broadens out, moving towards greater magnitudes of current as the fractal dimension increases. It means that more and more of the small clusters, which can trap the vortices best, are being involved in the game. Hence the number of vortices broken away from pinning centers by the Lorentz force is reducing, so the smaller part of a magnetic flux can flow. The shift of the critical current distribution towards higher currents can be described by dependencies of the mean critical current distribution on the fractal dimension, as it is shown in the inset of Fig.11. The mean critical current obeys the strong super-linear law specified by Euler gamma function:

$$\bar{i} = \left(\frac{2+D}{2}\right)^{\frac{2+D}{2}} \Gamma\left(1-\frac{D}{2}\right) \tag{22}$$

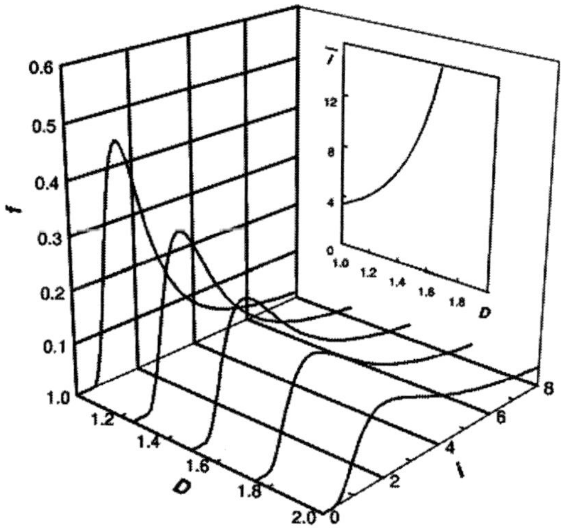

Fig.11. Influence of the fractal dimension of the boundary of the normal-phase clusters on the critical current distribution. The inset shows the dependencies of the mean critical current $\bar{i}$ on the fractal dimension.

Figure 11 clearly demonstrates that increasing the fractal dimension gives a growth of the contribution made by clusters of greater critical current to the overall distribution, resulting just in enhancement of the magnetic flux trapping.

7 Anomalous Statistical Properties of the Critical Current Fractal Distribution

In the general case of gamma distribution of the normal-phase cluster areas of Eq.(14) the probability density for the cluster areas has the form

$$w(A) = \frac{A^g \exp\left(-\frac{A}{A_0}\right)}{\Gamma(g+1)A_0^{g+1}} \tag{23}$$

For further consideration it is convenient to introduce the dimensionless area of the cluster $a \equiv A/\bar{A}$, for which the distribution function of Eq.(23) can be rewritten as:

$$w(a) = \frac{(g+1)^{g+1}}{\Gamma(g+1)} a^g \exp[-(g+1)a] \tag{24}$$

The mean dimensionless area of the cluster is equal to unity, whereas the variance is determined by one parameter g only: $\sigma_a^2 = 1/(g+1)$. The probability density of Eq.(24) is presented in Fig.12 for some characteristic values of g-parameter. The critical current distribution has the exponential-hyperbolic form of Eq.(16), for which the mean critical current, instead of Eq.(22), is

$$\bar{i} = G^{\frac{D}{2}} \frac{\Gamma\left(g+1-\frac{D}{2}\right)}{\Gamma(g+1)} \tag{25}$$

The critical current distribution has the skew bell-shaped form with inherent broad "tail" extended over the region of high currents (see Fig.11 as well as the inset of Fig.12). The mean critical current of Eq.(25), just as mean current of Eq.(22) in the special case of exponential distribution of the normal-phase clusters areas, is divergent in the range of fractal dimensions $D \geq 2(g+1)$. The region of the mean critical current divergence is shown in Fig.13,a. The divergent dependencies of the mean critical current on the fractal dimension for different values of g-parameters of gamma distribution are presented in Fig.14.

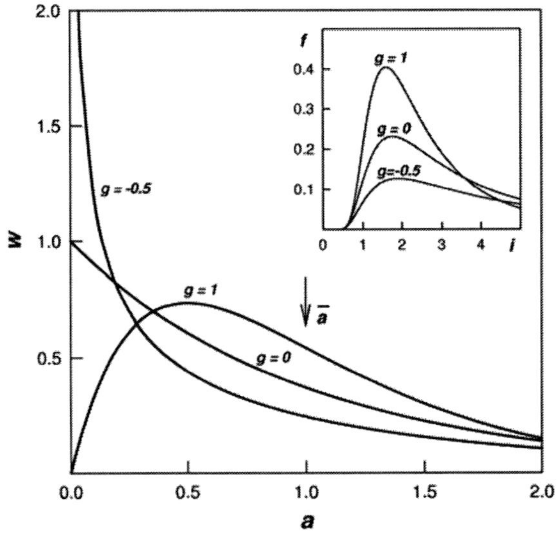

Fig.12. The distribution of the areas of normal-phase clusters at different values of g-parameter of gamma distribution. The inset shows the related critical current distributions for the fractal dimension of the cluster boundary $D = 1.5$.

As may be seen from the inset of Fig.12, in the case of gamma distribution, the critical current distribution spreads out with some shift to the right as g-parameter decreases. This broadening can be characterized by the standard deviation of critical currents

$$\sigma_i = G^{\frac{D}{2}} \sqrt{\frac{\Gamma(g+1-D)}{\Gamma(g+1)} - \left[\frac{\Gamma(g+1-D/2)}{\Gamma(g+1)}\right]^2} \tag{26}$$

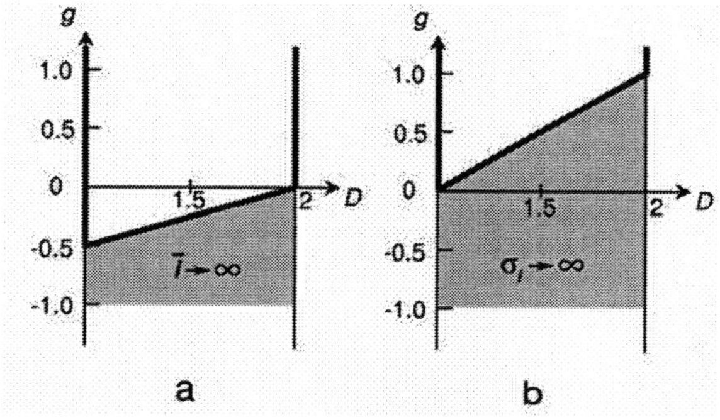

Fig.13. The regions of the divergence (a) of the mean critical current and (b) the standard deviation of critical currents of the fractal distribution of the critical currents.

The standard deviation grows nonlinearly with increase in the fractal dimension, as is illustrated in Fig.15. The peculiarity of the distribution of Eq.(16) is that its variance becomes infinite in the range of fractal dimensions $D \geq g+1$ (see Fig.13,b). At the same time, the mode of the distribution $mode f(i)=(G/\theta)^{D/2}$ remains finite for all possible values of the fractal dimension $1 \leq D \leq 2$. The distributions with divergent variance are known in probability theory - the classic example of that kind is Cauchy distribution [71]. However, such an anomalous feature of exponential-hyperbolic distribution of Eq.(16) is of special interest, inasmuch as the current-carrying capability of a superconductor would be expected to increase in the region of giant variance. Then the statistical distribution of critical currents has a very elongated "tail" containing the contributions from the clusters of the highest depinning currents.

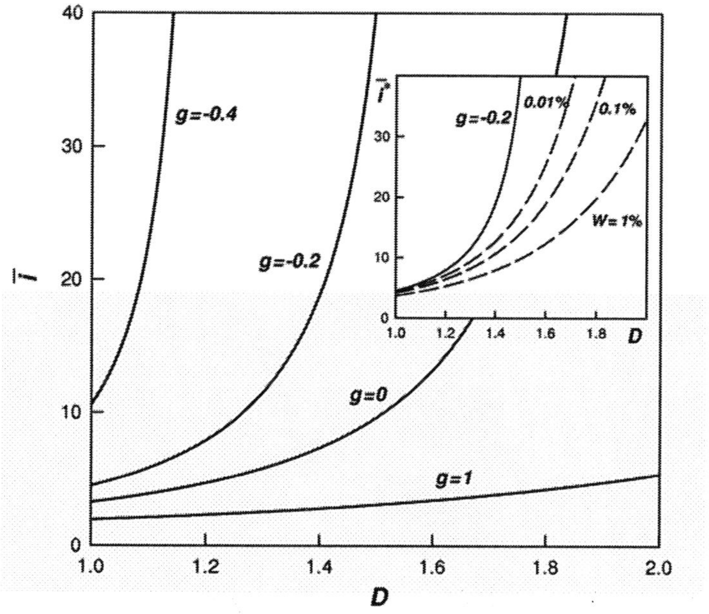

Fig.14. Influence of the fractal dimension of the cluster boundary on the mean critical current. The inset shows mean critical current graphs for truncated distribution at g = -0.2 with the different degree of truncation. The dotted lines are given for truncated distribution, the solid lines are for initial distribution.

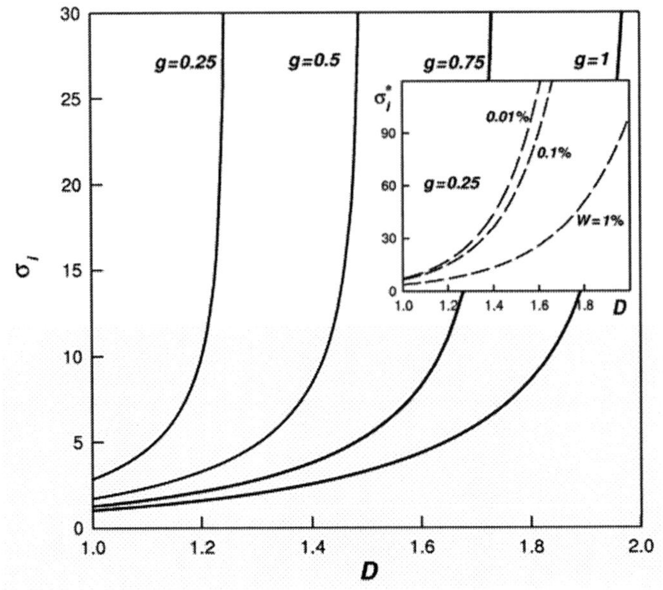

Fig.15. Influence of the fractal dimension of the cluster boundary on the standard deviation of critical currents. The inset shows standard deviation graphs for truncated distribution at g = -0.2 with the different degree of truncation. The dotted lines are given for truncated distribution, the solid lines are for initial distribution.

The reason for divergence of statistical moments of the exponential-hyperbolic critical current distribution of Eq.(16) consists in the behavior of the cluster area distribution of Eq.(24). As may be seen from Fig.12, the graph of this function takes essentially different shapes depending on the sign of g-parameter - from the skew unimodal curve at $g>0$ to the monotonic curve 3 with hyperbolic singularity at zero point at $g<0$. In the borderline case of $g=0$, which separates these different kinds of the functions, the distribution obeys the exponential law. It is just for negative values of g-parameter that the mean critical current diverges (Fig.13,a). The contribution from the clusters of small area to the overall distribution grows at $g<0$. Since the clusters of small size have the least number of weak links over a perimeter, they can best trap the magnetic flux. Therefore, an increase of the part of small clusters in the area distribution of Eq.(24) leads to a growth of the contribution with high depinning currents made by these clusters in the critical current distribution of Eq.(16). Just as a result of this feature the mean critical current diverges at $g<0$. Nevertheless, the total area between the curve of the probability density and the abscissa remains finite by virtue of the normalization requirement.

Obviously, the proper critical current cannot be infinitely high as well as the clusters of infinitesimal area do not really exist. There is the minimum value of the normal-phase cluster area A_m, which is limited by the processes of the film growth. So in YBCO based composites prepared by magnetron sputtering [12], the sample value of minimum area of the normal-phase cluster has been equal to A_m=2070 nm^2 at mean cluster area $\overline{A}$=76500 nm^2, that corresponds to the lower bound of the dimensionless area of the cluster $a_m \equiv A_m / \overline{A}$=0.027. In view of this limitation, we will describe the distribution of the normal-phase cluster areas by the truncated version of the probability density of Eq.(24):

$$w(a|a \geq a_m) = \frac{h(a - a_m)}{1 - W(a_m)} w(a) \tag{27}$$

where $h(x) \equiv \begin{cases} 1 & x \geq 0 \\ 0 & x < 0 \end{cases}$ is the Heaviside step function, and

$$W(a_m) \equiv \int_0^{a_m} w(a)da = \frac{\gamma[g+1,(g+1)a_m]}{\Gamma(g+1)} \tag{28}$$

is the truncation degree, which is equal to the probability $\Pr\{\forall a<a_m\}$ to find the cluster of area smaller than the least possible value a_m in the initial population. In the expression of Eq.(28) $\gamma(\nu, z)$ is the complementary gamma function.

The expression of Eq.(27) gives the conditional distribution of probability, for which all the events of finding a cluster of area less than a_m are excluded. Thus the truncation provides a natural way to fulfill our initial assumption that the cluster size has to be greater than the coherence and penetration lengths. New distribution of cluster areas gives rise to the truncated distribution of the critical currents:

$$f(i|i \leq i_m) = \frac{h(i_m - i)}{1 - W(a_m)} f(i) \tag{29}$$

where $i_m \equiv \left[\frac{G}{(g+1)a_m}\right]^{\frac{D}{2}}$ is the upper bound of the depinning current, which corresponds to the cluster of the least possible area a_m.

Then, instead of Eqs.(26) and (25), the standard deviation and mean critical current are

$$\sigma_i^* = G^{\frac{D}{2}} \sqrt{\frac{\Gamma[g+1-D,(g+1)a_m]}{\Gamma[g+1,(g+1)a_m]} - \left\{\frac{\Gamma\left[g+1-\frac{D}{2},(g+1)a_m\right]}{\Gamma[g+1,(g+1)a_m]}\right\}^2} \tag{30}$$

$$\bar{i}^* = \frac{\Gamma\left[g+1-\frac{D}{2},(g+1)a_m\right]}{\Gamma[g+1,(g+1)a_m]} G^{\frac{D}{2}} \tag{31}$$

For the truncated distribution of Eq.(29) the possible values of depinning currents are bounded from above by the quantity i_m, therefore the mean critical current as well as the variance do not diverge any more. Both of these characteristics are finite for any fractal dimensions, including the case of maximum fractality (D=2). The dotted lines in the insets of Figs.14, 15 show the corresponding graphs drawn for the same values of truncation degree. All the curves are calculated according to the formulas of Eqs.(30), (31) at g = -0.2 (see the corresponding curve in Fig.14). In this case the variance for initial distribution of the critical currents is infinite, so no graph for $W(a_m)$=0 is shown in the inset of Fig.15. The truncation degree is related to the least possible area of the cluster by the equation (28).

The degree of truncation gives the probability measure of the number of normal-phase clusters that have the area smaller than the lower bound a_m in the initial distribution. If the magnitude of $W(a_m)$ is sufficiently small (no more than several percent), then the procedure of truncation scarcely affects the initial shape of the distribution and still enables the contribution from the clusters of zero area (therefore, of infinitely high depinning current) to be eliminated. It is interesting to note that the very similar situation occurs in analyzing the statistics of the areas of fractal islands in the ocean [53]. The island areas obey the Pareto distribution, which also has the hyperbolic singularity at zero point that causes certain of its moments to diverge. For exponential distribution of the cluster areas, which is valid in the above-mentioned case of YBCO films [12], the probability to find the cluster of area smaller than a_m in the sampling is equal to $\Pr\{\forall a<a_m\}$=2.7% only. In principle, the truncation procedure could be made here, too, but there is no need for that, because the contribution of infinitesimally small clusters is finite at g=0 (and equal to zero at g>0).

Referring to Figs.14, 15 it can be seen that dependencies of the standard deviation and the mean critical current on the fractal dimension become smoother and smoother with increase in the degree of truncation. At the same time, the inherent tendency of initial distribution is still retained: as the fractal dimension increases, the critical current distribution broadens out (the variance grows, see the inset of Fig.15), moving towards higher currents (the mean critical current grows, see the inset of Fig.14). As may be seen from the inset of Fig.12, this trend is further enhanced with decreasing g-parameter. The most current-carrying capability of a superconductor should be achieved when the clusters of small size, which have the highest currents of depinning, contribute maximally to the overall distribution of the critical currents. Such a situation takes place just in the region of giant variance of critical currents. So far, the least magnitude of g-parameter (equal to zero) has been realized in YBCO composites containing normal-phase clusters of fractal dimension D=1.44 [12]. The critical currents of superconducting films with such clusters are higher than without them [8, 69, 70]. It would thus be expected to further improve the current-carrying capability in superconductors containing normal-phase clusters, which will be characterized by area distribution with negative magnitudes of g-parameter, especially at the most values of fractal dimensions.

8 Current-Voltage Characteristics and Vortex Dynamics near the Resistive Transition

By virtue of the fact that any motion of the magnetic flux causes the energy dissipation in superconductors, the question of how such a process could be prevented, or only suppressed, is of prime practical importance. The study of resistive state peculiarities leads to the conclusion that the cluster fractality affects the electric field induced by the flux motion. The found distribution of the critical currents allows us to find the electric field arising from the magnetic flux motion after the vortices have been broken away from the pinning centers. Inasmuch as each normal-phase cluster contributes to the total critical current distribution, the voltage across a superconductor can be represented as the response to the sum of effects made by the contribution from each cluster. Such a response can be expressed as a convolution integral

$$\frac{v}{r_f} = \int_0^i (i - i') f(i') di', \tag{32}$$

where v is the dimensionless voltage and r_f is the dimensionless flux flow resistance. The similar approach is used universally in all the cases where depinning currents obey a statistical law [68, 72, 73]. The subsequent consideration will be essentially concentrated on the consequences of the fractal nature of the normal-phase clusters specified by the distribution of Eqs.(15), (18), so all the problems related to possible dependence of the flux flow resistance r_f on a transport current will not be taken up here. The voltage across a superconductor V and flux flow resistance R_f are related to the corresponding dimensionless quantities v and r_f by the relationship: $V/R_f = I_c(v/r_f)$.

After substitution of the critical current distribution of Eq.(16) into the expression of Eq.(32), upon integration, we get the final expression for the current-voltage (V-I) characteristics in the general case of the gamma distribution of the cluster areas:

$$\frac{v}{r_f} = \frac{1}{\Gamma(g+1)} \left[i\Gamma\left(g+1, Gi^{-\frac{2}{D}}\right) - G^{\frac{D}{2}}\Gamma\left(g+1-\frac{D}{2}, Gi^{-\frac{2}{D}}\right)\right] \tag{33}$$

In the special case of exponential cluster area distribution (g=0) the general formula of Eq.(33) can be simplified:

$$\frac{v}{r_f} = i\exp\left(-Ci^{-\frac{2}{D}}\right) - C^{\frac{D}{2}}\Gamma\left(1-\frac{D}{2},\ Ci^{-\frac{2}{D}}\right) \tag{34}$$

where $C \equiv ((2+D)/2)^{2/D+1}$.

In extreme cases of Euclidean clusters and clusters of the most fractality the expression of Eq.(34) can be further transformed:

Clusters of Euclidean boundary (D=1 at g=0):

$$\frac{v}{r_f} = i\exp\left(-\frac{3.375}{i^2}\right) - \sqrt{3.375\pi}\,erfc\left(\frac{\sqrt{3.375}}{i}\right) \tag{35}$$

where $erfc(z)$ is the complementary error function.

Clusters of boundary with the maximum fractality (D=2 at g=0):

$$\frac{v}{r_f} = i\exp\left(-\frac{4}{i}\right) + 4Ei\left(-\frac{4}{i}\right) \tag{36}$$

where $Ei(z)$ is the exponential integral function.

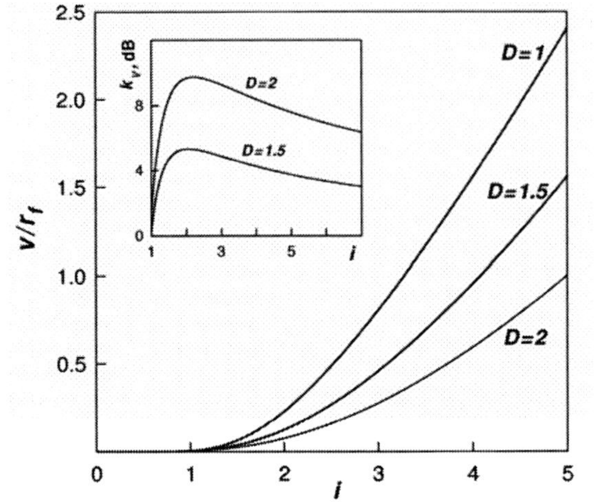

Fig.16. Current-voltage characteristics of superconductors with fractal clusters of a normal phase. The inset shows the voltage attenuation factor for the different fractal dimension of the cluster boundary.

The *V-I* characteristics calculated using the formulas of Eqs.(34) - (36) are presented in Fig.16. Two lines drawn for extreme case of $D = 1$ and $D = 2$ bound the region the *V-I* characteristics can fall within for any possible values of fractal dimension. Figure 16 shows that the fractality reduces appreciably an electric field arising from the magnetic flux motion. In the inset of Fig.16 the voltage attenuation factor $k_V \equiv 20\log_{10}(V(D=1)/V(D))$ is presented. This effect is especially strong in this range of the currents ($1<i<3$), where the pinning enhancement also has a maximum (see also Fig.10). Both these effects have the same nature, inasmuch as their reason lies in the peculiarity of the critical current distribution. The influence of the fractal dimension of the cluster boundary on the critical current distribution is demonstrated in Fig.11. An increase of fractality causes a significant broadening of the tail of the distribution $f = f(i)$. Therefore, more of small clusters, which have higher critical currents of depinning, are being involved in the process. Hence the smaller number of vortices can move, creating the smaller electric field. In turn, the smaller the electric field is, the smaller is the energy dissipated when the transport current passes through the sample. Therefore, the decrease in heat-evolution, which could cause transition of a superconductor into a normal state, means that the current-carrying capability of a superconductor containing such fractal clusters is enhanced.

Though the fractality of the clusters reduces the voltage arising from the vortex motion in the range of the currents $i>1$, the situation is quite different in the neighborhood of the resistive transition below the critical current. When $i<1$, the higher the fractal dimension of the normal-phase cluster is, the larger is the voltage across a sample and the more stretched is the region of initial dissipation in *V-I* characteristic. In Fig.17 this region is shown under magnification. As may be seen from this graph, the critical current i_c is preceded by some initial region of the finite voltage drop starting with i_{on}, so the resistive transition is not absolutely abrupt. The existence of this initial section on the *V-I* characteristic arises from the peculiarities of the fractal distribution in the range of small currents. It has been just a similar initial region of fractal dissipation has been observed in high-resolution measurements of dynamical resistance of HTS-normal metal composites [11]. The arrows in Fig.17 show the values of the onset current i_{on} at the voltage resolution of $10^{-5}v/r_f$. Dependence of the onset current on the fractal dimension of the cluster boundary is presented in the inset.

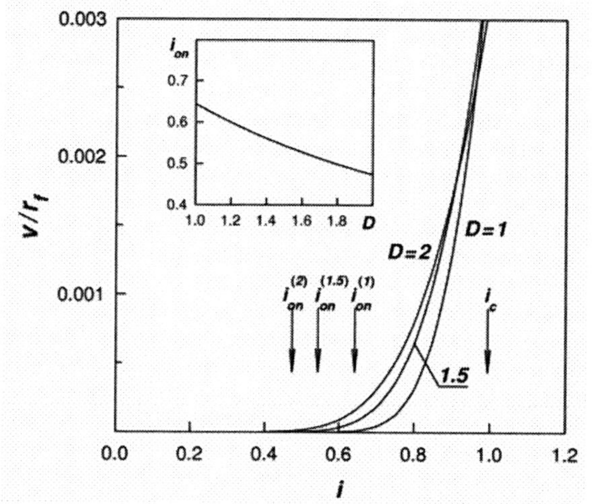

Fig.17. Initial region of current-voltage characteristic at different values of fractal dimension of the normal-phase clusters. The arrows show the values of the onset current i_{on} at the voltage resolution of $10^{-5}v/r_f$. Dependence of the onset current on the fractal dimension of the cluster boundary is presented in the inset.

The significant difference in *V-I* characteristic behavior below and above the resistive transition is related to dependence of free vortex density on the fractal dimension for various transport currents. The resistive characteristics provide additional information about the nature of a vortex state in superconductor with fractal clusters of a normal phase. Since the *V-I* characteristics of Eqs.(33), (34) are non-linear, *dc* (static) resistance is not constant and depends on the transport current. The more convenient parameter is the differential resistance $r_d = dv/di$, a small-signal *ac* parameter that gives the slope of the *V-I* characteristic. The corresponding dimensional quantity R_d can be found using the formulas $R_d = r_d R_f / r_f$. Figure 18 shows the graphs of the differential resistance as a function of transport current for superconductor with fractal normal-phase clusters. The curves drawn for the Euclidean clusters (D=1) and for the clusters of the most fractal boundaries (D=2) bound a region containing all the resistive characteristics for an arbitrary fractal dimension. As an example, the dashed curve shows the case of the fractal dimension D=1.5. The dependencies of resistance on the current shown in Fig. 18 are typical of the vortex glass, inasmuch as the curves plotted on a double logarithmic scale are convex and the resistance tends to zero as the transport current decreases, $r_d(i \to 0) \to 0$, which is related to the flux creep suppression [20, 68]. A vortex glass represents an ordered system of vortices without any long-range ordering. At the same time, the vortex configuration is stable in time and can be characterized by the order parameter of the glassy state [74, 75]. In the *H-T* phase diagram, mixed state of the vortex glass type exists in the region below the irreversibility line. The dashed horizontal line at the upper right of Fig.18 corresponds to a viscous flux flow regime ($r_d = r_f$=const), which can only be approached asymptotically.

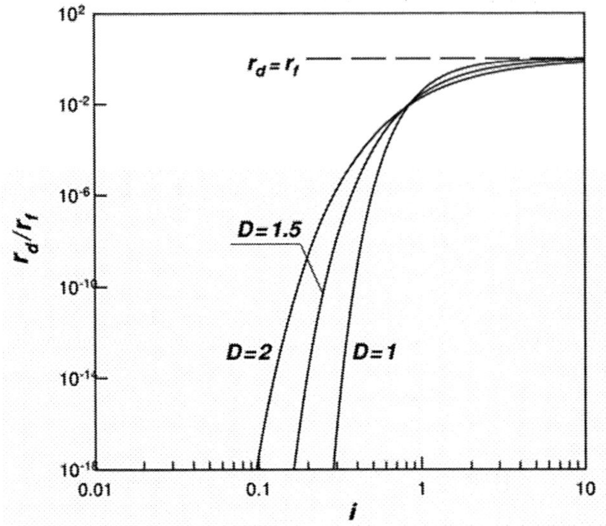

Fig.18. Dependence of the *dc* resistance on the transport current for superconductor with fractal normal-phase clusters. The dashed horizontal line $r = r_f$ at the upper right corresponds to the viscous flow of a magnetic flux.

The resistance of superconductor is determined by the density n of free vortices broken away from pinning centers of Eq.(20). In the special case of exponential cluster area distribution (g=0) the free vortex density is equal to

$$n == \frac{B}{\Phi_0} \exp\left(-Ci^{-\frac{2}{D}} \right) \qquad (37)$$

The more vortices are free to move, the stronger the induced electric field, and therefore, the higher is the voltage across a sample at the same transport current. If we differentiate the *V-I* characteristic of Eq.(34) and compare with formula of Eq.(37), we will see that the differential resistance is proportional to the density of free vortices: $r_d = (r_f \Phi_0 / B)n$. Resistance of a superconductor in the resistive state is determined by the motion just of these vortices.

Figure 19 demonstrates dependence of the relative density of free vortices $n(D)/n(D=1)$ (relatively to the value for clusters with Euclidean boundary) on the fractal dimension for different magnitudes of transport currents. The vortices are broken away from pinning centers mostly when $i>1$, that is to say, above the resistive transition. Here the free vortex density decreases with increasing the fractal dimension. Such a behavior can be explained by the peculiarity of the critical current distribution of Eq.(16). The curve of this distribution broadens out, moving towards greater magnitudes of current as the fractal dimension increases. It means that there are more clusters of high depinning current in superconductor. The smaller part of the vortices is free to move, the smaller the induced electric field. The fact that the fractality of cluster boundary enhances pinning is also demonstrated in Fig.18, where the resistance decreases with increasing fractal dimension above the resistive transition. The relative change in free vortex density depends on the transport current (see inset in Fig.19) and, in the limiting case of the most fractal boundary D=2 reaches a minimum for i=1.6875 (this curve should goes below others in Fig.19). That corresponds to the maximum pinning gain and the minimum level of dissipation.

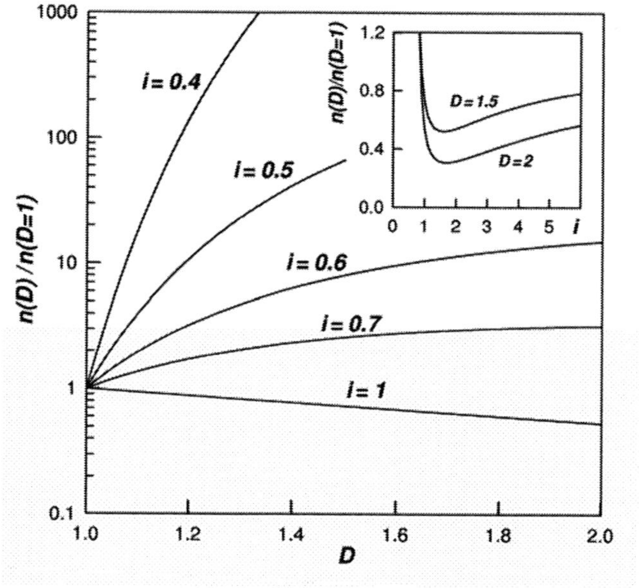

Fig.19. Dependence of the free vortex density on the fractal dimension of the normal-phase clusters for different values of a transport current. The inset shows the free vortex density versus current for two values of a fractal dimensions.

Below the resistive transition (i<1) the resistance as well as the free vortex density increase for the clusters of greater fractal dimension (see Figs.18 and 19). Such a behavior is related to the fact that the critical current distribution of Eq.(16) broadens out covering both high and small currents as the fractal dimension increases. For this reason, the breaking of the vortices away under the action of transport current begins earlier for the clusters of greater fractal dimension. In spite of sharp increase in relative density of free vortices (Fig.19), the absolute value of vortex density in the range of currents involved is very small (much smaller than above the resistive transition). So the vortex motion does not lead to the destruction of the superconducting state yet, and the resistance remains very low. The low density of vortices at small currents is related to the peculiarity of exponential-hyperbolic distribution of Eq.(16). This function is so "flat" in the vicinity of the coordinate origin that all its derivatives are equal to zero at the point of i=0: $d^k f(0)/di^k=0$ for any value of k. This mathematical feature has a clear physical meaning: so small a transport current does not significantly affect the trapped magnetic flux because there are scarcely any pinning centers of such small critical currents in the overall statistical distribution, so that nearly all the vortices are still pinned. This interval corresponds to the so-called initial fractal dissipation regime, which was observed in BPSCCO samples with silver inclusions as well as in polycrystalline YBCO and GdBCO samples [11].

9 Conclusion

So, in the present work the fractal nature of the normal-phase clusters is revealed, and relation between the fractal properties of the clusters and dynamics of the trapped magnetic flux is established. The fractal distribution of the critical current is obtained. It is found that the fractality of cluster boundary strengthens the flux pinning and thereby hinders the destruction of superconductivity by the transport current. *V-I* characteristics of fractal superconducting structures in a resistive state are obtained. It is revealed that the fractality of the boundaries of the normal-phase clusters reduces the electric field arising from magnetic flux motion, and thereby raises the critical current of a superconductor. The fractal properties of the normal-phase clusters significantly affect the resistive transition. This phenomenon is related to the features of the fractal critical current distribution. The resistance characteristics correspond to a mixed state of the vortex glass type. An important result is that the fractality of the normal-phase clusters enhances pinning, thus decreasing the resistance above the resistive transition and increasing the current-carrying capability of a superconductor.

This work is supported by the Saint Petersburg Scientific Center of the Russian Academy of Sciences.

References

[1] Tonomura A.; Kasai H.; Kamimura O.; Matsuda T.; Harada K.; Nakayama Y.; Shimoyama J.; Kishio K.; Hanaguri T.; Kitazawa K.; Sasase M.; Okayasu S. *Nature* 2001, 412, 620-622.

[2] Lidmar J.; Nelson D. R.; Gorokhov D. A. *Phys. Rev. B* 2001, 64, 144512(1-10).

[3] Dam B.; Huijbregtse J. M.; Rector J. H. *Phys. Rev. B* 2002, 65, 064528(1-8).

[4] Higuchi T.; Yoo S. I.; Murakami M. *Phys. Rev. B* 1999, 59, 1514-1527.

[5] Mezzetti E.; Gerbaldo R.; Ghigo G.; Gozzelino L.; Minetti B.; Camerlingo C.; Monaco A.; Cuttone G.; Rovelli A. *Phys. Rev. B* 1999, 60, 7623-7630.

[6] Beasley M. R. In *Percolation, Localization and Superconductivity*; Goldman A. M; Wolf S. A.; Ed.; NATO ASI Series, Ser.B; Plenum Press: New York, NY, 1984; Vol. 109, pp 115-143.

[7] Krusin-Elbaum L.; Blatter G.; Thompson J. R.; Petrov D. K.; Wheeler R.; Ullmann J. Chu C. W. *Phys. Rev. Lett.* 1998, 81, 3948-3951.

[8] Kuzmin Yu. I. *Tech. Phys. Lett.* 2000, 26, 791-793 [Translated from: *Pis'ma Zh. Tekh. Fiz.* 2000, 26, 81-87].

[9] Olson C. J.; Reichhardt C.; Nori F. *Phys. Rev. Lett.* 1998, 80, 2197-2200.

[10] Surdeanu R.; Wijngaarden R. J.; Dam B.; Rector J.; Griessen R.; Rossel C.; Ren Z. F.; Wang J.H. *Phys. Rev. B* 1998, 58, 12467-12477.

[11] Prester M. *Phys. Rev. B* 1999, 60, 3100-3103.

[12] Kuzmin Yu. I. *Phys. Lett. A* 2000, 267, 66-70.

[13] Kuzmin Yu. I. *Phys. Rev. B* 2001, 64, 094519(1-13).

[14] Pashitski A. E.; Polyanskii A.; Gurevich A.; Parrell J. A.; Larbalestier D. C. *Physica C* 1995, 246, 133-144.

[15] Fukumoto Y.; Li Q.; Wang Y. L.; Suenaga M.; Haldar P. *Appl. Phys. Lett.* 1995, 66, 1827-1829.

[16] Suenaga M.; Fukumoto Y.; Haldar P.; Thurston T. R.; Wildgruber U. *Appl. Phys. Lett.* 1995, 67, 3025-3027.

[17] Hlasnik I.; Jansak L.; Majoros M.; Kokavec J.; Chovanec F. *IEEE Trans. Magnetics* 1996, 32, 2806-2809.

[18] Smith A. W.; Jaeger H. M.; Rosenbaum T. F.; Kwok W. K.; Crabtree G. W. *Phys. Rev. B* 1999, 59, R11665-R11667.

[19] Indenbom M. V.; Konczykowski M.; Van Der Beek C. J.; Holtzberg F. *Physica C* 2000, 341-348, part 2, 1251-1252.

[20] Blatter G.; Feigel'man M. V.; Geshkenbein V. B.; Larkin A. I.; Vinokur V. M. *Rev. Mod. Phys.* 1994, 66, 1125-1388.

[21] Blatter G.; Geshkenbein V. B.; Koopmann J. A. G. *Phys. Rev. Lett.* 2004, 92, 067009(1-4).

[22] Yamafuji K.; Kiss T. *Physica C* 1996, 258, 197-212.

[23] Ziese M. *Physica C* 1996, 269, 35-45.

[24] Ziese M. *Phys. Rev. B* 1996, 53, 12422-12429.

[25] Scalapino D. J. *Phys. Reports* 1995, 250, 329-365.

[26] Kerchner H. R.; Norton D. P.; Goyal A.; Budai J. D.; Christen D. K.; Kroegeer D. M.; Paranthaman M.; Lee D. F.; List F. A.; Feenstra R.; Brandt E.H. *Phys. Rev. B* 1999, 60, 6878-6883.

[27] Pastoriza H.; Candia S.; Nieva G. *Phys. Rev. Lett.* 1999, 83, 1026-1029.

[28] Küpfer H.; Wolf Th.; Zhukov A. A.; Meier-Hirmer R. *Phys.Rev. B* 1999, 60, 7631-7639.

[29] Maggio-April I.; Renner C.; Erb A.; Walker E.; Fisher O. *Nature* 1997, 390, 487-489.

[30] Liu J. Z.; Jia Y. X.; Shelton R. N.; Fluss M. J. *Phys. Rev. Lett.* 1991, 66, 1354-1357.

[31] Oussena M.; De Groot P. A. J.; Porter S. J.; Gagnon R.; Taillefer L. *Phys. Rev. B* 1995, 51, 1389-1392.

[32] Oussena M.; De Groot P. A. J.; Deligiannis K.; Volkozub A. V.; Gagnon R.; Taillefer L. *Phys. Rev. Lett.* 1996, 76, 2559-2562.

[33] Duran C. A.; Gammel P. L.; Wolfe R.; Fratello V. J.; Bishop D. J.; Rice J.P.; Ginsberg D. M. *Nature* 1992, 357, 474-476.

[34] Duran C. A.; Gammel P.L.; Bishop D. J.; Rice J. P.; Ginsberg D. M. *Phys. Rev. Lett.* 1995, 74, 3712-3713.

[35] Welp U.; Gardiner T.; Gunter D. O.; Veal B. W.; Crabtree G. W.; Vlasko-Vlasov V. K.; Nikitenko V. I. *Phys. Rev. Lett.* 1995, 74, 3713-3714.

[36] Wijngaarden R. J.; Griessen R.; Fendrich J.; Kwok W.-K. *Phys. Rev. B* 1997, 55, 3268-3275.

[37] Hanisch J.; Attenberger A.; Holzapfel B.; Schultz L. *Phys.Rev. B* 2002, 65, 052507(1-4).

[38] Rykov A. I.; Tajima S.; Kusmartsev F. V.; Forgan E. M.; Simon Ch. *Phys. Rev. B* 1999, 60, 7601-7606.

[39] Schuster Th.; Kuhn H.; Koblischka M. R.; Theuss H.; Kromüller H.; Leghissa M.; Kraus M.; Saemann-Ischenko G. *Phys. Rev. B* 1993, 47, 373-383.

[40] Kilic A.; Kilic K.; Senoussi S. *J. Appl. Phys.* 1998, 84, 3254-3262.

[41] Sonier J. E.; Kiefl R. F.; Brewer J. H.; Bonn D. A.; Dunsiger S. R.; Hardy W. N.; Liang R.; Miller R. I.; Noakes D. R.; Stronach C. E. *Phys. Rev. B* 1999, 59, R729-R732.

[42] Spitzer F. *Principles of Random Walk*; Princeton University Press: Princeton, NJ, 1964.

[43] Gardiner C. W. *Handbook of Stochastic Methods*; Springer-Verlag: Berlin, 1983.

[44] Dorogovtsev S. N.; Kuzmin Yu. I. *Phys. Lett. A* 1992, 170, 245-247.

[45] Fangohr H.; Cox S. J.; De Groot P. A. J. *Phys. Rev. B* 2001, 64, 064505(1-10).

[46] Turchinskaya M.; Kaiser D. L.; Gayle F. W.; Shapiro A. J.; Roytburd A.; Vlasko-Vlasov V.; Polyanskii A.; Nikitenko V. *Physica C* 1993, 216, 205-210.

[47] Welp U.; Gardiner T.; Gunter D.; Fendrich J.; Crabtree G. W.; Vlasko-Vlasov V. K.; Nikitenko V. I. *Physica C* 1994, 235-240, 241-244.

[48] Feynman R. P.; Hibbs A. R. *Quantum Mechanics and Path Integrals*; McGraw-Hill: New York, NY, 1965.

[49] Kuzmin Yu. I. *Phys. Lett. A* 2001, 281, 39-43.

[50] Kuzmin Yu. I. *Phys. Solid State* 2001, 43, 1199-1206 [Translated from: *Fiz. Tverd. Tela* 2001, 43, 1157-1164].

[51] Mandelbrot B. B. *Science* 1967, 155, 636-638.

[52] Mandelbrot B. B. *Fractals: Form, Chance, and Dimension*; Freeman: San Francisco, CA, 1977.

[53] Mandelbrot B. B. *The Fractal Geometry of Nature*; Freeman: San Francisco, CA, 1982.

[54] Mandelbrot B. B. in *Fractals in Physics*; Pietronero L.; Tosatti E.; Ed.; North-Holland: Amsterdam, 1986, pp 3-28.

[55] Feder J. *Fractals*; Plenum Press: New York, NY, 1988.

[56] Sidorenko A.; Surgers C.; Trappmann T.; Von Lohneysen H. *Phys. Rev. B* 1996, 53, 11751-11756.

[57] Laibowitz R. B.; Voss R. F.; Alessandrini E. I. In *Percolation, Localization and Superconductivity*; Goldman A. M; Wolf S. A.; Ed.; NATO ASI Series, Ser.B; Plenum Press: New York, NY, 1984; Vol. 109, pp 145-160.

[58] Alexander S. *Phys. Rev. B* 1983, 27, 1541-1557.

[59] Hong D. C.; Stanley H. E.; Coniglio A.; Bunde A. *Phys. Rev. B* 1986, 33, 4564-4573.

[60] Pike R.; Stanley H. E. *J. Phys. A* 1981, 14, L169-L177.

[61] Gefen Y.; Aharony A.; Alexander S. *Phys. Rev. Lett.* 1983, 50, 77-80.

[62] Havlin S.; Ben-Avraham D. *Adv. Phys.* 1987, 36, 695-798.

[63] Baziljevich M.; Bobyl A. V.; Bratsberg H.; Deltour R.; Gaevski M. E.; Galperin Yu. M.; Gasumyants V.; Johansen T. H.; Khrebtov I. A.; Leonov V. N.; Shantsev D. V.; Suris R. A. *J. Phys.* (Paris) IV 1996, 6 (Colloque C3), 259-261.

[64] Surdeanu R.; Wijngaarden R. J.; Griessen R.; Rossel C.; Ren Z. F.; Wang J. H. *Physica C* 1997, 282-287, 2219-2220.

[65] Kuzmin Yu. I. *Phys. Lett. A* 2001, 281, 39-43.

[66] Kuzmin Yu. I. *Tech. Phys. Lett.* 2002, 28, 568-571 [Translated from: *Pis'ma Zh. Tekh. Fiz.* 2002, 28, 74-82].

[67] Kuzmin Yu. I. *Phys. Lett. A* 2002, 300, 510-517.

[68] Brown B. *Phys. Rev. B* 2000, 61, 3267-3269.

[69] Kuzmin Yu. I.; Plechakov I. V. *Tech. Phys. Lett.* 1999, 25, 475-477 [Translated from: *Pis'ma Zh. Tekh. Fiz.* 1999, 25, 30-36].

[70] Kuzmin Yu. I., Plechakov I. V.; Razumov S. V. *Phys. Solid State* 1999, 41, 1594-1597 [Translated from: *Fiz. Tverd. Tela* 1999, 41, 1739-1742].

[71] Hudson D. *Statistics*; CERN: Geneva, 1964.

[72] Warnes W. H.; Larbalestier D. C. *Appl. Phys. Lett.* 1986, 48, 1403-1405.

[73] Wördenweber R. *Phys. Rev. B* 1992, 46, 3076-3083.

[74] Fisher M. P. A. *Phys. Rev. Lett.* 1989, 62, 1415-1418.

[75] Fisher D. S.; Fisher M. P. A.; Huse D. A. *Phys. Rev. B* 1991, 43, 130-159.

In: New Frontiers in Superconductivity Research
Editor: Barry P. Martins, pp. 83-105

ISBN: 1-59454-850-1

Chapter 3

VORTEX ENTRY CONDITIONS IN TYPE-II SUPERCONDUCTORS: GIBBS-LONDON APPROACH VS GINZBURG-LANDAU THEORY

I.L. Maksimov[*]

Nizhny Novgorod University, 603950, Nizhny Novgorod, Russian Federation

Abstract

The problem of the vortex entry into type-II superconductors (bulk samples, thin films) is analysed theoretically. Two approaches for the problem are discussed:

i) the Gibbs-London theory, based on the concept of geometrical barrier, controlling vortex entry/exit into/out from superconductor, and
ii) the quantum-mechanical Ginzburg-Landau theory, employing the concept of the wave function of superconducting condensate.

The quantitative conditions for the entry of vortices into type-II superconductors being in the Meissner and/or mixed state, will be described in the framework of two different approaches. Specifically, a detailed theory of geometrical barrier will be constructed, on the basis of which the shape of the vortex fragment will be described, and the vortex-entry criterion will be obtained. The analytical and numerical study of the Ginzburg-Landau equations reveal the existence of a modulation instability of superconducting condensate. The instability threshold is reached when superconducting order parameter turns to zero at the sample surface and the supervelocity (or gauge-invariant vector potential) of the condensate reaches some a threshold value at the superconductor edge or surface. For superconductors with high value of Ginzburg-Landau parameter this criterion practically coincides with criterion based on the depairing current density. Due to this instability, vortices start to nucleate at the edge or surface and then penetrate deeper into the sample.

[*] E-mail address: ilmaks@phys.unn.ru

1 Introduction

Since the discovery of type-II superconductors until now, much interest was devoted to the effects that surface or edge barriers have on the magnetic and dissipative properties of these materials. One of the most intriguing problems is the problem of the finding conditions at which the Abrikosov vortices, being driven by the external magnetic field start penetration inside the superconductor. Indeed, to find the conditions at which the virgin Meissner state is destroyed, so that the magnetic topological defect emerges inside the sample, thus transferring superconductor to the mixed state, would be fundamentally important step in describing the magnetic behaviour of superconductors. In majority of research devoted to this problem the vortex-entry condition is formulated in terms of the so-called surface barrier which opposes an easy vortex penetration even for sufficiently high magnetic field exceeding the lower critical field H_{c1} . Starting from the pioneer work by De Gennes [1] the problem was formulated in terms of constructing the Gibbs free-energy for the test vortex (for details see [2]), the analysis of which makes it possible to describe the vortex-entry field H_S ($H_S>H_{c1}$) for which the surface barrier disappears. More detailed study of the surface barrier energetics in the context of the London theory of superconductors was made by Petukhov and Chechetkin [3], who analysed the conditions for the entry or the exit of the vortex semiloops into a superconductor (see also Galaiko [4]), in the framework of the London model combined with the Gibbs free energy approach.

However, the results obtained on the basis of such a purely phenomenological Gibbs-London approach does not seem entirely convincing since the quantum features of the both the superconducting condensate and the vortex itself were ignored. Following Kramer approach [5] the emphasis in the problem solution was directed to the analysis of the stability of the Meissner state with respect to both order parameter and electromagnetic fluctuations in a system. For example, a series of results obtained on the basis of the analytical studies of the linearized Ginzburg-Landau equations (GLE) were reported by [5-8]. A number of works was dedicated to the direct modeling of the vortex entry into type-II superconductors [9-11] by numerically solving the time-dependent Ginzburg-Landau equations. After discovery of the high-temperature superconductors (HTSC) the problem of the magnetic field penetration into superconductor received more close attention. This is caused by the specific geometry of the single-crystal samples, possessing rather rectangular shape with a flat edge surface. For the case of the HTSC samples the concept of the geometrical edge barrier was introduced [12] in order to describe quantitatively the magnetic features of low-dimensional superconducting samples.

In the present review a criterion of the vortex entry into type-II superconductor, being in the mixed state, is formulated on the basis of two complementary approaches – the Gibbs-London approach and the Ginzburg-Landau theory, based on a numerical solution of the time-dependent Ginzburg-Landau equations (TDGLE), supplemented by an analytical study of the linearized GLE. We show that the obtained criterion is valid for both bulk and for thin-film type-II superconductors and allows, in particular, to estimate the first-vortex entry field H_{en} for superconductors with an ideal surface.

The paper is organized as follows. In Sec. 2 the vortex entry conditions are studied in bulk superconductors on the basis of a Gibbs-London approach. Specifically, the theory of geometrical barrier will be presented, the vortex-entry field will be derived within the single-vortex approach. In Sec. 3 the vortex entry conditions are studied in bulk superconductors on the basis of a numerical (Sec. 3. 1-3, 5) and analytical (Sec. 3. 4) solution of the GLE. The situation in thin-film superconductors is analysed in Sec. 3. 6.

2 Geometrical Barrier, a Single-Vortex Approach

1. Geometrical barrier is observed in type-II superconducting non-ellipsoidal samples (slabs, films) possessing high demagnetizing factor, which are placed in perpendicular magnetic field. These samples demonstrate irreversible behavior of the magnetic flux penetration into pin-free superconductors (even in the absence of bulk pinning), and show a visible magnetization hysteresis which is exclusively caused by the shape of near-the-edges cross-section of a sample and was first discovered in type-I superconducting samples [13].

The concept of the geometrical barrier (GB) for the case of type-II superconductors has been introduced by Zeldov *et al.* in [12] This concept made it possible to investigate vortex structures arising inside thick HTSC single crystal/tape of rectangular cross-section [12] and thin-films with an edge barrier [14, 15] placed into perpendicular magnetic field or carrying a transport current [16, 17]. It was shown that GB results in a magnetization hysteresis of a sample even in the absence of bulk pinning. However, theoretical description of geometrical edge barrier is not as yet complete. For example, M.Benkraoda and J.Clem in their study [18] have theoretically calculated the magnetic moment of a thick strip, assuming elliptically rounded edges of the sample. Meanwhile, the presence of the corners in a real sample of rectangular cross-section makes this approach rather questionable.

In this section the geometrical barrier is theoretically considered for a thick superconducting sample (strip) with rectangular edges. Within the bounds of a single -vortex approach a single-vortex Gibbs potential is calculated and the conditions for the full vortex penetration into a superconductor are determined. Dependence of the “pilot-vortex” inclination angle as well as its position on an external magnetic field is found out at the initial penetration stage.

2. Let us consider a superconducting strip of rectangular cross-section, by width w ($|x| \le w/2$), thickness d <<w ($|z| \le d/2$), placed in a perpendicular magnetic field (see fig. 1). It is supposed, that on the account of the symmetry of a sample it is possible to consider its part (a “unit cell”) containing one vortex (per unit length) weakly interacting with nearest neighbors. Thus, magnetic response of entire sample may be represented as a superposition of the non-interacting multi-plane vortex structures, to describe which it is sufficient to consider only one of them, thus reducing the solution of a complex multi-vortex problem to the more simple case – a single- vortex one.

In the magnetostatic approach magnetic field intensity outside the strip can be presented as $\vec{H} = -\nabla\phi$, where ϕ — scalar magnetostatic potential. By setting $\vec{\boldsymbol{B}} = rot\vec{\boldsymbol{A}}$, where $\vec{\boldsymbol{B}}$ is the magnetic induction, $\vec{A} = \{0, A, 0\}$ — vector potential, we have $H_x = -\partial\phi / \partial x$, $B_z = H_z = \partial A / \partial x$. Here we have assumed for simplicity $\mu = 1$ outside the strip, and also

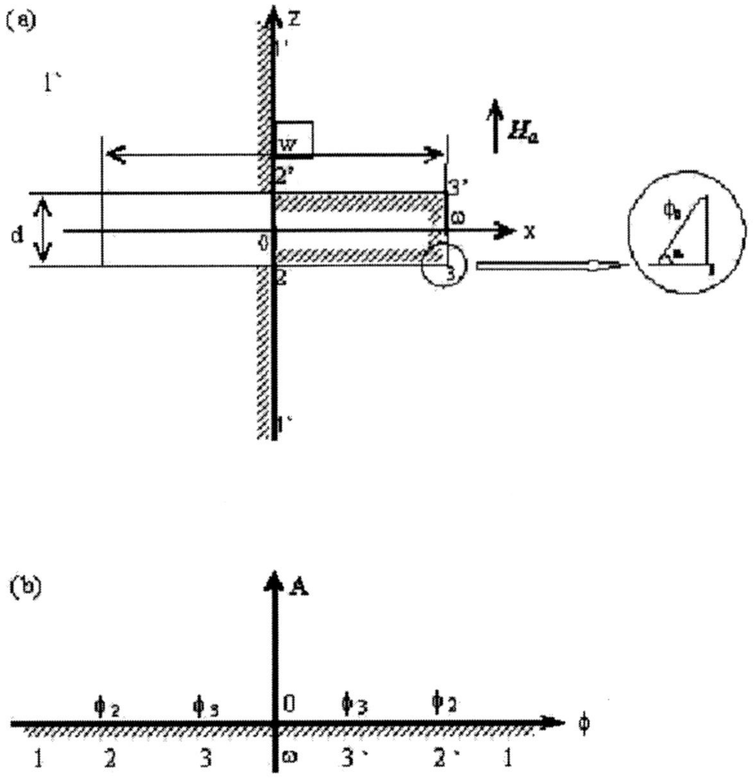

Fig. 1. (a) Cross-section y=0 of a superconducting strip in the plane (x, z), placed into perpendicular magnetic field H_a. The inset schematically represents vortex fragment. (b) The complex plane (φ, A): top half plane it is mapped with the help of Schwartz transformation onto unhatched area of the plane (x, z).

have used a thick-film (λ<<d) approach, where λ is the London penetration depth. In what follows we employ the Gibbs-London approach, relevant for the high-κ superconductors ($\kappa = \lambda/\varsigma$ being the Ginzburg-Landau parameter, and ξ being the coherence length [3]).

To define the magnetic field near a sample we perform a conformal mapping of the top of half-plane $\psi = \phi + iA$ on an external part of a polygon 1 ' 2 ' 3'ω123 in a complex plane $u = x + iz$ [19], taking into account the principle of area boundary [4], as shown in fig. 1a, 1b. Notice, that the zero on the plane (ϕ, A) is mapped into point ω (w/2,0) on the plane (x, z), but not into O (0,0).

The function $u(\psi)$ realizing required conformal mapping of a complex plane $\psi = \phi + iA$ on a complex plane $u = x(\psi) + iz(\psi)$, is represented by the Schwarz integral [19]:

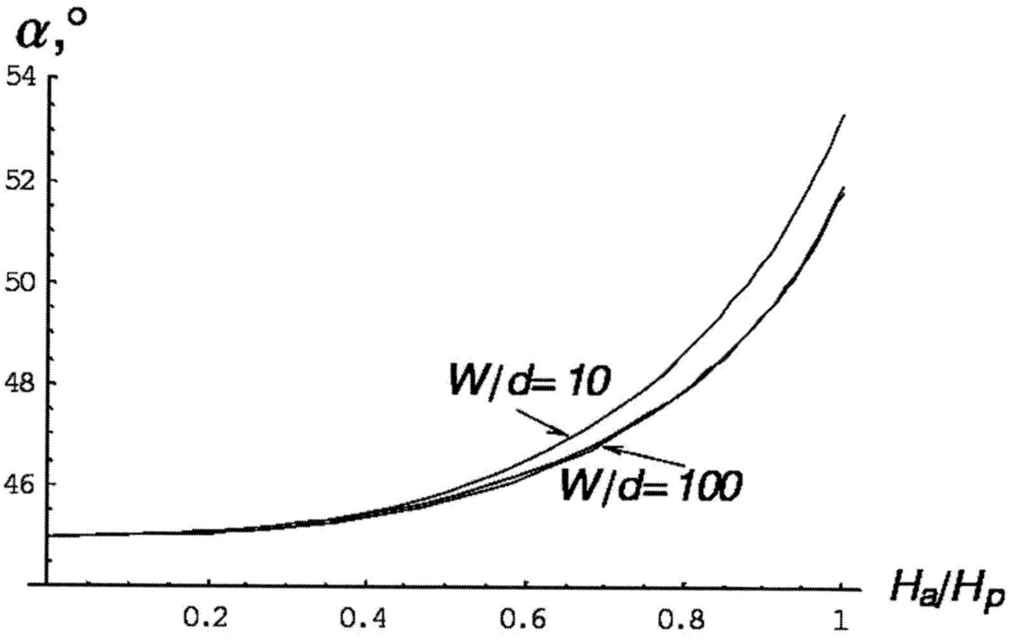

Fig. 2. Dependence of the vortex inclination angle on external magnetic field for the case w/d=10, 100.

$$u(\psi) = C\int_0^{\psi}(\xi-\phi_2)^{\alpha_2-1}(\xi-\phi_{2'})^{\alpha_3-1}(\xi-\phi_{3'})^{\alpha_{3'}-1}(\xi-\phi_2)^{\alpha_{2'}-1}d\xi + C_0 \qquad (2.1)$$

Constants C_1, C_0, $\phi_{2'}$, $\phi_{3'}$, ϕ_2, ϕ_3, in (1) should be determined from the conditions of a problem at the corners $\pi\alpha_K$ (k =2, 2`, 3 `, 3) (see fig. 2), which brings $\alpha_2=\alpha_{2'}=1/2$, $\alpha_3=\alpha_{3'}=3/2$ [13, 19]. Taking into account the symmetry of the problem, we have $\phi_2=-\phi_{2'}$, $\phi_3=-\phi_{3'}$. With the account $u(\psi=0)=w/2$, one finds $C_0=w/2$. Since distribution of the magnetic field is antisymmetric relative to the Ox axis, hereafter we consider only one of corners, specifically, the corner 3. Using asymptotic behaviour of the magnetic field components far from a sample (at $|u|\to\infty$) $H_x \to 0, H_z \to H_a, \dfrac{d\psi}{du} \to iH_a$, we find $C=-iH_a^{-1}$.

Thus, final expression for u(ψ) can be presented in the form

$$u(\psi) = \frac{1}{iH_a}\int_0^{\psi}\frac{(\omega^2-\phi_3^2)^{1/2}}{(\omega^2-\phi_2^2)^{1/2}}d\omega + \frac{w}{2} \qquad (2.2)$$

Constants ϕ_2 and ϕ_3 are determined from the following system of the equations:

$$0.5w = u(\varphi_3) - u(\varphi_2) = \frac{\phi_3}{H_a} a f_1(a) \tag{2.3a}$$

$$0.5iD = u(\varphi_3) - u(0) = \frac{\phi_3}{H_a} \frac{\pi}{4a} f_2(a) \tag{2.3b}$$

where $a \equiv \phi_2 / \phi_3$, and

$$f_1(a) \equiv \int_{\arcsin(1/a)}^{\pi/2} \sqrt{\sin^2 t - 1/a^2}\, dt, \quad f_2(a) \equiv \frac{4}{\pi} \int_0^1 \frac{(1-t^2)^{1/2}}{(1-(t/a)^2)^{1/2}} dt. \tag{2.4}$$

Using (3) we obtain the equation determining geometry-dependent parameter *a*:

$$\frac{w}{d} = \frac{4a^2}{\pi} \frac{f_1(a)}{f_2(a)}, \tag{2.5}$$

The constants ϕ_2 and ϕ_3 of the conformal mapping are defined as $\phi_3 = 2aH_a d / \pi f_2(a)$, $\phi_2 = a\phi_3$.

3. The components of the magnetic field intensity on a surface of a sample $\widetilde{H}_x(x) \equiv H_x(x, z = -d/2)$ and $\widetilde{H}_z(z) \equiv H_z(x = w/2, z)$ can be calculated from (2); it would be not instructive to present here quite cumbersome exact expressions for $\widetilde{H}_x(x)$ and $\widetilde{H}_z(z)$. Instead, for the further calculations it is rather convenient to take advantage of the approximation expressions for the quantities $\widetilde{H}_z(z)$ and $\widetilde{H}_x(x)$:

$$\begin{cases} H_z(x = \frac{w}{2}, \xi) = aH_a \dfrac{1 - [1 - \frac{1}{\beta}]\xi}{(1-\xi)^{1/3}}, & \dfrac{1}{\beta} = \left(\dfrac{4}{3\pi}\right)^{1/3}; \\ H_x(\eta, |z| = \frac{d}{2}) = H_a \left[\dfrac{a^2 f_1(a)}{a^2 - 1}\right] \dfrac{1 - (1-\eta)^{\gamma}}{\gamma(1-\eta)^{1/3}}, & \dfrac{1}{\gamma} = \left(\dfrac{a f_2(a)}{3 f_1(a)}\right)^{1/3} \left[\dfrac{a^2 - 1}{a^2 f_1(a)}\right] \end{cases} \tag{2.6}$$

which provide correct asymptotical behavior near the corner 3 and at the points 2 and ω (here we employ dimensionless variables $\zeta = 2|z|/d$ and $\eta = 2x/w$). Our study shows that the degree of an error of the approximation (6) near the sample edges ($w/2 - d/2 \le |x| \le w/2$) does not exceed 15 % for w/d >10.

4. Gibbs-free energy in the presence of a vortex fragment inside a sample can be calculated as a difference between the vortex energy ε_0 and the work A produced by the Meissner current while moving a vortex from the corner 3 into deep of a sample (see [12]): $G = \varepsilon_0 - A$. The vortex energy ε_0 is equal $\varepsilon_0 = (\phi_0 / 4\pi\lambda)^2 L \ln\kappa$ (in high κ limit one may neglect the energy of the vortex core) [20], here ϕ_0 is the flux quantum, κ is the Ginzburg-Landau parameter, L is length of the vortex.

Consider Abrikosov vortex fragment of length L (which is schematically represented in the inset of the fig.1.a), which has penetrated into the edge area of a sample and is inclined by the angle α to the plane 23. Inside a sample the vortex is rectilinear, since on the distances greater than $L_{eff} \sim \lambda$ (λ being London penetration depth) Meissner currents are exponentially damped and practically do not affect the vortex. Within the framework of our model, in the limit $\lambda \to 0$, Meissner currents are concentrated on the surface of a sample. Therefore, it is reasonable to assume that on the distance smaller than L_{eff} from the sample planes the vortex is bent, so that its ends are perpendicular to the planes 23 and 33`, thus satisfying the minimum- energy constraint for such configuration. Performing standard calculation on the basis of a method suggested in ref. [18], and adjusted according to the specifics of the present model, it is quite easy to find Meissner currents density along planes 23 and 33': $\vec{j}_{23} = -c/(4\pi L_{eff})\tilde{H}_x(x)\vec{e}_y$ and $\vec{j}_{33'} = -c/(4\pi L_{eff})\tilde{H}_z(z)\vec{e}_y$

Let's consider the work while moving the ends of a vortex along planes 23 and 33', separately. By definition it is possible to write the following expressions:

$$A_{23}(x_0) = \int_{w/2}^{x_0} \left|\vec{F}_{23}(x)\right| d(-x) \text{ and } A_{33'}(z_0) = \int_{-d/2}^{z_0} \left|\vec{F}_{33'}(z)\right| dz .$$

Here Lorentz force $\vec{F}_k$ on k-th surface may be expressed as $\vec{F}_k = \frac{1}{c}\left[\vec{j}_k \times \vec{\phi}_0\right] L_{eff}$, where L_{eff} — effective length of a vortex in the field of surface currents, $L_{eff} \sim \lambda$. It is more preferable to describe the vortex penetration near the edge introducing its inclination angle α with respect to the bottom surface 23 (see the inset in fig.1.a). By using the expression $\eta = 1 - \frac{d}{w}(1-\xi)ctg\alpha$ it is possible to calculate the currents-related total work $A(\xi,\alpha) = A_{23} + A_{33'}$ and finally the Gibbs-free energy of the vortex fragment

$$G(\xi,\alpha) = \frac{3}{16\pi}\phi_0 H_{c1} d \left[\frac{2}{3}\frac{1-\xi}{\sin\alpha} - \left(\frac{\pi}{36}\right)^{1/6} \frac{H_a}{H_{c1}}\left(\frac{w}{d}\right)^{1/2} g\left(\frac{w}{d}\right)(1-\xi)^{2/3} \times \right.$$
$$\left. \times \left\{\frac{2}{5}(\beta-1)(1-\xi) + 1 + \mathrm{ctg}^{2/3}\alpha - \frac{(1-\xi)^\gamma}{1+\frac{3}{2}\gamma}\left(\frac{d}{w}\right)^\gamma \mathrm{ctg}^{2/3+\gamma}\alpha\right\}\right], \tag{2.7}$$

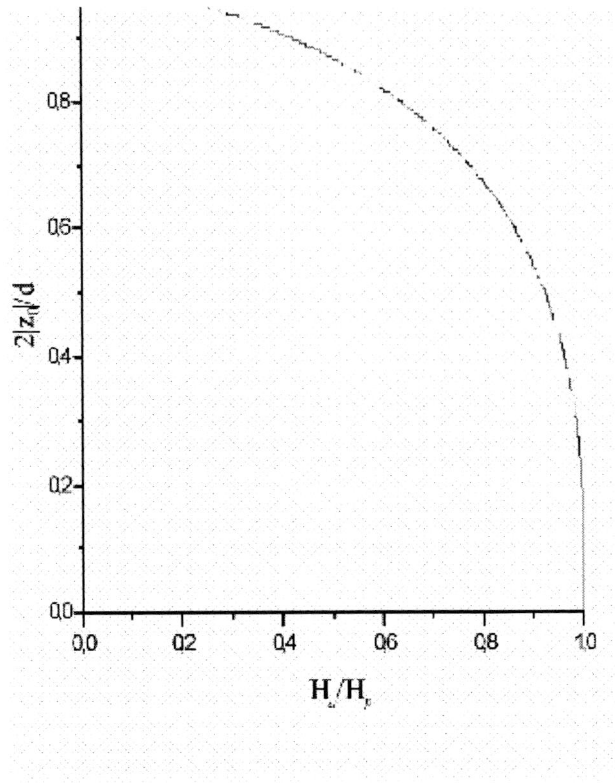

Fig. 3. Field-dependent vortex position near the sample edge for the case w/d= 100.

where H_{c1} is the first critical field for the bulk sample in longitudinal geometry [20]: $H_{c1} = \phi_0 /(4\pi\lambda^2)\ln\kappa$. To derive (2.7) the expressions for a and γ were used. Here the function $g(w/d)$ defined from the equation $a = \sqrt{\pi w/4d}\cdot g\,(\mathrm{w/d})$ is weakly dependent on the ratio w/d: numerical calculation shows that the function $g\in[1.1, 1]$ for $w/d\geq 10$.

By minimizing $G(\zeta, \alpha)$ with respect to both variables one comes to the equations system:

$$\begin{cases} (1-\frac{2|z|}{d})^{1/3} = \frac{H_a}{H_{c1}}\left(\frac{w}{d}\right)^{1/2}\frac{tg^{1/3}\alpha}{\cos\alpha}\left[1-(1-\frac{2|z|}{d})^{\gamma}\left(\frac{d}{\mathrm{w}}\right)^{\gamma} ctg^{\gamma}\alpha\right]\left(\frac{\pi}{36}\right)^{1/6}\cdot g\left(\frac{\mathrm{w}}{d}\right) \\ 1+(\beta-1)(1-\frac{2|z|}{d}) = [1-(1-\frac{2|z|}{d})^{\gamma}\left(\frac{\mathrm{d}}{\mathrm{w}}\right)^{\gamma}\mathrm{ctg}^{\gamma}\alpha]\cdot \mathrm{tg}^{4/3}\alpha \end{cases} \qquad (2.8)$$

Numerical solution of the system (8) makes it possible to find the dependence of the vortex inclination angle and its position on the applied field (see figs. 2,3). We should emphasize that our approach, unlike previous ones, contains no hypothesis on the configuration of the penetrating vortex, which is determined self-consistently.

Magnetic field, at which the vortex will cross equator ($z = 0$) of a sample, we define as the field of the vortex penetration through geometrical barrier, H_p. At further increase of the applied field, the vortex ultimately penetrates deep inside superconducting sample. The angle α_p corresponding to the value H_p is defined as the penetration angle.

5. Considering penetration of vortex into a sample, it is possible to mark out the following basic characteristics:

1) Vortices penetrate into the edge corner of a sample having inclination angle $\alpha = 45^\circ$ *(for $H_a \approx 0$)* (see Fig.1.a).This is due to the fact that H_x and H_z are practically identical in the close vicinity of the sample corner.
2) In a process of further increase of an external magnetic field the vortex orientation changes: the angle α increases for *$H_a > 0$*; (see Fig. 2). The field-dependent vortex position near the sample edge for the case w/d= 100 is shown in Fig.3. Physically, this behavior is caused by the variation difference of the magnetic field components H_x and H_z. Note that similar behavior demonstrates distribution of the magnetic lines, obtained numerically by Brandt [21] for pin-free superconductors with high demagnetization.
3) At the vortex-penetration field $H_a = H_p$ the vortex reaches the equator of the sample (z =0; see Fig. 3) while its inclination angle α reaches maximum value: $\alpha = \alpha_p$ (for example, $\alpha_p \approx 52^\circ$ for w/d =100; see fig. 2). It is worth noting that in the thin-film limit w/d $\to$1, α_p does not depend upon geometry anymore reaching the maximum value $\alpha_p^{\infty} = 57^\circ$ (see Fig. 2).
4) For $H_a > H_p$ two possibilities may take place:

i)the joining of symmetrical vortex fragments into one vortex occurs followed by its subsequent straightening and moving towards the center of a sample as a unit, in position, perpendicular to its surface [2, 3].

ii)the nearest vortex fragments do not collapse into one, but form a complex vortex structure in which a sporadic reconnection of the magnetic flux lines takes place.

The vortex-penetration magnetic field for the case i) is represented as $H_p = H_{c1}\left(\frac{d}{w}\right)^{1/2}\chi_p$. Here, the dependence of the factor χ_p on the ratio of w/d is extremely weak (χ_p changes by 9% when w/d increases by five orders of its magnitude: $\chi_p \in [0.824, 0.9]$, w/d$\in[10, 10^5]$).

6. Let us discuss finally the applicability of the single-vortex approach. It seems not quite realistic to analyse a sole vortex penetrated near the sample edges. It would be reasonable to expect the appearance of a series of vortices concentrated near the sample corner thus forming there a mixed-state quasi-continuum. However, as numerical solution [21] demonstrates, the multiple vortex penetration and the mixed state formation takes place predominantly near the corners, so that the "pilot-vortex" is separated by some distance l from the near-the-edge vortices. Simple estimate shows that the intervortex interaction is negligible until intervortex distance reaches λ . Therefore, in the limit $l \gg \lambda$, one may neglect exponentially small multivortex corrections to the single-vortex Gibbs-free energy.

We wish to point out that all above results are obtained in the absence of bulk pinning. It is evident that bulk imperfections will affect considerably both the condition of the "pilot-vortex" penetration inside the strip and the structure of the near-the-edge mixed-state, formed by the multiple vortex fragments. To account for the bulk-pinning effects in the framework of the present model would be a challenging task for future study.

7. This Section is devoted to a study of a geometrical barrier in low-dimensional type-II superconductors. This problem is quite actual in connection with the wide use of both conventional and HTSC high-demagnetization monocrystals in modern physical experiment [22, 23]. Here we succeeded to construct analytical expression for the Gibbs-free energy of the vortex fragment entering from the corners of a sample. Dependence of the vortex inclination angle as well as its position on an external magnetic field is found out at the initial penetration stage. As a result of the present work it is possible to define the GB suppression field at which near-the-edge vortices start penetrating deep into a sample. The results of this study could be useful at the analysis of experimental data related to the vortices penetration into superconductors possessing high demagnetizing factor.

3 Vortex-Entry Condition: Ginzburg-Landau Theory

In this Section we discuss the problem of the magnetic field penetration inside type-II superconductors with an ideal surface. This means that the defects of any type – bulk defects, which tend to pin vortices inside the sample, and surface defects, which control the current-density and order parameter distribution inside superconductor, will not be considered here.

1 General Equations

Let an infinite superconducting slab of width *w* ($0 < y < w$) be placed in a parallel magnetic field $\vec{H} = (0,0,-H)$. The time-dependent GLE in this case are two-dimensional because the problem is homogeneous along the coordinate *z*,

$$\tau \frac{\partial \Psi}{\partial t} = -\left[(-i\nabla - \vec{A})^2 \Psi + \Psi(|\Psi|^2 - 1)\right] + \chi \tag{3.1}$$

$$\frac{\partial \vec{A}}{\partial t} = \mathrm{Re}\left[\Psi^* (-i\nabla - \vec{A})\Psi\right] - \kappa^2 \nabla \times \nabla \times \vec{A} \tag{3.2}$$

Here the length is scaled in units of the coherence length $\varsigma(T)$, time in units of $\tau_0 = 4\pi\lambda^2\sigma_n / c^2$ (σ_n being the normal-state conductivity), the vector potential $\vec{A}$ is presented in terms of $\Phi_0/(2\pi\varsigma)$, $H_c = \Phi_0/(2\sqrt{2}\pi\varsigma\lambda)$, where $\Phi_0 = 0.5ch/e$ is the magnetic flux quantum, κ is the Ginzburg-Landau parameter, τ is the characteristic relaxation time [24], χ is a random ``force" which simulates fluctuations of the order parameter, [9] and Re means the real part. The boundary conditions for Eqs.(3.1,2) along the transverse coordinate y are:

$$[(curl\,\vec{A})_z\,|_{y=0,w} = -H_z\ ;\ (-i\,\partial/\partial y - A_y)\Psi\,|_{y=0,w} = 0 \qquad (3.3)$$

For $\vec{A}$ and Ψ we assume periodic boundary conditions along the longitudinal coordinate x in order to model a sample of infinite length.

2 Vortex Entry Condition - Numerical Study

Equations (1,2) are solved numerically by employing a method similar to that described in [9]. Namely, we introduce a link variable $U(j) = \exp(-i\int A_j dx_j)$, where $x_j = x, y$, that allows to satisfy the gauge-invariance condition on the grid introduced by discretizing the variables x and y. To solve this system of time-dependent equations we use the standard Euler method.

The parameters of the sample were chosen as follows: width w=25, length L=50 (in units of ξ). Our numerical analysis shows that the results of our study are not sensitive with respect to an increase of L; the reason for this is discussed below in this Section. The Ginzburg-Landau parameter κ=2 and the time constant τ=2 were chosen to minimize the computation time. From the microscopic theory it follows that $\tau^{-1} = 12$ (see [24], however, the review [25] contains additional arguments for the more optimal selection of τ . For each change of H in steps of $\Delta H = 0.01$ (in units of H_{c2}) we compute the solution of Eqs. (3.1,2). A series of iteration steps is performed until the magnetic moment becomes independent of time. The random ``force" χ is distributed uniformly within the range $-s \le \chi \le s$; the obtained results do not depend on the noise amplitude s provided $s << 1$. In our simulations we chose $s=0.001$.

During numerical solution of the Eqs. (3.1,2) we studied the condition at which the entry of vortices into the superconductor starts. It turns out that the vortex nucleation is controlled by one single quantity: the supervelocity of the condensate $\vec{\Pi} = \nabla\varphi - \vec{A}$ (in dimensionless units) at the sample surface. As soon as this quantity at the sample edge, Π_x^{edge}, reaches the critical value Π_{cr}, a modulation instability of the order parameter of superconducting condensate emerges at the edge. The quantity $\Pi_{cr} \approx 1$, practically does not depend upon the external field for our selection of parameters. In what follows we consider vortex entry/exit conditions near the left ($y=0$) edge of the sample; from the symmetry of the problem it follows that these conditions near the opposite edge ($y=w$) are completely analogous. The analytical theory of the modulation instability emergence will be described below (see subsection 4). Here we point out that the crucial role, which plays the supervelocity of the

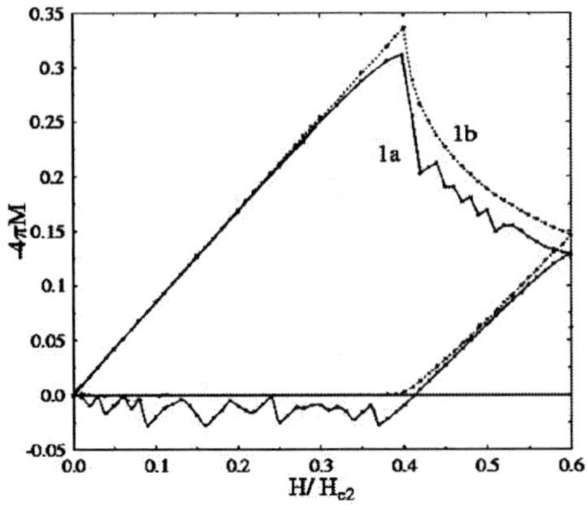

Fig. 4. Magnetization curve of a bulk superconductor. Curve 1a is obtained by numerical solution of the TDGLE, and curves 1b is calculated on the basis of a phenomenological model of the surface barrier [26].

condensate as the main quantity controlling the instability onset (and evolution), reflects the |validity of the Landau-Feynman scenario for the vortex entry into superfluid bose-liquid with the specification for the superconductors.

From our numerical solution of the GLE we calculate the magnetization curve of a type-II superconductor with ideal surface (see Fig. 4). It is seen that the magnetization curve has a nonmonotonic serrated shape resulting from the threshold character of the vortex entry. Indeed, let at some value of the external magnetic field the inequality $\Pi_x^{edge} > \Pi_{cr}$, be fulfilled at the edge of the superconductor. Then a chain of vortices with a period l starts to nucleate at the slab edge, as it is shown in Fig. 5. Our numerical calculation reveals that the spatial period of the vortex chain depends on the quantity $\Delta\Pi = \Pi - \Pi_{cr}$. The greater is $\Delta\Pi$, the less is l, i.e. the more vortices per unit length enter a superconductor at one step. The entering vortices reduce the value of Π_x^{edge} since the part Π_x^H of total supervelocity, $\Pi_x = \Pi_x^H + \Pi_x^v$, generated by an external magnetic field, and the part Π_x^v, due to the penetrated vortices, have opposite sign at the slab edge. The more vortices enter the slab at a single step of the magnetic field, the less becomes Π_x^{edge}. It is therefore necessary to further raise the magnetic field (in order to increase the component Π_x^H) to fulfill the critical condition $\Pi_x^{edge} > \Pi_{cr}$ again. Then, the modulation instability starts again, and another vortices chain appears near the each edge, penetrating inside the sample.

Each instability results in the appearance of distinct serrations not only in the magnetization curve but also in the dependence Π_x^{edge} on H (see Figs. 4 and 6). Particularly large serrations of the magnetization can be obtained for a mesoscopic superconductor (of size comparable with the vortex core radius ξ), as it was discussed in [11]. As displays our numerical analysis, an increase of the specimen size L only, brings in small corrections to the

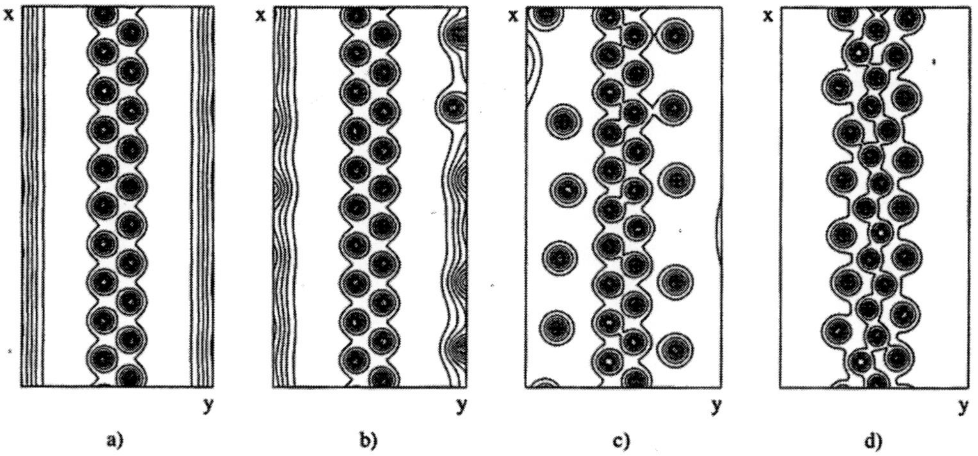

Fig. 5. Time evolution of the vortex nucleation process at *t=15* τ_0 *(a), t=24* τ_0 *(b), t=30* τ_0 *(c), t=180* τ_0 *(d)* after fixing the magnetic field $H = 0.42H_{c2}$. Shown are the contour lines of the order parameter $|\Psi|$ in a bulk superconductor.

obtained results; this is due to the fact that vortices enter the superconductor in form of a chain, the space period of which is much less than *L*, see Fig. 5. Besides, it turns out that the higher is $\Delta\Pi$, the less is the time of the vortex entry.

3 Vortex Entry Criterion: Numerical Results *vs* London Theory

Let us discuss the criterion for the vortex entry in more details. In the Gibbs-London theory it is assumed that the vortex overcomes the surface barrier and thus enters superconductor when the current density near the edge reaches the maximal possible value J_{GL}: $j_{edge} = J_{GL}$, where $J_{GL} = \sqrt{2}/(6\pi\sqrt{3})cH_S/\lambda$. According to the advanced GL theory vortex enters the superconductor when superconducting order parameter is locally suppressed at the sample edges, i.e. when $\Pi_x^{edge} = \Pi_{cr}$. Therefore, it would be useful to compare phenomenological and field-theoretical predictions. For this reason we compare figures 6 and 7, which show how $\vec{\Pi}_{edge}$ and j_{edge} depend on the external field *H*. From Fig.6 it is seen that Π_x^{edge} increases *monotically* up to the maximum value, reached at $\mathrm{H} = 0.4\,\mathrm{H}_{c2} \equiv H_S$; beyond this field visible serrations of $\vec{\Pi}_{edge}(H)$ appear. On the contrary, Fig. 7 demonstrates *nonmonotonic* behavior of j_{edge} : this quantity reaches its maximum before the modulation instability of the Meissner state starts. Indeed, at increasing of magnetic field the edge current

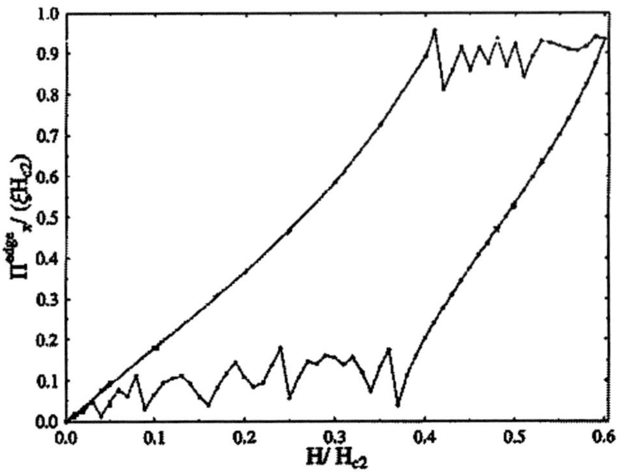

Fig. 6. The dependence of the length-average of Π_x^{edge} on the applied magnetic field.

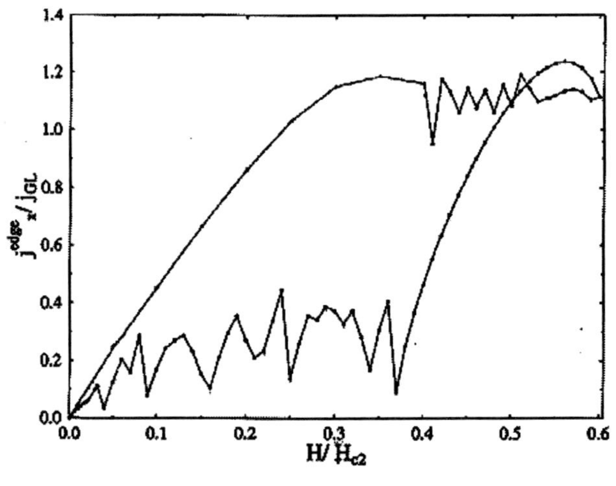

Fig. 7. The dependence of the length-average of j_x^{edge} on the applied magnetic field.

density reaches maximum value at *H=0.35* and then decreases monotonically up to the barrier suppression field $H = H_S = 0.41$. Thus, we may conclude that the true criterion of vortex entry is controlled by the value of the supervelocity $\vec{\Pi}_{edge}$ at the edge (a supervelocity condition), rather than by the current density j_{edge} at the sample edge (a current density condition). However, since the latter criterion is more easy to apply, one may neglect minor details of the current density variation near the sample edge and (with some reservation) employ this criterion in practical estimates.

4 Vortex Entry Criterion - Analytical Study

To find the quantity $\vec{\Pi}_{cr}$ analytically, we consider the stationary Ginzburg-Landau equations

$$\Delta F + \kappa^2(1 - F^2 - \vec{\Pi}^2)F = 0, \tag{3.4}$$

$$\nabla \times \nabla \times \vec{\Pi} + F^2 \vec{\Pi} = 0 \tag{3.5}$$

in which we introduced new variables F=| Ψ | and $\vec{\Pi}$. In this case it is more convenient to measure distances in units of λ The supervelocity $\vec{\Pi}$ is expressed in units of $\Phi_0/(2\pi\varsigma)$, and the magnetic field in units of $\Phi_0/(2\pi\varsigma\lambda) = \sqrt{2}H_c$, where $H_c = \Phi_0/(2\sqrt{2}\pi\varsigma\lambda)$, is the thermodynamic critical field [3]. We seek the solution of these equations in the form $F = F_0 + \delta f$ and $\vec{\Pi} = \vec{Q} + \delta\vec{q}$, where δ*f* and $\delta\vec{q}$ are small $(|\delta f/F_0| << 1, |\delta\vec{q}|/|\vec{Q}| << 1)$ perturbations of the stationary values of the order parameter F_0 and momentum $\vec{Q}$, correspondingly. Disregarding terms of the order higher than linear in δ*f* and $\delta\vec{q}$, we obtain a system of three linear differential equations for the functions δ*f* and $\delta\vec{q}$,

$$\Delta\delta f + \kappa^2(1 - 3F_0^2 - \vec{Q}^2)\delta f - 2\kappa^2 F_0(\vec{Q}\delta\vec{q}) = 0, \tag{3.6}$$

$$-\frac{\partial^2 \delta q_y}{\partial x^2} + \frac{\partial^2 \delta q_x}{\partial x \partial y} + F_0^2 \delta q_y + 2F_0 Q_y \delta f = 0, \tag{3.7}$$

$$-\frac{\partial^2 \delta q_x}{\partial y^2} + \frac{\partial^2 \delta q_y}{\partial x \partial y} + F_0^2 \delta q_x + 2F_0 Q_x \delta f = 0. \tag{3.8}$$

Performing the Fourier transform with respect to the longitudinal coordinate *x* with wavevector *k* (physically, this means that we are searching for the solution representing vortex chain of a period $\propto k^{-1}$) we obtain the following system of equations:

$$\kappa^{-2}\frac{\partial^2 f}{\partial y^2} + (1 - 3F_0^2 - \vec{Q}^2 - (k/\kappa)^2)f - 2F_0(\vec{Q}\vec{q}) = 0, \tag{3.9}$$

$$ik\frac{\partial q_x}{\partial y}+(F_0^2+k^2)q_y+2F_0Q_yf=0, \tag{3.10}$$

$$-\frac{\partial^2 q_x}{\partial y^2}+ik\frac{\partial q_y}{\partial y}+F_0^2 q_x+2F_0Q_xf=0, \tag{3.11}$$

supplemented by the boundary conditions: $df/dy\,|_{y=0,w}=0$ and $(dq_x/dy-ikq_y)\,|_{y=0,w}=0$. Hereafter, f and $\vec{q}$ denote the Fourier-transforms of the perturbations δf and $\delta\vec{q}$.

As shows our numerical analysis of the equations (3.1,2) (with κ=2) the "most dangerous" perturbations are localized at the slab surface over a length of about ξ. This feature will be analytically confirmed below for the case $\kappa>>1$, at which perturbations are also localized near the surface over a length of several ξ (i.e. much less than λ). For such a distance from the surfaces we may reasonably assume that $Q_y\approx 0$ (this assumption follows from our boundary condition $Q_y^{edge}=0$, which reflects the absence of vortices near the surface), taking into account the fact that both F_0 and Q_x do not depend on the longitudinal coordinate *x*.

It seems quite constructive to expand $F_0(y)$ and $Q_x(y)$ near the edge surface of the slab into a Taylor series keeping only the linear terms with respect to $y\to 0$. Hereafter, for convenience, we choose the reference surface at *y=0*. It is also useful to take advantage from the relation $F_0^2=1-\vec{Q}^2$ which follows from the GLE for the order parameter F_0 (ignoring the term with second derivative in Eq. (3.4)) and is typical for high-κ materials. Inserting q_y from Eq. (3.10) into Eq. (3.11) we obtain a system of two second-order differential equations, valid near the reference slab surface *y=0*:

$$\kappa^{-2}\frac{\partial^2 f}{\partial y^2}+(1-3F_0^2-Q_x^2-(k/\kappa)^2)f-2F_0Q_xq_x=0, \tag{3.12}$$

$$-\frac{F_0^2}{F_0^2+k^2}\frac{d^2q_x}{dy^2}+\frac{d}{dy}\left(\frac{k^2}{F_0^2+k^2}\right)\frac{dq_x}{dy}+F_0^2q_x+2F_0Q_xf=0. \tag{3.13}$$

Note that in the Meissner state Eqs. (3.12,13) are valid not only on the slab surface, but also in the bulk, since in the Meissner state the condition $Q_y=0$ holds everywhere in the slab due to the absence of vortices. Equations Eqs.(3.12,13) have been derived and studied numerically in Refs. [5,6] to check the stability of the Meissner state in bulk superconductors.

In order to pursue more analytical approach we shall concentrate further on the physically most interesting limit κ>>1. Keeping the interest in the solution of Eqs.(3.12,13) which vanishes deep inside the slab, we may neglect the term, containing the first derivative in Eq. (3.13) since the function $k^2/(F_0^2+k^2)$ depends on *y* rather smoothly (in the limit k~κ>>1 and $F_0 \le 1$). In addition, the choice κ>>1 allows us to disregard the term with the second derivative in Eq. (12) at the critical wavenumber $k = k_c$ (see below). As a result we arrive at the Airy equation for the perturbation q_x, only

$$\frac{d^2 q_x}{dy^2} + (b + ay)q_x = 0\,. \tag{3.14}$$

It is worth noting here that Eq. (3.14) with proper boundary conditions is applicable to study the stability of both Meissner state and mixed state in bulk superconductors with respect to the vortex entry, as well. In the latter case the validity region of Eq.(3.14) is $\Delta y << w_{vf}$ where Δy is the size of perturbation in the *y*-direction and w_{vf} is the width of the vortex-free zone near the superconductor surface. In Eq.(3.14) the coefficients *a* and *b* are expressed through the constants *P* and *R*, which appear while expanding the order parameter $F_0^2 \approx 1 - P$ and the vector potential related quantity $Q_x^2 \approx P + Ry$ near the reference superconductor edge *y=0*,

$$a = -4R(1 + k^2 - P)(3P - 2 - (k/\kappa)^2)/(2P - 2 - (k/\kappa)^2)^2, \tag{3.15a}$$

and

$$b = -(1 + k^2 - P)(6P - 2 - (k/\kappa)^2)/(2P - 2 - (k/\kappa)^2)\,. \tag{3.15b}$$

The coefficient *R* may be expressed as a direct function on the magnetic field near the edge H_{edge}: $R = -2H_{edge}\sqrt{P}$. Note that for bulk superconductors H_{edge} coincides with the external field *H*; for the case of a thin-film some geometry-related factor appears: $H_{edge} \approx \gamma H$. The solution of the Eq.(3.14) is the well-known Airy function

$$q_x = D\Phi\{-a^{1/3}(y + b/a)]\,, \tag{3.16}$$

where *D* is an integration constant. From the boundary condition $q_y(y=0) = 0$, which reduces to $\partial q/\partial y\,|_{y=0} = 0$, a simple relation follows

$$b = -\gamma_1 a^{2/3}, \tag{3.17}$$

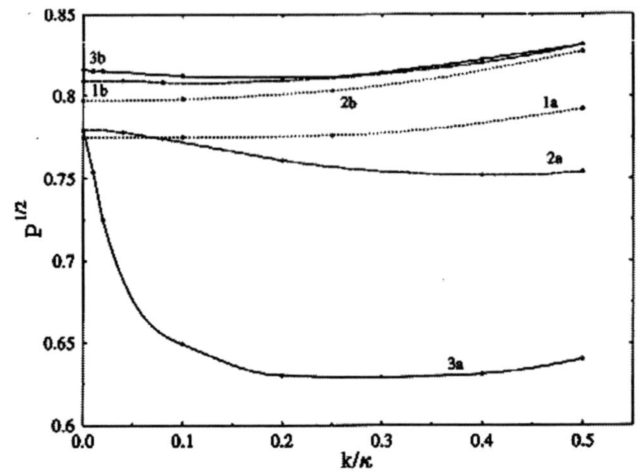

Fig. 8. The dependence of $\Pi_{cr} = P^{1/2}$ on the wave-number *k* of the disturbance q_x for different values of magnetic field *H* and GL parameter κ: (1) κ=2, (2) κ=5, (3) κ=100. (a) $H = H_S$, (b) $H \approx 0.9 H_{c2}$.

where $\gamma_1 \approx -1.02$ is the first zero of the first derivative of the Airy function. Equation (3.17) defines the value of the supervelocity Π_{edge} on the slab surface for which a periodic, spatially modulated disturbance $q_x \neq 0$ exists for a given value of the perturbation wave vector *k*. Inserting the expressions for *a* and *b* into Eq. (3.17) one obtains the dispersion relation in the form

$$(P-1-k^2)(6P-2-(k/\kappa)^2)^3(2P-2-(k/\kappa)^2) = gP(3P-2-(k/\kappa)^2)^2 \qquad (3.18)$$

that contains only one field-dependent control parameter $g = 64 |\gamma_1|^3 H_{edge}^2$. The given equation relates the value of the supervelocity Π_{edge} at the slab surface to the wavenumber $k = k_c$ for which a nontrivial solution q_x becomes possible. Numerical solution of the dispersion equation Eq. (3.18) defines the critical wavenumber $k = k_c$, for which the quantity *P(k)* reaches its minimum. In Fig.8 the dependence $\Pi_x^{edge} = \sqrt{P(k)}$ is shown for different values of κ and *H*. It is interesting that the value $\Pi_{cr} \equiv \sqrt{\Pi(k = k_c)}$ only slightly depends on the magnetic field *H*.

It is worth noting that for the case κ>>1, the perturbation is localized at the superconductor surface; this is reflected by the exponential decay of the Airy function $\Phi(\theta)$ in the positive θ region θ>0. The estimate for the characteristic scale Δy of the perturbation decay deep into the film yields $\Delta y \approx b^{-0.5}$. The numerical solution of Eq. (3.13) shows that for large values of κ at the critical perturbation is localized on a scale of several ξ in the *y*-direction. For more details the reader is referred to the original paper [26].

5 Vortex Exit

Let us discuss now the conditions for the vortex exit which takes place when the external field is decreased. As can be seen from Figs. 3.4 and 3.6, the dependences *M(H)* and $\Pi_x^{edge}(H)$ are nearly linear functions of the (initially) decreasing magnetic field. Physically this is explained by the fact that during the initial stage of the field decrease the vortices do not exit from the superconductor. At some value of the magnetic field (in our case at $H = 0.36H_{c2}$) the vortices start to exit from the sample. At $H < 0.36H_{c2}$ both dependences *M(H)* and $\Pi_x^{edge}(H)$ become non-monotonic because not one row of the vortices but several vortex rows may leave the sample at once. Our study shows that the process of the vortex exit is controlled by the behavior of Π_x^{edge} as well. As soon as Π_x^{edge} changes sign with the decrease of *H* (see Fig.6), the vortices leave the defect-free superconductor [26]. In the opposite case the vortices do not exit.

We should emphasize that when the field decreases below some value $H = H^*$ (in our case $H^* = 0.41H_{c2}$), both the magnetization and magnetic field coincide in sign. Such a ``paramagnetic" response, resembling the result reported for the first time in [27] for the bulk pin-free superconductors, is essentially due to surface pinning, which is a generic feature of type-II superconductors with an ideal surface.

6 Thin Film

It is easy to extent the above results for the case of thin supercondicting film (of thickness *d* less then London penetration length). In this geometry $\vec{\Pi}$ inside the film is practically independent of *z*; together with the boundary conditions $\partial\Psi/\partial z\,|_{z=\pm d/2} = 0$ this gives a *z*-independent order parameter Ψ. By averaging equation (3.1) over the coordinate *z* one arrives at the two-dimensional GL equation (valid for superconducting domain $z = 0; 0 \le y \le w; -\infty \le x \le \infty$.

$$\Delta F + \kappa_{eff}^2 (1 - F^2 - \vec{\Pi}^2)F = 0, \tag{3.19}$$

$$\nabla \times \nabla \times \vec{\Pi} = -F^2 \vec{\Pi}\,\delta(z)\theta(y)\theta(w - y) \tag{3.20}$$

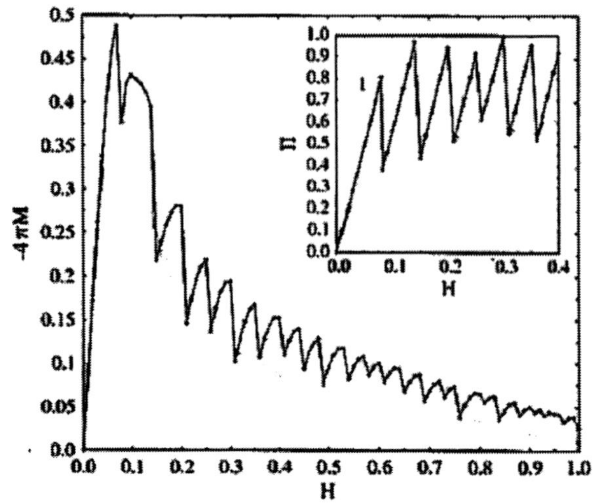

Fig. 9. Magnetization curve of a superconducting film. The applied field is expressed in units of H_{c2}^{film}, magnetization M is expressed in units H_{c2}^{film} $0.5H_{c2}^{film}/\kappa_{eff}^2$. The inset shows the field-dependent Π_x^{edge} for the film.

where $\kappa_{eff} = \lambda_{eff}/\varsigma$, and $\lambda_{eff} = \lambda^2/\varsigma$ and the distance now is measured in units of λ_{eff}. $\vec{\Pi}$ is expressed in units of $\Phi_0/(2\pi\varsigma)$, and the magnetic field in units of $\Phi_0/(2\pi\varsigma\lambda_{eff})$. Similarly to bulk case, by performing standard linearization procedure one arrives to the equations system analogous to Eqs. (3.12,13) (with corresponding modifications). By solving the dispersion equation one finds the criterion for the modulation instability emergence in thin films [28]. The corresponding magnetization curve supplemented by the dependence $\Pi_x^{edge}(H)$ is shown in Fig. 9. It seen that qualitative behavior of both *M(H)* and $\Pi_x^{edge}(H)$ in thin films is similar to those in bulk samples. This fact supports the conclusion that the physical picture of the vortex entry in type-II superconductors is quite universal. Indeed, Fig.10 demonstrates similar scenario of the vortex penetration inside thin film for the case when external field is switched to the value $H = 0.21H_{c2}^{film} > H_S^{film}$. For this case two additional vortex rows are generated near the film edge, demonstrating modulation instability development. Thus, we see that for both superconducting film and bulk superconductor, the criterion for the vortex entry is met when the kinematic momentum at the sample edge (or the surface) Π^{edge} reaches the critical value Π_{cr}: $\Pi^{edge} = \Pi_{cr}$ defined by the corresponding dispersion equation. Therefore, the quantitative parameters of the vortex entry threshold will appreciably depend on the sample geometry and the structure of the edge surface. For example, it would be rather useful to define the vortex penetration field from the Ginzburg-Landau approach. However, this would require appropriate modification of the governing equations (see e.g. [28]).

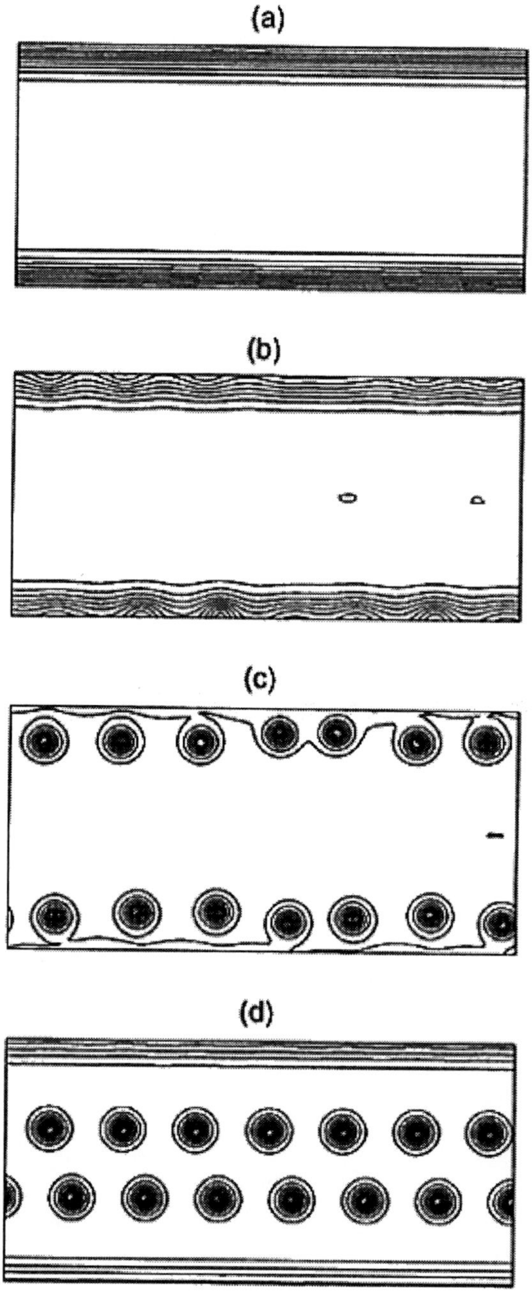

Fig. 10. Time evolution of the vortex nucleation process after *1500 (a)*, 2600 *(b)*, 3000 *(c)*, and 4000 *(d)* time steps after fixing the magnetic field $H = 0.21 H_{c2}^{film} > H_{S}^{film}$. Shown are the contour lines of the order parameter $|\Psi|$.

4 Summary

The problem of the vortex entry into type-II superconductors (bulk samples, thin films) was studied theoretically. Two approaches for the problem were employed:

i) the Gibbs-London theory, based on the concept of geometrical barrier, controlling vortex entry/exit into/out from superconductor, and
ii) the quantum-mechanical Ginzburg-Landau theory, employing the concept of the wave function of superconducting condensate.

Employing phenomenological approach for the problem of a geometrical barrier in type-II superconductors, we succeeded to construct analytical expression for the Gibbs-free energy of the vortex fragment entering from the corners of a sample. Dependence of the vortex inclination angle as well as its position on an external magnetic field is found out at the initial penetration stage. As a result we make it possible to define the GB suppression field at which near-the-edge vortices start penetrating deep into a sample. The results of this study could be useful at the analysis of experimental data related to the vortices penetration into superconductors possessing high demagnetizing factor.

By performing both numerical and analytical solution of the time-dependent Ginzburg-Landau equations, the vortex entry condition for bulk and thin-film type-II superconductors is studied. A universal "supervelocity criterion" for the vortex entry into type-II superconductors is formulated, which is shown to be more precise than the conventional "supercurrent criterion"'.

Acknowledgments

The author is grateful to Dr. E.H. Brandt, Prof. J.R. Clem, Dr. A. A. Elistratov, Dr. V. Jeudy, Dr. G.M. Maksimova, Dr. D.Yu. Vodolazov and to Mr. O. Bobrikov for cooperation and for helpful discussions of the obtained results. This work was supported by the Russian Ministry of Science, Project N_0 107-1(00), by the Russian Foundation for Basic Researh, Grant N_0 01-02-16593, by the Russian Ministry of Education, Grant N_0 E00-3.4-331, and the COBASE Foundation Grant N_0 INT-0002341.

References

[1] P.G. de Gennes, *Sol. St. Comm.* v.3, p. 127 (1965).

[2] P.G. de Gennes, *Superconductivity of Metals and Alloys*, W.A. Benjamin, Inc (New York-Amsterdam) (1996).

[3] B.V.Petukhov and V.R. Chechetkin, *Zh. Eksp. Teor. Fiz.*, v. 65, p. 1653 (1973). [*Sov. Phys. JETP*, v. 38, p. 827, (1974)].

[4] V.P. Galaiko, *Zh. Eksp. Teor. Fiz.*, v. 50, p. 1322 (1966). [*Sov. Phys. JETP*, v. 23, p. 878, (1966)].

[5] L. Kramer, *Phys. Rev.* v. 170, p.475 (1968).

[6] H.J.Fink, A.G.Presson, *Phys. Rev.* v. 182, p.498 (1969).

[7] L.G.Aslamazov, S.V.Lempickii, *Zh. Eksp. Teor. Fiz.*, v. 84, p. 2216 (1983).

[8] I.Aranson, M.Gitterman, and B.Y.Shapiro, *Phys. Rev. B*, v. 51, p.3092 (1995).

[9] R.Kato, Y.Enomoto, and S.Maekawa, *Phys. Rev. B*, v. 47, p.8016 (1993).

[10] R.Kato, Y.Enomoto, and S.Maekawa., *Physica C*, v. 227, p.387 (1994).

[11] C.Bolech, G. C. Buscaglia, and A.Lopez, *Phys. Rev. B*, v. 52, p. R15719 (1995).

[12] E. Zeldov, A. I. Larkin ,V.B. Geshkenbein, M. Konczykowski, D. Majer, B.Khaykovich, V.M. Vinokur and H.Shtrikman, *Phys. Rev. Lett.* v. 73, p. 1428 (1994).

[13] A. Fortini and E. Paumier, *Phys. Rev. B*, v. 14, p. 55 (1976).

[14] I. L. Maksimov and A. A. Elistratov, *JETP Lett.* v. 61, p. 208 (1995).

[15] I. L. Maksimov and A. A. Elistratov, *Appl. Phys. Lett.* v. 72, p. 1650 (1998).

[16] M. Yu. Kupriyanov and K. K. Likharev, *Sov. Phys. Solid State* v. 16, p. 1835 (1975)).

[17] I. L. Maksimov and A. A. Elistratov, *Appl. Phys. Lett.* v. 72, p. 1650 (2002).

[18] M. Benkraouda and J. R. Clem, *Phys. Rev. B*, v. 53, p. 5716 (1996).

[19] M.A. Lavrentjev and B.V. Shabat, Methods of complex variable functions theory, Moscow, *Nauka*, 1973.

[20] V.V. Shmidt, Introduction into physics of superconductivity, Moscow, *Nauka*, 2001.

[21] E.H. Brandt, *Phys. Rev. B,* v. 59, p. 3369 (1999).

[22] V. Jeudy and D. Limagne, *Phys. Rev. B*, v. 60, p. 9720 (1999).

[23] N. Morozov, E. Zeldov, M.Konczykowski and R.A. Doyle, *Physica C* v. 291, p. 113 (1997).

[24] L.P.Gor'kov, N.B.Kopnin, *Usp. Fiz. Nauk*, v.116, p.413 (1975). [*Sov. Phys. Uspekhi*, v. 18, p.496 (1975)].

[25] Ivlev B.I. and Kopnin N.B., *Usp. Fiz. Nauk*, v.142, p.435 (1984). [*Sov. Phys. Uspekhi*, v. 27, p.206 (1984)].

[26] D.Yu. Vodolazov, I.L. Maksimov, E.H. Brandt, *Physica C*, v. 384, p.211 (2003).

[27] F.F.Ternovskii and L.N.Shekhata, *Zh. Eksp. Teor. Fiz.*, v.622, p.2297 (1972). [*Sov. Phys. JETP*, v. 35, p. 1202, (1972)].

[28] D.Yu. Vodolazov, I.L. Maksimov, E.H. Brandt, *EPL*, v. 48, p.313 (1999).

In: New Frontiers in Superconductivity Research
Editor: Barry P. Martins, pp. 107-143
ISBN 1-59454-850-1

Chapter 4

AMORPHOUS NB-GE THIN FILMS AS A MODEL SYSTEM FOR EXPERIMENTS ON FUNDAMENTAL PROPERTIES OF VORTEX TRANSPORT

Dinko Babić*
Department of Physics, Faculty of Science,
University of Zagreb, Bijenicka 32,
HR-10000 Zagreb, Croatia

Abstract

Amorphous Nb_xGe_{1-x} thin lms have recently attracted the attention of researchers interested in basic properties of vortex dynamics. These samples are conventional, isotropic superconductors in the dirty limit, with a high Ginzburg-Landau parameter $\kappa \sim 80 - 100$, the transition temperature $T_c \sim 2.7 - 3$ K and the upper critical magnetic eld B_{c2} at zero temperature of $4 - 5$ T. Their amorphous structure and homogeneity lead to a low density of defects and consequently to a very weak pinning. Together with the experimental accessibility of the entire magnetic eld vs temperature phase diagram of the mixed state, this makes them suitable for investigations of those aspects of vortex motion which are general rather than sample speci c. I shall highlight several experimental results the understanding of which has bene ted from this unique combination of the simplicity of vortex matter and marginal role of pinning in the studied amorphous $Nb_{0.7}Ge_{0.3}$ microbridges. The phenomena under discussion are of a fundamental character and are not manifested in such a clear form in most other type II superconductors, where they are usually hindered by other mechanisms - such as strong pinning in classical superconductors or uctuation phenomena inherent in the depinned vortex state of high-T_c compounds. The experimental methods used in the studies presented are measurements of the time-averaged voltage (magnetoresistance, voltage-current characteristics) and the related voltage noise, whereas the analysis relies on the Larkin-Ovchinnikov theory of ux ow in its original or modi ed form. Another advantageous property of amorphous Nb-Ge lms is that they maintain their superconducting properties when deposited onto a substrate through a mask produced by standard electron-beam lithography. Thus the samples can be made in various shapes containing features as small as a few hundred nanometres. Since we

*E-mail address: dbabic@phy.hr

already understand well how vortices in amorphous Nb-Ge lms move, it is perhaps not too bold to predict an upcoming development in fabricating mesoscopic devices which would involve vortex motion as the principle of operation.

Keywords: Vortex, Flux Flow, Dissipation, Nonequilibrium, Fluctuations

1 Introduction

Almost half a century since they took part in a breakthrough that elucidated the phenomenon of superconductivity, V. L. Ginzburg and A. A. Abrikosov were in 2003 nally awarded the Nobel prize for physics. Together with Landau, Ginzburg introduced a complex macroscopic wave function $\psi = |\psi| \, e^{i\varphi}$ as the superconducting order parameter, the super uid density being $n_s = |\psi|^2$ and the phase φ accounting for the quantum nature of superconductivity. This was a foundation of extending the Landau theory of phase transitions to the case of superconductivity, which resulted in two famous Ginzburg-Landau (GL) equations as the basis for our understanding of various superconducting properties.

The GL theory approached superconductivity in a general manner, and it was obvious that different situations required different applications of the GL formalism. One of the most intriguing ndings was that there were two sorts of superconductors, today known as type I and type II. In the former class the superconductivity was destroyed by applying a small magnetic eld, whereas the latter ones endured much larger magnetic elds, exhibiting a weakened diamagnetic response but not the complete loss of superconductivity. The weakened and still diamagnetic magnetisation implied a co-existence of normal and superconducting domains, therefore the name the *mixed state*, but the understanding of this phenomenon had to wait for the concept of the *supercurrent vortex*, introduced by Abrikosov in 1957 [1].

By analysing what would happen if the magnetic- eld penetration depth λ were larger than the coherence length ξ he found that the energy of interface between the superconducting and normal domains favoured the contact area as large as possible. Consequently, the magnetic eld penetrates a type II superconductor in the form of normal-metal tubes of a radius $\sim \xi$. The superconducting order parameter increases from zero in the centre of the tube to a constant value outside, whereas the magnetic- ux density decreases away from the centre with a decay length $\lambda > \xi$.

As the distance from the tube centre increases, the local magnetic eld decreases because it is screened by the supercurrent that circulates around the normal tube core and forms a *vortex*. The continuity of ψ along a circular path around the core sets the requirement for each vortex carrying an integer number of the magnetic ux quantum $\phi_0 = h/2e = 2.07 \times 10^{-15}$ Tm^2, where h is the Planck constant and $2e$ the charge of a Cooper pair. Another picture introduced by Abrikosov was the *vortex lattice*. If a vortex is exposed to a current density J perpendicular to its axis, it will feel a force (per unit length) $J\phi_0$, so a vortex collective will be in a static equilibrium only if $J = 0$ at the position of each vortex. This is realised through a regular arrangement, i.e. the vortex lattice, in which a vortex occupies a position where the supercurrents of all other vortices cancel out. The validity of this concept was demonstrated by experiments which made use of a magnetic decoration technique combined with electron microscopy [2].

From the point of view of technological applications of superconductivity, type II superconductors have for decades been in the focus of interest due to their capability of sustaining large currents without showing any resistance. The central problem in high-current applications, e.g. for strong magnets, has always been how to immobilise vortices. If moving with a velocity u, vortices generated by a magnetic field B induce an electric field $E = uB$, which results in power dissipation. Vortex motion is impeded by *flux pinning*, present in all real materials. Areas where the pairing in a superconductor is less favourable then elsewhere attract vortex cores - a normal metal, thus a vortex at such a place is in a potential well and energy is required for setting it in motion. Pinning is hence associated with imperfections in crystal and chemical structure, and, as a rule of thumb, less "perfect" superconducting materials dissipate less power than their more ordered counterparts with the same average chemical composition. The question of how to increase the pinning in a particular material that may be promising for high-current applications belongs equally to the domains of fundamental and applied research.

The purpose of this Chapter is to address a different, actually an opposite issue. Namely, despite enormous efforts made in the past to understand how vortices move under the influence of an applied current, we have only recently began to test the existing theories and to develop new ones - in order to meet the challenge of upcoming experimental findings. Flux pinning, mentioned above, has for decades been a serious obstacle to our understanding of *basic rules* of vortex dynamics. These basic rules refer to vortex systems with no pinning whatsoever, so that already a vanishingly small J causes a vortex motion which we then call *flux flow* (FF).

In classical (poly)crystalline superconductors the pinning is usually strong, sample dependent, and dominates the power dissipation almost up to the upper critical magnetic field B_{c2}, which causes difficulties in studying the generic properties of FF. For example, since FF in these samples does not occur over a wide range of B and temperature T, there has been a common misconception in using the Bardeen-Stephen model [3] uncritically and too often in vague, extrapolated forms. This model predicts FF resistivity $\rho_f \approx \rho_n(B/B_{c2})$, where ρ_n is the normal-state resistivity, but it is valid only in the limit $T \ll T_c$ and $B \ll B_{c2}$, as noted later in Section 2.3. On the other hand, pinning in this regime is appreciable and a verification of $\rho_f \propto B/B_{c2}$ dependence is not straightforward.

When the high-T_c superconductors (HTS's) appeared in 1986, and after it became obvious that these compounds had a quite low irreversibility line $B_{irr}(T)$ - a magnetic field above which pinning is negligible, there was a hope that these materials were suitable for investigating FF. However, later it turned out that the state for $B > B_{irr}$, often termed "vortex liquid", was dominated by peculiar fluctuations [4] linked possibly to the nature of superconducting pairing in HTS's [5]. Although ohmic, as expected for FF, in HTS's the dissipation for $B > B_{irr}$ cannot be unequivocally explained by mechanisms related to vortex motion. For example, in $Bi_2Sr_2CaCu_2O_{8+x}$ - which is the most two-dimensional of all HTS's, for both large [6] and small [7] magnetic fields the magnetoresistance follows remarkably well the B dependence characteristic of a depairing that occurs in two-dimensional superconductors *above* T_c. Although a qualitative explanation of such a decoherence phenomenon in the mixed state has been given [8], we still lack its microscopic description. Hence, the intricate properties of the mixed state of HTS's set restrictions to their suitability for research on fundamental properties of vortex dynamics.

Samples of another kind have appeared as candidates for such studies. These are thin lms of amorphous alloys of certain transition metals (e.g. Nb, Mo) with some elements of the IV group (Si, Ge). The idea behind these superconducting systems is rather simple: there are no crystallographic defects in amorphous materials, so if we can produce stable, homogeneous lms they should exhibit a weak pinning. This eventually turned out to be correct. They had been known before the discovery of HTS's, but, as the whole research in superconductivity virtually revolved around HTS's for more than a decade, they were somewhat put aside. The interest for them has recently been renewed, for several reasons.

First, after incredibly extensive work on vortex transport in HTS's it has become rather clear that we have attempted to deal with extremely complex vortex systems without previously gaining enough knowledge about fundamental principles of vortex-motion power dissipation. Such a topic requires a simple, weakly-pinned vortex matter - which is ful lled in the case of the lms mentioned. Second, their production - done routinely by magnetron sputtering, is compatible with standard resists for electron-beam lithography. This has offered a new opportunity: using these lms in nanoscale science and technology, for instance as parts of hybrid superconducting samples [9], superconductors with locally controlled pinning sites [10], the smallest superconducting wires [11], etc.

In this Chapter I shall concentrate on a research carried out on one type of samples mentioned above: amorphous $Nb_{1-x}Ge_x$ microbridges (a-$Nb_{1-x}Ge_x$) with $x \approx 0.3$, i.e. narrow wires between two large contact pads. Interestingly, the crystalline counterpart of almost the same stoichiometry (Nb_3Ge) had until the era of HTS's the highest $T_c \sim 23$ K of all known superconductors. Amorphous $Nb_{0.7}Ge_{0.3}$ thin lms have a low T_c, in the range $2.7 - 3$ K, and the GL parameters are $\xi(0) \sim 7$ nm, $\lambda(0) \sim 1$ μm, $\kappa \sim 80 - 100$. These values may vary slightly from sample to sample, but one property stays the same: the weak pinning, especially if the lms are in two-dimensional limit (lm thickness d is of the order or less than 2ξ). This permits clean investigations of FF over a large part of the magnetic eld vs temperature (B, T) phase diagram of the mixed state, with the background pinning being either irrelevant or accountable.

The studies presented refer to vortex transport close to equilibrium, when vortex cores maintain essentially their static properties, as well as to the dynamic effects that, for instance, lead to dramatic nonlinearities in the voltage-current characteristics. The latter phenomena are related to the situations when the distribution function of normal excitations is no longer set by the bath temperature only. In turn, the usual two-dimensional (B, T) phase diagram becomes three-dimensional, with the induced electric eld as the third axis.

Before proceeding I would like to comment on the units and notation used. I shall insist on SI units for all physical quantities and expressions, and shall try to stay as much as possible with the notation usual in superconductivity research. However, I shall use *tesla* as the unit for both magnetic- ux density and applied magnetic eld, as ampere per metre - strictly speaking appropriate in the latter case, is less commonly used. This choice is justi ed reasonably by the fact that a-$Nb_{1-x}Ge_x$ is a high-κ superconductor, so the magnetic- ux density in most cases of interest does not depart much from the applied magnetic eld.

2 Overview of the Relevant Theory

In this Section the theoretical models used to discuss the experiments on vortex motion in a-$Nb_{0.7}Ge_{0.3}$ microbridges will be presented. I shall restrict myself to highlighting the physical pictures and main results, leaving to the reader to consult the cited literature for details.

2.1 Determination of Superconducting Parameters

Amorphous $Nb_{1-x}Ge_x$ thin lms are weak-coupling, s-wave superconductors in the dirty limit, having the mean free electron path l much smaller than the BCS coherence length ξ_0. The basis of theoretical approach to this category of superconductors has been given by Gor'kov [12], and used later by Kes and Tsuei who derived practical expressions for calculating the values of superconducting parameters directly from transport measurements [13], as follows.

The GL coherence length of a superconductor with $l \ll \xi_0$ and a diffusion constant D is close to T_c given by

$$\xi(t) = \sqrt{\frac{e\phi_0 D}{8k_B T_c(1-t)}} \ , \tag{1}$$

where $t = T/T_c$, k_B is the Boltzmann constant, and (for l much smaller than the sample dimensions) $D = v_F l/3$ links the diffusion constant and the Fermi velocity v_F. This expression agrees numerically very well with a coherence length $\tilde{\xi}(t) = 0.855[\xi_0 l/(1-t)]^{1/2}$ calculated using the BCS expression $\xi_0 = \hbar v_F/\pi\Delta_0$, where $\Delta_0 \approx 1.764 k_B T_c$ is the superconducting gap Δ at $T = 0$. We can therefore, without a loss of generality, use Eq.(1) in connection with the GL solution for the upper critical magnetic eld $B_{c2}(T) = \phi_0/2\pi\xi^2(T)$. We de ne

$$S = -\left(\frac{dB_{c2}}{dT}\right)_{T=T_c} = \frac{B_{c2}(0)}{T_c} = \frac{\phi_0}{2\pi T_c \xi^2(0)} \ . \tag{2}$$

The above equations are valid only close to T_c, thus one should not forget that the actual value of $B_{c2}(T=0)$ is around 70 % of $B_{c2}(0)$ [14]. The superconducting condensation energy density is given by $U_c = B_c^2(0)/2\mu_0 = N(0)\Delta_0^2/2 = 1.56N(0)(k_B T_c)^2$, where $N(0)$ is the density of electron states at the Fermi level per one direction of spin, $B_c(0)$ the thermodynamic critical magnetic eld at $T = 0$, and $\mu_0 = 4\pi/10^7$ H/m. Since the resistivity of a diffusive conductor at $T = 0$ is given by $\rho_0 = 1/2e^2 D N(0)$, we can use this and Eqs.(1),(2) together with

$$-\left(\frac{dB_c}{dT}\right)_{T=T_c} = 1.73\frac{B_c(0)}{T_c} \tag{3}$$

and

$$\kappa = \frac{1}{\sqrt{2}}\left(\frac{dB_{c2}/dT}{dB_c/dT}\right)_{T=T_c} = \frac{1}{1.73\sqrt{2}}\frac{B_{c2}(0)}{B_c 0} \tag{4}$$

to calculate (in SI units, i.e. ρ_0 in Ωm, S in T/K)

$$\kappa = 3.54 \times 10^4 \sqrt{\rho_0 S} \ . \tag{5}$$

For most situations the temperature dependence of κ can be disregarded. To nd the GL magnetic- eld penetration depth $\lambda(0)$ we use the BCS expressions $\lambda(0) = \lambda_L(0)(\xi_0/l)^{1/2}$ and $\kappa = 0.715\lambda_L(0)/l$, where λ_L is the London penetration depth, and obtain

$$\lambda(0) = 1.63\kappa\xi(0) \ . \tag{6}$$

For a-$Nb_{1-x}Ge_x$ lms the errors in experimentally determined S and ρ_0 are acceptably small, and the superconducting parameters can be found with suf cient accuracy. This is demonstrated in Fig.1, where the resistivity $\rho(T)$ of an a-$Nb_{0.7}Ge_{0.3}$ microbridge is shown together with the $B_{c2}(T)$ (inset). Far above $T_c \approx 2.75$ K (determined as the midpoint of the 10 % - 90 % transition curve), i.e. where there is no rounding of $\rho(T)$ due to the superconducting uctuations, the resistivity is almost insensitive to temperature ($\rho^{-1}(d\rho/dT) \sim 10^{-4}$ K^{-1}) and can hence be used for accurate determination of ρ_0 by extrapolation.

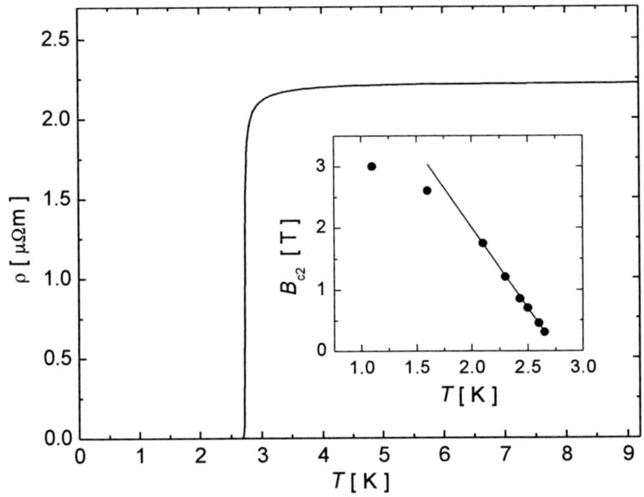

Figure 1: The resistivity $\rho(T)$ of an a-$Nb_{0.7}Ge_{0.3}$ microbridge close to the superconducting transition at $T_c \approx 2.75$ K. Inset: The values of $B_{c2}(T)$ found from the magnetoresistivity measurements (circles). The solid line is a t of $B_{c2}(T) = B_{c2}(0)(1 - T/T_c)$ to the points close to T_c.

The $B_{c2}(T)$ values shown in the inset to Fig.1 by the circles originate from measurements of the ohmic magnetoresistivity in the mixed state, carried out using a small applied current. The methods of nding B_{c2} will be discussed later, but their results agree within 10 %, which is suf cient for deducing S by tting the expression $B_{c2}(T) = B_{c2}(0)(1-T/T_c)$ to the data close to T_c. Such a t, with $B_{c2}(0) = 7.2$ T and $S = 2.6$ T/K, is shown in the inset to Fig.1 by the solid line.

In conclusion, to determine the superconducting parameters of a superconductor in the dirty limit it is suf cient to nd T_c, ρ_0, S, and insert these into formulae summarised in Table 1. The values typical of a-$Nb_{1-x}Ge_x$ lms are $T_c \sim 2.8$ K, $\rho_0 \sim 2.5$ $\mu\Omega$m and $S \sim 2.5$ T/K.

Table 1: Summary of the expressions for determination of superconducting parameters of dirty superconductors from transport measurements. To calculate a parameter use the given expression, insert ρ_0 in Ωm, S in T/K and T_c in K, and multiply with the corresponding multiplier. The values typical of a-Nb$_{1-x}$Ge$_x$ lms are given in the bottom row.

	κ	$\xi(0)$ [m]	$\lambda(0)$ [m]	$B_c(0)$ [T]	$N(0)$ [J^{-1}m^{-3}]
expression	$(\rho_0 S)^{1/2}$	$(T_c S)^{-1/2}$	$(\rho_0/T_c)^{1/2}$	$T_c(S/\rho_0)^{1/2}$	S/ρ_0
multiplier	3.54×10^4	1.81×10^{-8}	1.05×10^{-3}	1.15×10^{-5}	1.78×10^{41}
typical of a-Nb$_{1-x}$Ge$_x$ lms	90	7×10^{-9}	1000×10^{-9}	30×10^{-3}	1.8×10^{47}

2.2 Equilibrium (B,T) Phase Diagram

Equilibrium (B,T) diagram describes the nature of a vortex state for a given combination of B and T. If there is no pinning, three phase boundaries are relevant: the lower critical magnetic eld B_{c1} - separating the Meissner phase and the mixed state with the vortex lattice, B_{c2} - where a second-order transition to the normal state occurs, and the *melting* □*eld* B_m - where the vortex lattice undergoes a transition to a state conceptually similar to a liquid. Since $B_{c1} \sim B_{c2} \ln\kappa/\kappa^2$, in high-$\kappa$ superconductors $B_{c1} \ll B_{c2}$ and the $B < B_{c1}$ region is small compared to the whole superconducting part of the (B,T) plane. Moreover, the $B_{c1}(T)$ line is in many cases considerably below those which mark important changes in vortex transport. In particular, it is of little relevance for this topic and will not be discussed further.

The vortex lattice may have a square or triangular symmetry, the latter being favoured energetically. For theoretical calculations, such as those in Sections 2.3 and 2.5, the round-cell approximation is sometimes used, i.e. the Wiegner-Seitz cell of the lattice is replaced by a circle of the same area. At a nite temperature vortices oscillate around their equilibrium positions, which may lead to melting into a state termed "vortex liquid" - where no long-range order exists and vortices are essentially individual entities. According to the Lindemann criterion [15] this occurs when the mean vortex displacement $\langle X^2\rangle^{1/2}$ increases beyond a certain fraction $c_L < 1$ of the vortex-lattice constant a. Therefore, B_m is de ned by the condition $c_L^2 a^2 = \langle X^2\rangle$. Other possible causes of vortex-lattice melting are the unbinding of vortex-antivortex pairs that may exist in two-dimensional superconductors [16], and superconducting-order-parameter uctuations of $3DXY$ class [17, 18] found in certain HTS's. The transition at B_m can be detected by measurements of thermodynamic quantities, for instance magnetisation [19] or speci c heat [20], but can also leave its signature in transport properties [21].

Pinning de nes additional lines in (B,T) diagram. The most general of these is the *irreversibility* □*eld* $B_{irr}(T)$, which separates a vortex state with nite pinning $(B < B_{irr})$

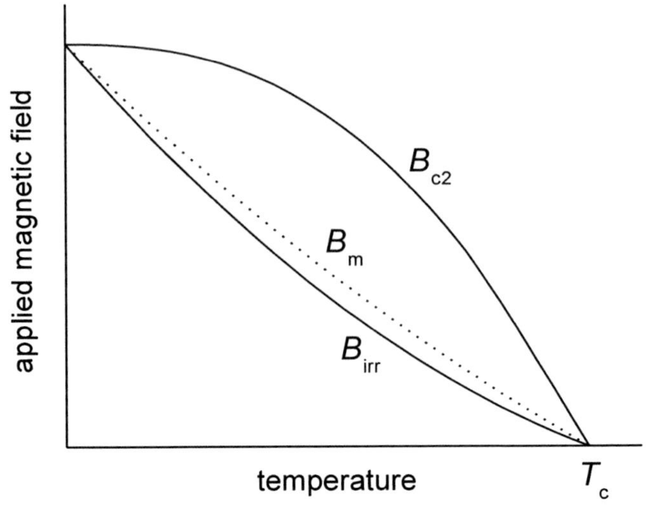

Figure 2: Schematic of equilibrium (B, T) phase diagram of superconductors similar to a-Nb$_{1-x}$Ge$_x$ lms. Only the boundaries relevant for vortex transport are shown, these are B_{c2}, B_{irr} and B_m. For $B < B_{irr}$ pinning is important and vortices form a solid which can be a vortex lattice over the entire sample, a system of crystalline domains, or a glass. In the range $B_{irr} < B < B_{c2}$ pinning is immeasurably small, thus already a vanishingly small current would initiate ux ow. A possible position of B_m is shown by the dotted line.

from a depinned vortex state ($B > B_{irr}$). Its name originates from magnetisation experiments, as the magnetisation is irreversible for $B < B_{irr}$ and reversible for $B > B_{irr}$. Sometimes B_m and B_{irr} can coincide, but more generally B_m is considered to be the upper limit of B_{irr} [4]. Just as above B_{irr} vortices do not necessarily form a regular array, below B_{irr} pinning may reduce the long-range crystalline order and cause a breakdown of the lattice into regions with a short-range order [22], or even a completely glassy vortex structure.

Two boundaries in (B, T) diagram can be found relatively easily from transport measurements - B_{c2} and B_{irr}, while determination of B_m in principle requires a more detailed comparison with theory. Although B_{c2} and B_{irr} can also be inferred from theoretical models, they can be estimated with acceptably small error by following these rules: at a given temperature, $B_{c2}(T)$ is the magnetic eld at which the magnetoresistivity approaches the normal-state resistivity, whereas at $B_{irr}(T)$ every measurable indication of pinning in the voltage-current characteristics disappears. The transitions at these boundaries are smooth, hence in practice there are transition widths which appear as errors in the determined values.

A (B, T) diagram characteristic of samples considered here is depicted in Fig.2. The shapes and relative ratios of B_{c2} and B_{irr}, shown by the solid lines, are similar to those for an 18-nm-thick a-Nb$_3$Ge lm studied in Ref.[23] (lms with $d \gg \xi$ were found to have B_{irr} closer to B_{c2}). The dotted line represents a possible B_m, which, in principle, can coincide with B_{irr}. A theoretical prediction for B_m, at intermediate temperatures positioned similarly as shown, has been calculated in Ref.[24] and compared to the experimental results of Ref.[23].

From the point of view of standard treatments of B_{irr}, B_m and B_{c2}, at $T = 0$ they all should merge at a single point, i.e. $B_{c2}(T = 0)$. However, there are experimental indications suggesting a separation of B_{c2} and B_m at $T = 0$ due to a quantum effect [23, 24], which is outside the scope of this Chapter.

2.3 Flux Flow Close to Equilibrium

Let us suppose that a magnetic eld $\mathbf{B} = B_z\hat{\mathbf{z}}$, where $|\hat{\mathbf{z}}| = 1$, is applied perpendicularly to a sample which stretches over a length L in the x direction and a width W in the y direction. If a vortex enters the sample at $y = 0$, travels across it with a uniform velocity u_y in the y direction and leaves out at $y = W$, classically one expects induction of a voltage $V = \phi_0 u_y/W$ over the x direction. The classical approach is, however, not without ambiguities, which one can verify by applying the Faraday's law to a square magnetic- ux pulse of a duration W/u_y.

On the other hand, superconductivity is a quantum phenomenon, thus the failure of classical physics to describe FF dissipation is not entirely surprising. A potential difference generated over a superconductor is generally related to the time derivative $\dot{\varphi}$ of the phase of ψ via the Josephson's equation $V = (\phi_0/2\pi)\dot{\varphi}$, which should hold for FF as well. A vortex moving with a velocity $u_y = \dot{y}$ causes a phase growth $\varphi(y) = 2\pi y/W$, so its travel over the whole sample shifts φ by 2π. This gives $V = \phi_0 u_y/W$ in a natural way. Since vortex density is given by $n = B_z/\phi_0$, for $N = nLW$ vortices we obtain the induced electric eld in the x direction $E_x = -u_y B_z$. This result can be written more generally as $\mathbf{E} = \mathbf{B} \times \mathbf{u} = \phi_0 n(\hat{\mathbf{z}} \times \mathbf{u})$.

The motion of vortices is essential in producing the dissipation also from the point of view of applying a current. An applied current will simply avoid static vortex cores, whereas the current density in a core will be nite, actually exactly equal to the applied current density J, if the vortex moves [25]. This associates a power-dissipation density of the order of $\rho_n J^2$ with each core. The interaction of J with the magnetic ux results in a force $\phi_0 \mathbf{J} \times \hat{\mathbf{z}}$ (per unit length) on vortices. This force is opposed by a viscous damping which is for slow vortices linear in $\mathbf{u}$ and proportional to a viscosity coef cient η. We can therefore write the equation of motion for vortex trajectories uniform in space:

$$\phi_0 \mathbf{J} \times \hat{\mathbf{z}} = \eta \mathbf{u} \ . \tag{7}$$

If vortex trajectories are not spatially uniform the inertial vortex mass m [26] should be taken into account as well, i.e. a term $m\dot{\mathbf{u}} = m(\mathbf{u} \cdot \nabla)\mathbf{u}$ should be added to Eq.(7). By nding $\mathbf{u}$ one can calculate the induced electric eld $\mathbf{E}$.

The physics of FF resistivity is related to the properties of normal electrons in the cores. The dissipation mechanisms are conveniently expressed through the viscosity coef cient η that enters Eq.(7) as a phenomenological parameter. There are two main contributions to FF resistivity [27]. One is related to Joule heating of electrons in the cores, which is described by the Bardeen-Stephen model [3] and used above to depict the simplest aspect of FF dissipation. The second mechanism, inferred by Tinkham [28], corresponds to a time dependence of ψ arising from vortex motion, since electrons must pair and depair at a rate u/ξ as vortices move. The time $\tau \sim \hbar/\Delta$ required for nucleation of superconductivity is nite, which causes a lagging of ψ behind the instantaneous equilibrium of a translatory vortex system. This results in a power-dissipation density of the order of $(u/\xi)^2 \tau U_c$, i.e. in each pairing-depairing cycle a fraction $(u/\xi)\tau$ of the superconducting condensation energy density U_c is dissipated as electrical resistance in the cores.

Larkin and Ovchinnikov (LO) quanti ed the above pictures by carefully taking into account the dynamic effects on ψ and the electron distribution function in the mixed state. They calculated FF conductivity σ_f for all cases where a static solution for $\psi(\mathbf{r})$ was known [27]. Derived for dirty superconductors, these results are appropriate for a-Nb$_{1-x}$Ge$_x$ lms. The LO expressions for σ_f close to equilibrium, i.e. when vortex cores keep their equilibrium properties, are summarised below.

- $T \ll T_c$, $B \ll B_{c2}$
  In this regime vortices are far from each other, their contributions to the resistivity simply add and the temperature dependence of σ_f corresponds to that of B_{c2}:

$$\sigma_f = 0.9\sigma_n/b \ , \tag{8}$$

  where $\sigma_n = 1/\rho_n$ is the normal-state conductivity and $b = B/B_{c2}$ a reduced magnetic eld representing the fraction of the normal state. Note that, apart from the factor 0.9, Eq.(8) reproduces the Bardeen-Stephen model. This is indeed the simplest situation which permits the usual visualisation of FF resistivity $\rho_f = 1/\sigma_f$ as ρ_n times the fraction of the normal state.

- B close to B_{c2}, any T
  Close to B_{c2} vortex cores almost overlap and ψ is small over the entire sample, almost resembling that of a gapless superconductor [29]. In this case σ_f is not much larger than σ_n, taking a form

$$\sigma_f = \sigma_n[1 + \alpha(T)(1 - b)] \ , \tag{9}$$

  where $\alpha(T)$ is a temperature-dependent constant with the values between 2 and 4.

- T close to T_c, any B
  The GL formalism can be used close to T_c to nd the solution for ψ, but the form of the electron distribution function imposes a necessity for a numerical calculation of σ_f, which is no longer a simple linear function of either b^{-1} or $(1 - b)$ (as in Eqs.(8) and (9), respectively). The result is

$$\sigma_f = \sigma_n \left[1 + \frac{g(b)}{b(1-t)^{1/2}}\right] \ . \tag{10}$$

  The function g can be approximated by the interpolation formulae $g(b) = 4.04 - b^{1/4}(3.96 + 2.38b)$ for $b < 0.315$, and $g(b) = 0.43(1 - b)^{3/2} + 0.69(1 - b)^{5/2}$ for $b > 0.315$.

2.4 Vortex Motion in the Presence of Pinning

As already noted, pinning sets up a force that tends to slow down or completely immobilise vortices. The simplest way to describe this effect is to introduce a critical current density J_c

as a depinning threshold, such that $u = 0$ for $J < J_c$ and $u = \phi_0(J - J_c)/\eta$ for $J \geq J_c$. If J_c in a given (B, T) point is a constant, then one expects to measure $\rho_f = dE/dJ$, i.e. the *differential resistivity* is predicted to agree with FF resistivity given by Eqs.(8)-(10). Such a "kink" structure of the voltage-current curves implies an instantaneous depinning of the whole lattice and a collective vortex motion due to the long-range correlations within the lattice. Signatures of threshold depinning are sometimes observed [30], but more frequently depinning starts below the threshold and progresses relatively smoothly. In other words, there is a certain current range between the rst appearance of E and the linear $E \propto (J - J_c)$ behaviour, suggesting that vortices may rst move over short distances locally before the depinning occurs at a global scale. This is commonly ascribed to a thermally-activated hopping vortex motion (HVM), which is sometimes also called *□ux creep*.

A vortex experiencing a pinning potential oscillates at a constant temperature around its equilibrium position, and if the pinning is weak enough a vortex jump to a neighbouring pinning site may occur. A homogeneous pinning potential can be in a given direction of space modelled as a rippled "washboard" with the minima positioned at the pinning centres and the height of the ripples representing the magnitude U_0 of the potential barrier for the jumps. In equilibrium ($J = 0$) the vortex jumps occur over a distance l_j in all directions equally, at a rate $\nu_0 \propto \exp(-U_0/k_BT)$. If $J \neq 0$ the "washboard" potential is tilted along the driving force $\mathbf{F} \propto \mathbf{J} \times \mathbf{B}$, which can be represented by a splitting $U_\pm = U_0 \mp U_F$ of the barrier height for the jumps in the direction (+) and opposite (-) to $\mathbf{F}$. The imbalance in the jump rates $\nu_\pm \propto \exp[-(U_0 \mp U_F)/k_BT)]$ causes a net vortex velocity $u = l_j(\nu_+ - \nu_-)$ perpendicular to J and thus an electric eld in the direction of J, i.e. a longitudinal dissipation. Because $u \propto \sinh(U_F/k_BT)$ and $U_F \propto J$, for $J \to 0$ we obtain $u \propto J$, which leads to a linear $E = \rho J$ dependence. This ohmic magnetoresistivity should not be confused with ρ_f - which is the highest resistivity that vortices can produce close to equilibrium. Even the smallest pinning results in a reduction of ρ below ρ_f.

There are numerous versions and re nements of the above two models of vortex motion in the presence of pinning, and we shall not go further with these details. Instead, it is instructive to address a question of how vortices move if the whole lattice does not undergo a collective translatory motion or vortices do not jump individually. The possibility of a situation in between these two extremes leads to the picture of the *vortex bundle*. This notion assumes that vortices can more easily overcome a potential barrier if they form bundles and act together, since the force on a bundle comprising $\mathcal{N}$ vortices is $\mathcal{N}$ times larger than that on a single vortex. The usual argument for the validity of this concept is the result of an early experiment on vortex-motion shot noise [31], the interpretation of which was later shown to still be open to debate (see e.g. Ref.[32]).

2.5 Flux Flow out of Equilibrium

As mentioned in Section 2.3, FF dissipation depends on the distribution of normal excitations in the mixed state. The electric eld generated in the cores by vortex motion introduces changes to the distribution function - which is for large E no longer set by the bath temperature only. These changes lead to a different, nonequilibrium FF, which results in strongly nonlinear voltage-current characteristics. The nonlinearities occur for J larger than that corresponding to ohmic σ_f of Section 2.3 but still much below $\sim en_s\Delta/p_F$ that marks depairing of Cooper pairs by a large superfluid velocity (p_F is the Fermi momentum).

The strongest effect of electric field on the distribution function takes place close to T_c, where a nonequilibrium correction of the order of $k_B(T_c - T)$ can be significant. By accounting for the effect of electric field, LO found that in nonequilibrium Eq.(10) transforms into

$$J = \sigma_n \left[1 + \frac{g(b)}{b(1-t)^{1/2}} \frac{1}{1 + \left(\frac{E}{E_i}\right)^2} \right] E \ . \tag{11}$$

The second term in the square parentheses contains a factor $[1+(E/E_i)^2]^{-1}$ that reflects a decrease in η and thus a vortex motion faster than in equilibrium. The characteristic electric field E_i corresponds to a vortex velocity $u_i = E_i/B$, which depends on T and microscopic sample parameters.

Namely, if an external perturbation drives electrons out of equilibrium, their collisions with phonons at a rate $\tau_{e,ph}^{-1}$ (at low temperatures, $\tau_{e,ph}^{-1} \propto (T/\theta_D)^3$, where θ_D is the Debye temperature) set $\tau_{e,ph}$ as the relaxation time for reaching the equilibrium after the perturbation has been turned off. This relaxation time is independent of whether the metal is in the normal or superconducting state. In the superconducting state not every electron-phonon collision is effective, so close to T_c the relaxation occurs at a rate $\tau_\epsilon^{-1} \sim (\Delta/k_B T_c)\tau_{e,ph}^{-1} < \tau_{e,ph}^{-1}$ [33]. If vortices in a diffusive superconductor are so fast that they travel a distance $l_\epsilon = \sqrt{D\tau_\epsilon}$ during the time τ_ϵ, the electrons excited by E will no longer be able to relax within the cores and will escape into the surrounding superfluid area. This defines $u_i \sim l_\epsilon/\tau_\epsilon = \sqrt{D/\tau_\epsilon}$, and if we take into account that $\Delta \propto (1-t)^{1/2}$ we obtain $u_i \propto [D(1-t)^{1/2}/\tau_{e,ph}]^{1/2}$. The precise expression for u_i is given by [27]

$$u_i = \sqrt{\frac{D[14\zeta(3)(1-t)]^{1/2}}{\pi \tau_{e,ph}}} \ , \tag{12}$$

where ζ is the Riemann zeta function. The above mechanism results in a shrinking of vortex cores according to

$$\xi^2(t,E) = \frac{\xi^2(t,0)}{1 + \left(\frac{E}{E_i}\right)^2} \ , \tag{13}$$

which continues until $\xi(t,E) = \xi(t,0)(1-t)^{1/4}$.

The $J(E)$ curve defined by Eq.(11) may have an "N" shape if E_i is small, i.e. a maximum of $J(E)$ around E_i followed by a region of $dJ/dE < 0$ and then again by a J increasing with increasing E. The region of negative differential conductivity is forbidden in current-biased measurements of voltage-current curves and is thus expected to result in a steep jump of the measured voltage, which is termed *flux-flow instability* (FFI).

LO discussed certain limitations to the applicability of the above model. First, they argued that it might not be valid if B is not small compared to B_{c2}. Namely, because electrons are heated by the electric field in the cores, and this heat is removed by phonons in the whole volume, the requirement for the electron system to remain in thermal equilibrium with the phonons is that the fraction of the normal state is small ($B \ll B_{c2}$). Hence, for

B sufficiently close to B_{c2} LO predicted a heating of electrons above the bath temperature, which was later expressed more quantitatively by Bezuglyj and Shklovskij who set the upper limit for the validity of the pure LO mechanism, described by Eq.(11), to $B \sim 0.4B_{c2}$ [34]. The second limitation to the appearance of vortex core shrinking is that it is not likely to occur far below T_c because for $T \ll T_c$ the distribution function is only weakly sensitive to nonequilibrium perturbations unless they are rather strong.

Another scenario of nonlinear flux flow and FFI was not discussed by LO, that is, a possibility of the electron heating for bath temperatures well below T_c. Although this idea, presented very recently [35, 36], has not obtained a rigorous microscopic description yet, it is physically clear. If the number of thermal phonons is small it may happen that electrons become heated to a temperature T^* above the phonon temperature T_0, which can occur for any B because it is the small number of phonons, and not the fraction of the normal state, which prevents an efficient heat removal. This nonequilibrium effect on the electron distribution function can be expressed through $T_0 \rightarrow T^*$, i.e. a dynamic equilibrium is established within the electron system at a temperature T^* and with a certain relaxation time τ_ε that is not related only to electron collisions with thermal phonons. This affects FF dissipation via

$$b = \frac{B}{B_{c2}} \quad \Longrightarrow \quad b^* = \frac{B}{B_{c2}(T^*)} = \frac{B}{B_{c2}(E)} \quad . \tag{14}$$

Inserted into Eq.(8) the above replacement may result in an N-shaped $J(E)$ and consequently an FFI, and if it is used together with Eq.(9) it gives a new line in the phase diagram of the mixed state: a transition to the normal state at a certain induced electric field $E_c(B,T)$.

3 Experiments on Vortex Motion in $Nb_{0.7}Ge_{0.3}$ Microbridges

The central topic of this Chapter are experiments on vortex motion in a selected type of amorphous a-$Nb_{1-x}Ge_x$ films, i.e. microbridges with 70 % of Nb and 30 % of Ge. These are $d = 20$ nm thick wires, of a width $W = 1 - 5$ μm and a length $L = 10 - 210$ μm, connecting two large contact pads. The thickness has been chosen such that the range $B_{irr} < B < B_{c2}$ is large enough for studies of vortex-motion phenomena not only in the presence of weak pinning but also in the depinned state (see Section 2.2). The microbridge geometry is suitable for addressing both the fundamentals of vortex transport and the local effects related to artificially created features which affect vortex motion. After ultrasonic bonding of the voltage probes close to a bridge, typically 95 % of the measured voltage drop occurs in the wire, whereas the additional voltage generated in the contact pads can be easily accounted for by geometrical arguments.

An example of such a sample [37] is shown in Fig.3. Although the experiments discussed in the rest of the Chapter refer to plain microbridges, i.e. simple a-$Nb_{0.7}Ge_{0.3}$ wires between contact pads, it is worthwhile to show a sample which illustrates possibilities of using a-$Nb_{1-x}Ge_x$ as an integral part of mesoscopic devices. The main geometrical relations, e.g. the size of the contact pads, the bonding scheme, the directions of $\mathbf{J}$, $\mathbf{B}$ and $\mathbf{u}$ (right inset), are the same for all the samples studied - with and without additional features on the bridges.

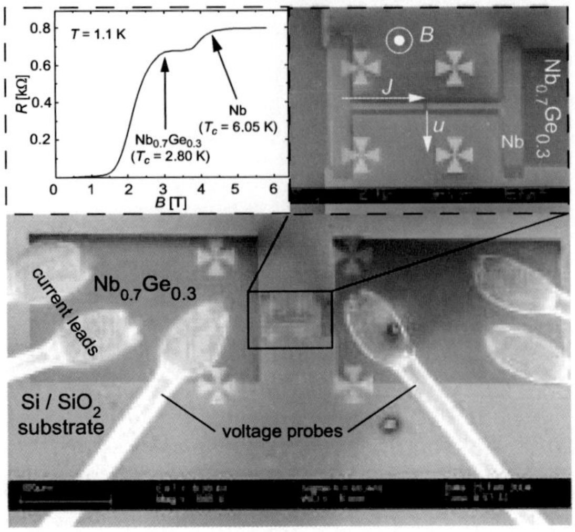

Figure 3: A hybrid sample comprising an a-Nb$_{0.7}$Ge$_{0.3}$ microbridge ($L = 50$ μm, $W = 5$ μm, $d = 20$ nm, on an Si/SiO$_2$ substrate) on top of which is a longitudinal Nb line (800 nm wide, 50 nm thick) with an opening (500 nm wide) in the middle. The crosses are alignment marks for multiple lithography steps. Right inset: The microbridge enlarged, with the indicated directions of $\mathbf{J}$, $\mathbf{B}$ and $\mathbf{u}$ used in the experiments presented. Left inset: The ohmic magnetoresistance at $T = 1.1$ K of the sample shown. Note the two transitions at the B_{c2} of a-Nb$_{0.7}$Ge$_{0.3}$ and Nb, as indicated by the arrows.

The samples are produced by combining the masks made by electron-beam lithography and magnetron sputtering as the film-deposition technique. Since a-Nb$_{1-x}$Ge$_x$ films remain stable at several hundreds of degrees centigrade [38] one can safely bake a resist spun on top of a previously deposited a-Nb$_{1-x}$Ge$_x$ film, which permits multiple lithography steps. For large samples one can use mechanical masks, but it is more advantageous to employ electron-beam lithography. The resist can be either poly(methyl methacrylate) (PMMA) [10, 36] or a double-layer combination of PMMA/MA [poly(methyl methacrylate / methacrylic acid)] underneath PMMA [32]. PMMA/MA provides a larger undercut and thus a larger distance between the resist and the deposition area, but for a-Nb$_{1-x}$Ge$_x$ this seems to be of little relevance. Various substrates give good results, for instance oxidised Si [10, 36, 32], sapphire [38], etc. Thus prepared masks can be used for the film deposition by magnetron sputtering.

This deposition method provides alloy films of good homogeneity and constant composition [39]. In general, Nb-Ge films can be obtained by using either a single Nb-Ge alloy target of appropriate composition or two elemental targets of Nb and Ge which are simultaneously sputtered. In the latter case, films of reproducible composition can be deposited in subsequent processes as long as the relationship between the individual magnetron power and the corresponding sputtering rate has been determined once. In our case, a-Nb$_{1-x}$Ge$_x$ films ($x \approx 0.3$) were obtained by simultaneous dc (Nb) and rf (Ge) magnetron sputtering from elemental targets in Ar atmosphere with the substrate held at room temperature. Prior to film deposition a presputtering of Nb in a high-vacuum system is beneficial in order to reduce further the background pressure to below the 10^{-5} Pa range by utilising the strong adsorption properties of Nb.

By the method described one obtains samples of high quality, homogeneous and with nearly the same values of superconducting parameters. For example, of around twenty samples only one showed a weak signature of inhomogeneity detected in the measured transport properties, and even then it was at a level of a few percent. Furthermore, we have produced several samples with small perforating holes in the microbridge area as arti cial pinning centres [10], nding no change whatsoever in the superconducting properties when compared to simultaneously produced plain microbridges. They differed only with regard to the pinning - which was a behaviour expected qualitatively and then studied in detail, but had the same $T_c = 2.75$ K and other superconductivity parameters.

Besides patterning perforating holes as pinning centres, one can think of making hybrid samples consisting of a superconductor as a medium for vortex motion and a material that pins a magnetic ux. The latter constituent of such a sample can be a magnetic material [40] or a strong-pinning superconductor, both of which can localise vortices. However, the superconducting order parameter may be affected by additional pair breaking in the proximity of a local permanent magnetic moment, whereas it remains a global property in the case of a superconducting addition in the sense outlined above. The sample in Fig.3 has been produced under these assumptions, i.e. that, without altering the generic superconducting properties, vortex motion in amorphous a-$Nb_{0.7}Ge_{0.3}$ could be controlled by adding regions of low vortex mobility. In our case the strong-pinning superconductor is a dirty niobium lm, but it is possible to use other materials as well, for instance NbN [9].

As shown in the right inset to Fig.3, vortices are created by applying a magnetic eld perpendicular to the lm plane; they are set in motion by applying a current along the bridge, so the vortex velocity is directed perpendicularly to the main bridge axis. In the sample shown the current density is localised around the opening in the Nb line - where it is also nonuniform in space. So, the vector $\mathbf{u}$ may depart locally from the direction indicated by the arrow in the right inset. The same applies whenever there are any other mesoscopic elements which modify the current paths, for example perforating holes discussed in detail in Ref.[10]. This complicates Eq.(7) in the sense of an additional term which contains the inertial vortex mass (see the discussion in Section 2.3). If no current-modifying elements are present, i.e. for simple bridges, the vortex velocity is uniform and the relations between $\mathbf{B}$, $\mathbf{J}$ and $\mathbf{u}$ are over the whole bridge as depicted in Fig.3 by the arrows (apart from insigni cant deviations close to the contact pads).

In the left inset to Fig.3 the ohmic magnetoresistance of the sample at $T = 1.1$ K is shown [37]. The zero-B resistance vs temperature curve (not shown) displays a steep decrease at $T = 6.05$ K and a superconducting transition of a-$Nb_{0.7}Ge_{0.3}$ at $T_c = 2.80$ K, below which the resistance is zero. The former step is attributed to the superconducting transition of the Nb line, although it occurs at around two thirds of T_c of bulk Nb (9.5 K), while the resistance drop at 2.80 K implies that the quality of a-$Nb_{0.7}Ge_{0.3}$ is not diminished by the presence of Nb. If we now look at the magnetoresistance at 1.1 K, we again observe a two-step behaviour, now corresponding to the vortex-motion dissipation. The lower transition at ~ 3.1 T agrees very well with that of a simultaneously measured plain microbridge, whereas the upper step at ~ 4.3 T is interpreted as the transition at the B_{c2} of Nb. This high value of the upper critical eld of Nb is probably related to its high level of disorder, which can also explain the decrease of the corresponding superconducting transition temperature. The reduction of the T_c of the Nb line, deposited by *magnetron sputtering*, is moderate in comparison to that which can occur for mesoscopic Nb wires produced by *evaporation technique* in combination with a PMMA mask, where T_c may drop to $\sim 2-3$ K.

The properties of the sample shown in Fig.3 imply that a-Nb$_{0.7}$Ge$_{0.3}$ lms can be combined with other superconducting materials to form mesoscopic hybrid samples consisting of regions of high and low vortex mobility. In our case the second, strong-pinning superconductor is particularly simple - a disordered Nb lm produced by magnetron sputtering. Having this technological problem mainly solved, one can design circuits for controlled vortex motion, which has already been started in several research laboratories. However, to deal with such an issue one rst has to understand well the basic properties of vortex motion in a-Nb$_{0.7}$Ge$_{0.3}$. This will be discussed in the remainder of this Chapter by means of selected examples of vortex-motion experiments on plain microbridges, i.e. samples similar to that in Fig.3 but without the Nb part. It will be shown that a number of these experimental ndings can be explained within the framework of the LO theory, which certainly adds to our current understanding of vortex dynamics.

3.1 Types of Vortex Motion Dissipation

Before turning to details of vortex dynamics in a-Nb$_{0.7}$Ge$_{0.3}$ microbridges, it is useful to discuss the types of the related dissipation semiquantitatively. This can be done by addressing the voltage-current curves in characteristic (B, T) points. The terminology used is given in Section 2.

We rst have to distinguish between the *close-to-equilibrium* and *nonequilibrium* conditions. In the former case vortex motion is governed by the properties set by a particular point in the equilibrium (B, T) phase diagram, with J being a weak perturbation, thus vortices undergo either HVM or FF - which both result in a linear $E(J)$. In a-Nb$_{0.7}$Ge$_{0.3}$ microbridges this occurs for $J \sim 0.1 - 1$ MA/m^2. The nonequilibrium conditions refer to the situations where the applied current plays an important role and leads to nonlinearities in the $E(J)$. At moderate J, for a-Nb$_{0.7}$Ge$_{0.3}$ microbridges typically 10-100 MA/m^2, the nonlinearities appear due to a competition between the driving force and the pinning force. In this process, originating from similar strengths of the two forces, vortex cores still preserve the equilibrium properties, therefore we refer to these conditions as to *weak nonequilibrium*. By increasing the applied current further a strong electric eld develops in the cores, which transforms them into entities beyond the equilibrium description and results in nonlinearities in the $E(J)$ as discussed in Section 2.5. For a-Nb$_{0.7}$Ge$_{0.3}$ microbridges this *strongly nonequilibrium* state is fully established above $J \sim 100-200$ MA/m^2.

The measured $E(J)$ is indeed as varied as anticipated above, which is demonstrated in Figures 4 and 5 for two characteristic temperatures - close to and well below T_c. The results refer to a ($W = 5$ μm, $L = 210$ μm) microbridge with $T_c = 2.75$ K, $\rho_0 = 3.3$ $\mu\Omega$m and $S = 2.6$ T/K. The two characteristic temperatures are $T = 2.5$ K ($t = 0.91$) and $T = 1.1$ K ($t = 0.40$), and the corresponding B_{c2} values are 0.65 and 3.0 T, respectively, as determined from the ohmic magnetoresistivity (see Section 3.2). A set of experimental $E(J)$ measured at $t = 0.91$, over a range of applied B as a parameter, is shown in Fig.4A, whereas in Fig.4B the same curves are plotted in a log-log scale. The results for $t = 0.40$ are displayed in Fig.4C and again plotted in a log-log scale in Fig.4D. For convenience, the scales in Figs.4A and 4C, and Figs.4B and 4D, are chosen to be the same. Two typical $E(J)$ curves in the regime where the strong nonequilibrium is not yet reached are displayed in Fig.5 in an expanded scale, i.e. this is the $E(J)$ for $t = 0.91$ and $b = 0.26$ (Fig.5A), and for $t = 0.40$ and $b = 0.30$ (Fig.5B).

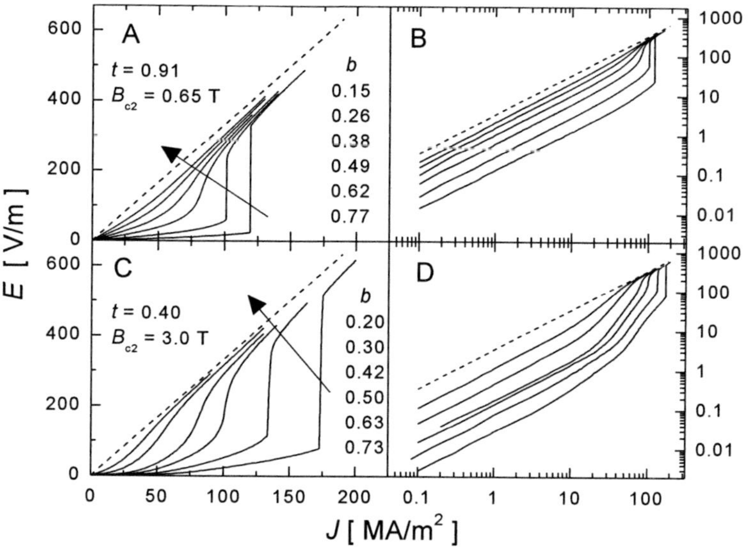

Figure 4: $E(J)$ of a microbridge with $T_c = 2.75$ K, $\rho_0 = 3.3$ $\mu\Omega$m and $S = 2.6$ T/K, for two characteristic temperature regimes: (A,B) close to T_c ($T = 2.5$ K, $t = 0.91$, $B_{c2} = 0.65$ T), and (C,D) well below T_c ($T = 1.1$ K, $t = 0.40$, $B_{c2} = 3.0$ T). The applied magnetic eld increases as shown by the arrows in A,C. The data in A and B are the same, but plotted differently, and the same applies to C and D. The dashed lines represent the normal-state electric eld $E_n = \rho_n J$.

We may rst focus on temperatures close to T_c, i.e. $1 - t \ll 1$, where B_{irr} is very low (e.g., for $t \approx 0.90$ less than 10 % of the B_{c2}). Hence, for $J \to 0$ one expects FF over a wide range of the applied B, with $E \propto J$ and dE/dJ equal to ρ_f given by Eq.(10). It is also expected that this region crosses-over directly into the strongly nonequilibrium regime, according to Eq.(11) for the LO mechanism of vortex-core shrinking. Such a structure of the $E(J)$ is really observed. The dissipation at low-to-moderate J, corresponding to the close-to-equilibrium conditions, is followed by a nonlinear change to E close to $E_n = \rho_n J$ (shown by the dashed line) at high J. For low b this change occurs in the form of an FFI *jump* predicted by Eq.(11) for a suf ciently small electric eld E_i, as seen for $b = 0.15$ and $b = 0.26$. For higher b the jump is smeared out but the measured $E(J)$ still follows Eq.(11), which will be discussed in more detail in Section 3.3.

The ohmic character of $E(J)$ prior to the onset of the strong equilibrium is demonstrated in Fig.4B, where the linearity extends over more than two decades in J. Another evidence for the validity of ascribing this linear $E(J)$ to FF is presented in Fig.5A, where a selected $E(J)$ (solid line) is plotted together with $E = \rho_f J$ (dashed line) calculated from Eq.(10) by inserting the independently determined values of T_c, B_{c2} and ρ_n. Obviously, below ~ 40 MA/m^2 the agreement of the experiment and theory is excellent, whereas the deviations at higher J indicate the appearance of the nonlinearity described by the term $[1+(E/E_i)^2]^{-1}$ in Eq.(11).

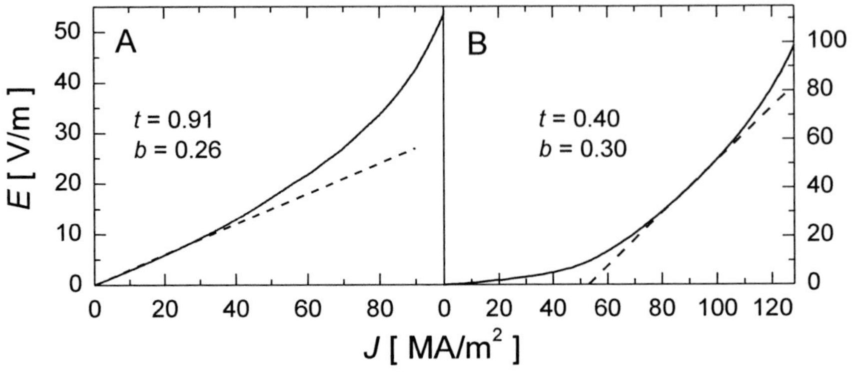

Figure 5: Selected $E(J)$ from Fig.4 in the regime where the strong nonequilibrium is not yet fully established. A: The measured curve for $t = 0.91$ and $b = 0.26$ (solid line) and a plot of Eq.(10) calculated using the independently determined B_{c2}, T_c and ρ_n (dashed line). B: The measured curve for $t = 0.40$ and $b = 0.30$ (solid line) and $E = \rho_f(J - J_c)$ plotted using ρ_f given by Eq.(8) and $J_c = 53$ MA/m^2 as the only adjustable parameter (dashed line). In both cases the curving-up of the measured $E(J)$ at high J corresponds to approaching the strong nonequilibrium.

Another temperature limit of interest is the situation well below T_c, i.e. $t \ll 1$, where all superconducting parameters are only weakly temperature dependent and pinning is no longer negligible - so that B_{irr} is closer to B_{c2} than at high temperatures discussed above. For $B < B_{irr}$ one expects a more complex $E(J)$, resembling HVM for $J \to 0$ (close to equilibrium), translatory motion of the vortex lattice against the pinning force at moderate J (weak nonequilibrium), and a signature of a strongly nonequilibrium regime at high J provided such a state exists. According to the discussion in Section 2.5, one could expect a strong nonequilibrium corresponding to a heating of electrons above the phonon temperature.

These three regimes, successive in J, imply $E \propto J$ for very small currents (with $dE/dJ < \rho_f$), then $E \propto (J - J_c)$ at moderate currents (with $dE/dJ \approx \rho_f$) and nally a nonlinear behaviour in the sense of Eq.(14) provided the absolute T is low enough (so that the heat removal by phonons is suf ciently ineffective). Such a structure of the $E(J)$ is indeed observed. The overall look of the $E(J)$ is shown in Fig.4C over a similar range of b as in Fig.4A. Although the $E(J)$ appears similar to that close to T_c, the differences are actually signi cant. These are revealed in Figs.4D and 5B. In the log-log plot of Fig.4D it can be seen that there is a linear part of the $E(J)$ at low J, but the transition to the strongly nonequilibrium $E(J)$ is not direct (as in Fig.4B). Furthermore, the slopes of the linear part are far below ρ_f.

This is also shown in Fig.5B, where the measured $E(J)$ for $b = 0.30$ (solid line) is compared to $E = \rho_f(J - J_c)$ (dashed line) plotted using $J_c = 53$ MA/m^2 and ρ_f calculated in terms of Eq.(8). The $E \propto (J - J_c)$ region is relatively small because the $E(J)$ curves up by approaching the strong nonequilibrium, which introduces an error in determining the slope dE/dJ. However, since the dashed line in Fig.5B - plotted by taking the theoretical $dE/dJ = \rho_f$, ts the measured $E(J)$ in a range of J it is reasonable to conclude that there is a collective translatory vortex motion against the pinning potential, bridging the regions of HVM and strong nonequilibrium. A detailed study of $E(J)$ in this regime [41], as well as the result of a vortex-motion-noise experiment [32], are in support of this interpretation.

Finally, although there is an FFI jump for low b, a quantitative analysis shows that neither the transition to the strong equilibrium nor the dissipation in this state can be described by Eq.(11) or its modi cations, whereas applications of the electron-heating approach seem to account for the experimental results adequately [35, 36]. This issue will be discussed in Section 3.4.

Hence, already at this level of analysis it is rather clear that several predictions of the theory of Sections 2.3-2.5, which are of general relevance, comply with the experimental results for a-$Nb_{0.7}Ge_{0.3}$ microbridges. This demonstrates clearly the usefulness of these samples in investigating the validities of a variety of theories and using thus acquired understanding in dealing with vortex motion in different experimental realisations.

3.2 Magnetoresistivity

In Section 3.1 it has been shown that the samples have an extended region of linear $E(J)$ at J of the order of 1 MA/m^2 or less. This allows us to de ne the ohmic magnetoresistivity $\rho = E/J$, originating in either HVM or FF. Signi cant deviations from the linearity occur for $B \to 0$ (in a very narrow B range) at low temperatures, where the pinning is strong enough to cause nonlinearities due to the onset of depinning.

The close-to-equilibrium regime can be therefore investigated by measuring ρ at a constant small J, performing either temperature or magnetic- eld sweeps. The maximum ρ, corresponding to FF resistivity ρ_f, is given by Eqs.(8)-(10), hence $\rho < \rho_f$ signi es a slowing-down of vortices by pinning. The $\rho(B,T)$ characteristics supplement the $E(J)$ results and provide a more complete quantitative understanding of vortex motion.

A measured $\rho(B)$ at a constant $T = 1.1$ K ($t = 0.40$) is shown in Fig.6 by the solid line. This result refers to the sample discussed in Section 3.1, so the corresponding $E(J)$ is displayed in Figs.4C, 4D and 5B. Qualitatively, it is obvious that the normal state resistivity ρ_n (dotted line) is approached around $B \sim 3$ T, while the initial part of $\rho(B)$ is far below the prediction of Eq.(8), plotted by the dashed line for $B < 1.5$ T. In this plot, $B_{c2} = 3.0$ T is assumed but any choice of B_{c2} around this value would clearly lead to the same conclusion. The above two observations can be analysed further.

The determination of B_{c2} in practice depends on a criterion used, so a comparison of ρ with theory in principle suffers form a related uncertainty. Nonetheless, as shown below, different criteria usually agree within 10 %, which is in many cases acceptable. The second-order transition at B_{c2} is characterised by smooth changes of measurable quantities within the transition width, which imposes the necessity of using extrapolation methods in determining the value of B_{c2}.

An example of this method is plotting Eq.(9) for ρ not much below ρ_n by choosing and varying the parameters α and B_{c2} within physically acceptable ranges until a reasonable agreement with the experimental ρ is obtained. This procedure is demonstrated in Fig.6, as indicated, where the theoretical curve (dashed line) corresponds to $\alpha = 3$ and $B_{c2} = 3.0$

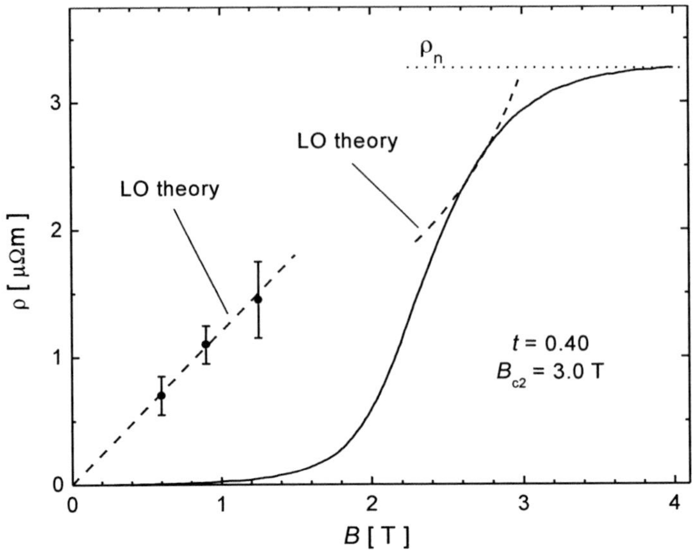

Figure 6: Measured magnetoresistivity of the sample discussed in Section 3.1 (solid line), obtained at $T = 1.1$ K ($t = 0.40$) by sweeping B . The dotted line represents the normal-state resistivity ρ_n, whereas the dashed lines are predictions of Eqs.(8) and (9), plotted using $B_{c2} = 3.0$ T and [for Eq.(9)] $\alpha = 3$. The solid circles are the slopes of some of the $E(J)$ in Fig.4C, inferred from the $E \propto (J - J_c)$ plots demonstrated in Fig.5B.

T. If no theory were available one could simply use a linear extrapolation of the upper part of $\rho(B)$, which would result in a not-much-different value of B_{c2}. Such simpli ed linear extrapolations are sometimes also used for estimating B_{c2} from other observables, e.g. magnetisation [42].

Below ~ 2.5 T the magnetoresistivity is clearly too small to correspond to FF, hence in this region the origin of $\rho(B)$ is in HVM. We can use the value of B_{c2} determined above and carry on with the plots $E = \rho_f(J - J_c)$, such as that shown in Fig.5B, to explore the applicability of Eq.(8). The quasilinear range between the HVM and strong nonequilibrium is de ned suf ciently well for B below ~ 1.3 T ($b \sim 0.45$), which allows us to determine the (approximately) constant slopes of experimental $E(J)$ in this regime. These are shown in Fig.6 by the solid circles and corresponding error bars, and compared to Eq.(8) (dashed line). The agreement is good enough for con rming the validity of Eq.(8) in terms of differential resistivity.

At temperatures closer to T_c, FF resistivity is given by Eq.(10), thus B_{c2} can be determined either by the above-described extrapolation or a t of Eq.(10) to the measured ρ. The latter approach is shown in Figs.7A and 7B, where for $T = 2.0$ K ($t = 0.70$) and $T = 2.5$ K ($t = 0.91$) the measured ρ is represented by the solid lines and the plots of Eq.(10) by the dashed lines.

The lower temperature, $t = 0.70$, is just at the border between the low-t (superconducting parameters depend weakly on t) and the high-t [superconducting parameters follow power laws in $(1 - t)$] regimes; see for example the inset to Fig.1. By a reasonable choice of B_{c2} as the only adjustable parameter in Eq.(10) - in this case 2.0 T, we obtain a good t

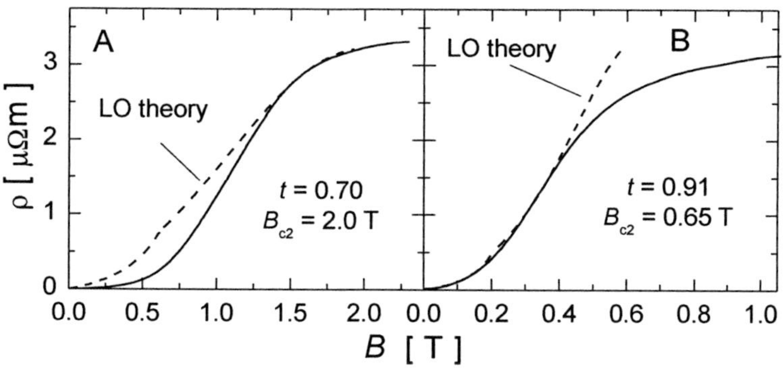

Figure 7: Measured magnetoresistivity (solid lines) closer to T_c, for $t = 0.70$ (A) and $t = 0.91$ (B). The dashed lines are plots of Eq.(10) using the interpolation formulae for the function g and the B_{c2} values of 2.0 T (A) and 0.65 T (B). The discrepancy between the experiment and theory are for $t = 0.70$ ascribed to a pinned state below $B_{irr} \sim 1.3$ T. For $t = 0.91$, B_{irr} is negligible but above $B \sim 0.4 - 0.5$ T the theory departs from the measured ρ presumably due to a widened uctuation region around the B_{c2}.

above $B \sim 1.3$ T. This also implies $B_{irr} \sim 1.3$ T, which agrees with the $E(J)$. A similar result, with $B_{c2} \approx 1.15$ T and $B_{irr} \sim 0.65$, was found for another sample ($L = 50$ μm, $W = 5$ μm) at $t = 0.82$ and con rmed independently by measurements of the voltage noise [32].

Even closer to T_c the irreversibility eld drops to a smaller percentage of B_{c2} and the magnetic- eld region of FF extends down to a quite low B. This situation is shown in Fig.7B for $t = 0.91$, where the deviation of Eq.(10), plotted using $B_{c2} = 0.65$ T, from $\rho(B)$ is very small from $B \to 0$ up to ~ 0.4 T, implying FF over this whole range of B. However, above $\sim 0.4 - 0.5$ T the magnetoresistivity departs signi cantly from Eq.(10), suggesting a wide uctuation range around the transition at B_{c2} when $T \to T_c$. The importance of non-Gaussian uctuations around B_{c2}, and their consequent in uence on observables in thus widened transition region, has been discussed in several theoretical works [43, 44].

3.3 Vortex-Core Shrinking Close to T_c

A number of experiments in the last decade have identi ed the existence of a jump in voltage-current characteristics in the mixed state of several type II superconductors, for instance $YBa_2Cu_3O_{7-\delta}$ [35, 45, 46, 47, 48, 49], Mo_3Si [50], 2H-$NbSe_2$ [51], Ta-Ge [52], Ti-N [53], etc., as well as in a-$Nb_{0.7}Ge_{0.3}$ [32, 36]. These jumps are dramatic, i.e. in a very narrow range of the applied current the measured voltage may rise for an order of magnitude, acquiring a value not much smaller than that corresponding to the normal state. Furthermore, the current density corresponding to such a jump is invariably several orders of magnitude below that expected for a depairing at the level of a single Cooper pair, and after the jump occurs the normal-state voltage is approached gradually rather than in an abrupt way.

These qualitative features, shown already in Fig.4, bring up a possibility of a vortex-motion phenomenon being the cause of the observed behaviour - therefore called FFI.

Several mechanisms have been ascribed to FFI, for example LO vortex-core shrinking [36, 45, 46, 48, 49, 50, 52, 53], depinning phenomena [51], electron heating [35, 36], hot spots [49], dynamic vortex-lattice crystallisation [54], with the applicability expected to depend on particular (B, T) conditions. A general requirement for the appearance of FFI is the existence of a region where the differential resistivity dE/dJ is negative, so that in a current-biased measurement of $E(J)$ the voltage avoids this forbidden region by jumping to the value permitted by $dE/dJ > 0$. This also implies a *hysteresis* in the $E(J)$, i.e. a difference between the ascending and descending current branches, which has indeed been observed [35, 50].

From a theoretical point of view, the most elaborate mechanism which can cause FFI is the LO vortex-core shrinking described in Section 2.5, since Eq.(11) implies that $J(E)$ may have a maximum at $E = E_i$. Experimentally, however, an unambiguous identi cation of the core-size reduction requires a careful analysis of the measured $E(J)$. Namely, the $E(J)$ predicted by Eq.(11) can also be smooth - exhibiting no jump, but if the underlying physics is correct it should hold for these smooth $E(J)$ as well. Most analyses of the LO model have relied on simply ascribing the position of the jump to E_i and subsequently calculating the characteristic velocity $u_i = E_i/B$. The latter quantity sometimes even showed a B dependence [46, 53] - in contrast to the prediction by LO, which led to extensions and modi cations of the original theory. The overall validity of the original LO approach as described by Eq.(11) has recently been tested by making use of the suitability of a-$Nb_{0.7}Ge_{0.3}$ microbridges for such an investigation [36].

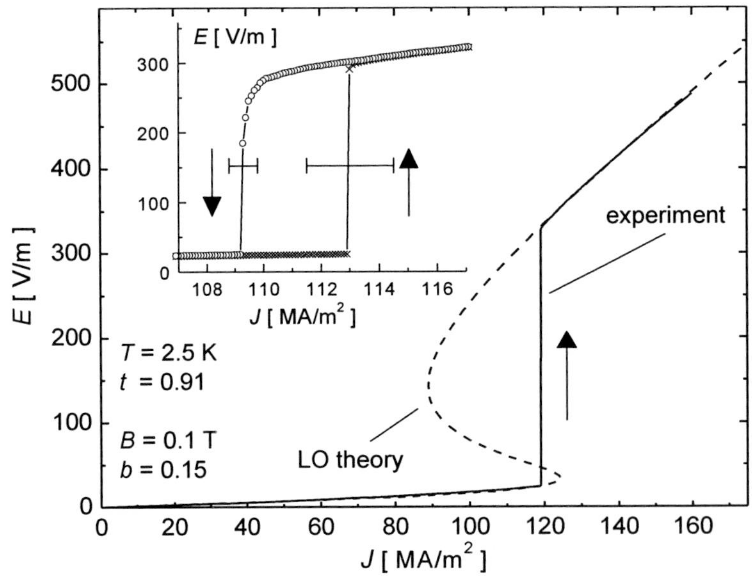

Figure 8: Comparison of a measured $E(J)$ with the FFI jump (solid line) from Fig.4A ($t = 0.91$ and $b = 0.15$), obtained by sweeping the applied current up, and the prediction of Eq.(11) shown by the dashed line. The region of $dE/dJ < 0$ is experimentally forbidden, resulting in a steep jump at $E_i \approx 30$ V/m and a hysteresis shown in the inset. The arrows point the current-sweep directions, and the horizontal error bars indicate the spread of the FFI in repeated measurements.

This check of the viability of vortex-core shrinking as the mechanism underlying the strongly nonequilibrium $E(J)$ and FFI jump has been performed by analysing the $E(J)$ in Fig.4A. These results refer to a temperature close to T_c, i.e. $t = 0.91$, and are therefore expected to be appropriate for such a task. To reduce the number of adjustable parameters in Eq.(11) the values of B_{c2}, T_c and σ_n have been determined independently as described in Sections 3.1 and 3.2. This means that only E_i has been left as a free parameter. The simplest way to nd E_i from a measured $E(J)$ is to extract $\tilde{J} = J - \sigma_n E$, which has a maximum at $E = E_i$ irrespectively of the shape of $E(J)$ (with or without FFI).

An $E(J)$ from Fig.4A ($b = 0.15$) with the FFI jump is shown in Fig.8 by the solid line. This curve was measured by increasing the applied current. The theoretical curve representing Eq.(11), with $E_i \approx 30$ V/m, is displayed by the dashed line. Obviously, when the maximum in $J(E)$ (exchange the axes to obtain this N-shaped curve) is reached the measured E cannot follow the $dE/dJ < 0$ part of the theoretical curve and hence jumps to the allowed $dE/dJ > 0$ branch. The $E(J)$ above the jump again agrees well with the theoretical prediction, which means that Eq.(11) can account for the FF dissipation on the both sides of the instability. This quantitative description of the $E(J)$ extends beyond previous analyses in terms of identifying the position of the jump only.

Besides the numerical procedure presented above, one can carry out an additional test of the structure of the FFI, that is, measurements of the $E(J)$ for increasing and decreasing applied current in order to detect a possible hysteresis. It is advisable to perform several short and slow current sweeps around the FFI in order to rule out a possibility of the hysteresis originating from artefacts such as, e.g., a temperature oscillation that may trigger the FFI. This is important especially when increasing J, since in this case the sensitivity to small perturbations is extreme as expected from the sharpness of the maximum of the theoretical $J(E)$. Hence, in different ascending current sweeps the currents corresponding to the FFI may be slightly different, since a small increase of E above E_i triggers the jump immediately. On the other hand, the sensitivity for a decreasing J should not be so critical, as the minimum of $J(E)$ is relatively broad.

The hysteretic behaviour was for our samples studied by repeating the current sweeps around the FFI at a rate ~ 0.1 MAm^{-2}s^{-1} and during a time period when the temperature stability, as given by the sensor reading, was within a few mK. The result of these measurements is shown in the inset to Fig.8. The directions of the current sweeps are indicated by the arrows, and the horizontal error bars show the ranges of the spread of the FFI in subsequent measurements. Note that the jump can occur over a wider range if J is increased, i.e. in this case the FFI is sharper - as expected from Eq.(11). In fact, in a single sweep the jump is restricted to within 0.1 MA/m^2 - which is the closest separation of two measurement points. In contrast to this situation, for decreasing J the FFI in different measurements spreads over a smaller range of J and one can also nd data points in the transition region, which implies a smoother transition - again as expected. If one compares the theoretical line in Fig.8 and the size of the hysteresis one nds that the hysteresis is considerably narrower than predicted theoretically, which has also been observed in other experiments of this kind [35, 50]. The precise reason for this reduction of the hysteresis width at the moment remains obscure.

If E_i is suf ciently large, Eq.(11) predicts a smooth $E(J)$ and no hysteresis. Since $E_i \propto B$ this is expected at higher B, which is the case indeed - as can be seen in Fig.4A.

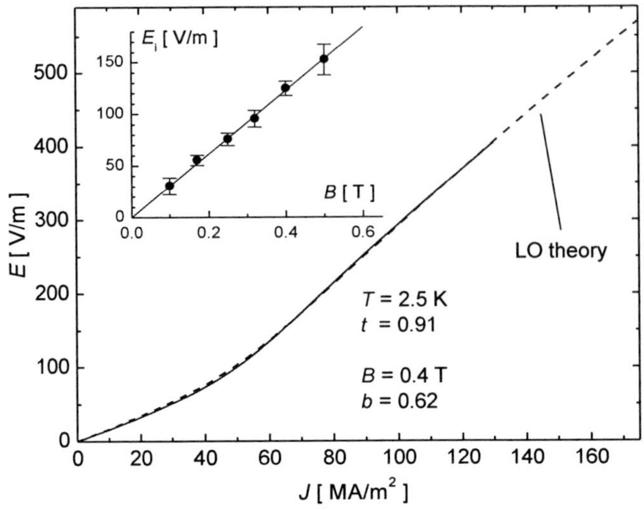

Figure 9: A measured smooth $E(J)$ (solid line) from Fig.4A ($t = 0.91$ and $b = 0.62$), compared to Eq.(11) calculated using $E_i \approx 125$ V/m (dashed line). Inset: The values of E_i inferred from the $E(J)$ in Fig.4A and plotted vs B. The solid line represents $E_i = u_i B$, where $u_i \approx 305$ m/s.

We can analyse the smooth $E(J)$ using the same procedure as described above to extract E_i and check the agreement with Eq.(11) by comparing the theoretical and measured $E(J)$.

A result typical of the smooth $E(J)$ is shown in Fig.9, where the measured $E(J)$ (for $b = 0.4$) is again represented by the solid line and the calculated curve (with $E_i \approx 125$ V/m) by the dashed line. As before, the agreement is remarkable, which suggests the validity of the LO mechanism even at high B. We can do the same for all the $E(J)$ in Fig.4A and plot the inferred E_i vs B, which is presented in the inset to Fig.9. The error bars indicate the range of E_i within which the t can be considered satisfactory given the errors in the used values of T_c, B_{c2} and σ_n. We see that the linear $E_i \propto B$ dependence is satis ed all the way up to $B = 0.5$ T ($b \approx 0.77$), i.e. for all the curves measured. The characteristic velocity, determined by the slope of the linear t (solid line), is $u_i \approx 305$ m/s. From Eq.(12) this gives the inelastic electron-phonon scattering time $\tau_{e,ph} \approx 0.18$ ns, which is of the order of magnitude expected for amorphous metals at low temperatures.

In the presented study the LO model holds up to at least $b \sim 0.75$, i.e. considerably above 0.4 which was in Ref.[34] regarded as the upper limit of the validity of the LO mechanism without any heating effect. The heating, of course, eventually must appear, but for the range of B and E of the curves in Fig.4A it seems to be of a secondary importance. Therefore close to T_c the vortex-core shrinking in its pure form has a range of occurrence wider than thought before.

3.4 Electron Heating at Low Temperatures

The jump in the $E(J)$ well below T_c - as presented in Fig.4C, seems to have the same shape as that in Fig.4A for T close to T_c, which would in turn suggest an extension of the vortex-core shrinking to unexpectedly low temperatures. This assumption is, however, far from being correct. Being initially misled with it, the author carried out an extensive numerical analysis in terms of Eqs.(8)-(11), checking several combinations of the close-to-equilibrium part and the decrease in vortex-motion viscosity due to the vortex-core shrinking given by the term $[1+(E/E_i)^2]^{-1}$. In short, the result of this procedure is: no numerical tting can result in any reasonable agreement between the mechanism of vortex-core shrinking and the $E(J)$ measured at low t. Quantitatively, this is a strong argument in favour of a difference between the strong equilibrium for $(1-t) \ll 1$ and $t \ll 1$. However, to explain the remarkable low-t nonlinearities in the $E(J)$ a different intuitive picture is required before one addresses the measured data numerically.

Such a model has been pointed out by Kunchur relatively recently [35] in his measurements and analysis of the $E(J)$ of several $YBa_2Cu_3O_{7-\delta}$ microbridges. He found a reasonable agreement of the (J, E) points that mark the FFI with a model of *electron heating* at low temperatures. As noted in Section 2.5, this point of view assumes an internal equilibration of the electron system at a temperature T^* above the phonon temperature T_0, in conditions of infrequent inelastic electron-phonon scattering. Since at the moment we lack a microscopic elaboration of this mechanism, we shall approach its manifestations by supposing that all observables in such a situation change their temperature dependences by a replacement $T_0 \rightarrow T^*$. Although this approximation may seem too crude, it is shown below that it nonetheless can account for the measured data surprisingly well.

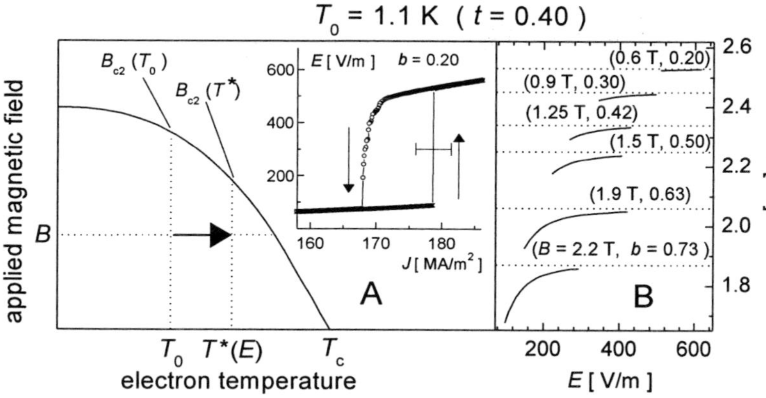

Figure 10: The model of electron heating by E in strongly nonequilibrium FF, with the experimental data obtained at the bath temperature $T_0 = 1.1$ K ($t = 0.40$), see Fig.4C. A: Schematic of the idea, assuming that the superconducting properties are governed by the electron temperature $T^*(E) > T_0$, reached as indicated by the arrow. Inset to A: A hysteretic $E(J)$ (for $b = 0.20$) implying an FFI at low b. B: The electron temperature calculated from the equilibrium $B_{c2}(T)$ characteristics.

The idea of the electron-heating approach is depicted in Fig.10A. A magnetic eld B is applied to a sample at a temperature T_0 which also sets the number of thermal phonons available for the heat removal from the electron system. The vortex motion produces power dissipation which, if the rate of electron collisions with thermal phonons is low, heats the electrons and thus shifts the electron temperature along the arrow shown. At the electron temperature $T^* > T_0$ all superconducting parameters acquire temperature dependences set by T^* rather than T_0, in particular the value of B_{c2} drops from $B_{c2}(T_0)$ to $B_{c2}(T^*)$.

This renormalised B_{c2} then governs the FF dissipation. Since T^* depends on E, the FF conductivity given by the formulae in Section 2.3 becomes nonlinear by a change $b \to b^*(E)$. Note that the decrease of B_{c2} implies a vortex-core *expansion* rather than contraction essential in the LO description.

There are two obvious impacts of this mechanism on the $E(J)$. The rst is the possibility of an FFI at low applied B, observed experimentally (see Fig.4C). Namely, $b^*(E)$ increases with increasing E, thus $J \propto E/b^*(E)$ may have a maximum de ned by $db^*/dE = b^*/E$. The second is a destruction of superconductivity when $B = B_{c2}(T^*)$, at E corresponding to a certain $E_c(B)$. The details of the FFI have been elaborated by Kunchur who has recorded a set of the FFI data appropriately dense in B. The main conclusion of this analysis is that the vortex velocity corresponding to the FFI is proportional to $1/\sqrt{B}$, i.e. it is not B independent as in the case of the LO mechanism. The Kunchur's picture implies again a hysteresis in the low-B $E(J)$, which was for our samples checked by measurements carried out in the same way as described in previous Section. The measured hysteresis (for $t = 0.40, b = 0.20$) is shown in the inset to Fig.10A. All remarks in Section 3.3 on the experimental approach for measuring the hysteresis apply here as well.

Instead of continuing with analysing the FFI we shall address in more detail the second issue - the destruction of superconductivity by the heating, a topic which was experimentally inaccessible to Kunchur due to the large B_{c2} in HTS's - *estimated* to around 100 T over most of the (B, T) diagram. For a-$Nb_{1-x}Ge_x$ this problem is absent, thus the $E \to E_n = \rho_n J$ data are available. Since we are dealing with a high-dissipation regime, where b^* is not much smaller than unity, the appropriate expression for analysing the $E(J)$ is given by Eq.(9) modi ed by a replacement $b \to b^*(E)$, so that $J = \sigma_f(E)E$. The parameter α in Eq.(9) is temperature dependent, and therefore changes with $T^*(E)$, but it has a restricted range of possible values (between 2 and 4) which gives a reason to believe that this variation may not affect the modelling substantially.

If the picture of a destruction of superconductivity by the electron heating is correct, it is plausible to assume that close to the dynamic phase boundary $E = E_c$ all physical variables are functions of $(1 - E/E_c)$. This approach is similar to assigning dependences on $(1 - T/T_c)$ in the vicinity of T_c, or on $(1 - B/B_{c2})$ close to B_{c2}. Since the way how the heating occurs depends on the applied magnetic eld, one expects that E_c is a function of B. With these notions we can invert Eq.(9) to calculate $b^* = 1 - (1/\alpha)[J/\sigma_n E - 1]$ for all the high-E data in Fig.4C. If α and $E_c(B)$ are chosen properly, then it should be possible to collapse all the data onto a single b^* vs $[1 - E/E_c(B)]$ curve.

The result of this procedure is presented in Fig.11A, where the scaled b^* vs $(1 - E/E_c)$ curve is shown. Only the high-b^* part is taken into account, because this is the range where the validity of Eq.(9) is expected. In this calculation, $\alpha = 3$ is taken for all the $E(J)$, which is an approximation the correctness of which is estimated when the analysis is completed. Note that the same value of α has been used in Fig.6 to analyse the corresponding close-to-equilibrium magnetoresistivity. Once α has been chosen, the obtaining of the scaling in Fig.11A corresponds to a determination of $E_c(B)$, which sets the horizontal axis.

The inferred values of E_c are plotted in Fig.11B by the symbols against equilibrium $(1 - b)$, exhibiting a linear dependence. Therefore, it is indeed possible to relate the $E(J)$ from Fig.4C to a change in B_{c2} by the heating, with the characteristic electric eld E_c exhibiting a simple relation to the initial conditions $[T_0, B/B_{c2}(T_0)]$. Finally, we can use

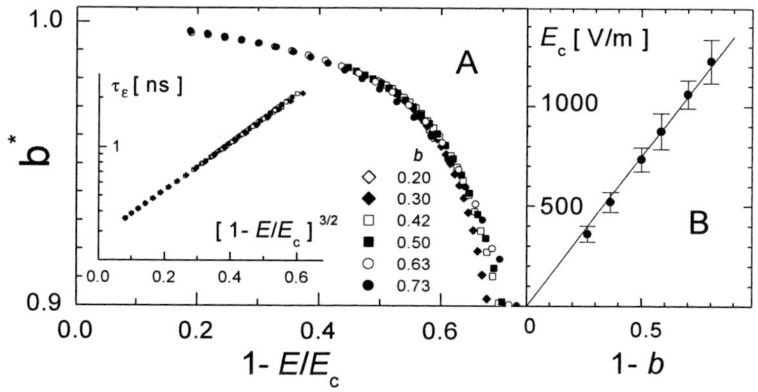

Figure 11: A: Scaling of the nonequilibrium reduced magnetic eld b^* with $[1 - E/E_c(B)]$ for the data in Fig.4C. In B, E_c is plotted against $(1 - b)$ and displays a linear $E_c = E_{c0}(1 - b)$ dependence. Inset to A: The nonequilibrium relaxation time τ_ε calculated for the data in A and plotted against $(1 - E/E_c)^{3/2}$ in a log-lin scale.

the equilibrium $B_{c2}(T)$ phase boundary to calculate $T^*(E)$, shown by the solid lines in Fig.10B for each b, by equating $B_{c2}[T^*(E)] = B/b^*(E)$. The $T_c(B)$ values de ned by the equilibrium condition $B = B_{c2}$ are shown by the dotted lines.

We can now turn more closely to the result $E_c = E_{c0}(1 - b)$, shown in Fig.11B by the solid line with a slope $E_{c0} \approx 1500$ V/m. This dynamic phase boundary can be related to thermodynamic properties of a superconductor in the mixed state, as follows. In the electron heating model the stationary state at T^* is established according to $JE\tau_\varepsilon = G_s(T_0) - G_s(T^*)$, since a part of the superconducting contribution G_s to the Gibbs free energy density is lost due to the "evaporation" of the super uid. The relaxation time τ_ε of the heated state has the same general meaning as the relaxation time τ_ϵ of the LO nonequilibrium state (see Section 3.3) but not necessarily the same origin.

To check whether the $E_c \propto (1 - b)$ behaviour can be derived using the simple argument given above we recall that for a high-κ superconductor close to B_{c2} we can write $G_s \approx U_c(1 - b)^2$ [29]. At the phase boundary E_c the above energy-balance equation takes a form $\sigma_n E_c^2 \tau_0 = U_c(T_0)(1 - b)^2$, where $\tau_0 \approx 0.25$ ns is the relaxation time at the destruction of superconductivity. This expression reproduces the $E_c \propto (1 - b)$ relation determined experimentally. We have here disregarded the change in the normal part of the Gibbs free energy density, but the obtained agreement justi es this approximation.

A step further is to study τ_ε, which provides information on the nature of the heated state. Using the approximate relationship $B_{c2} \approx B_c/\kappa\sqrt{2}$ and the obtained $B_{c2}(E)$ result we can nd $\tau_\varepsilon = [G_s(T_0) - G_s(T^*)]/JE$ from the data in Fig.11A, and plot it as a function of $(1 - E/E_c)$. We again obtain a scaling, of a de nite functional form $\tau_\varepsilon(E, B) = \tau_0 \exp\{3.5[1 - E/E_c(B)]^{3/2}\}$, as shown in the inset to Fig.11A. Note that the scaling of τ_ε is not a simple consequence of the scaling of b^*, because $B_{c2}(E)$ enters the expression for τ_ε independently of E_c.

Having such a well-de ned τ_ε we can discuss its underlying physical reason. Since no microscopic approach to the electron-heating model is available at the moment, we must use plausible assumptions as a groundwork for explaining the form of τ_ε. When heat is delivered to the electron system some super uid evaporates, which creates nonequilibrium quasiparticles. These quasiparticles live for a time τ_ε, after which they recombine into Cooper pairs and release energy, and this is what the nonequilibrium stationary state is supposed to look like. Since the collisions with thermal phonons are occasional only, there must be another mechanism responsible for the recombination. A good candidate for such a process is the emission of *nonequilibrium phonons*, known to be characteristic of the response of a superconducting lm to energy deposition [55].

Let us now try to set up a simple plausible model for the observed form of τ_ε, keeping in mind that such a simpli cation can be tested only by a rigorous microscopic calculation. Imagine that by reducing $E > E_c$ the superconducting gap $\Delta(E, B)$ starts to grow at $E = E_c$ from zero. As Δ increases, the number n_q of the nonequilibrium quasiparticle decreases, and if this decrease is exponential with Δ, which is not unreasonable for excitations over an energy gap, the differential equation for this process follows $d\Delta \propto -dn_q/n_q$. For each recombination we need two quasiparticles, hence the corresponding scattering rate is $\tau_q^{-1} \propto n_q^2$, resulting in $d\Delta = Cd\tau_q/2\tau_q$ with C being a constant. Since we cool the system down from $E_c(B)$ to a certain (E, B) point we can integrate $\int_0^{\Delta(E,B)} d\Delta = (C/2)\int_{\tau_0}^{\tau_\varepsilon(E,B)} d\tau_q/\tau_q$ to obtain $\tau_\varepsilon(E, B) = \tau_0 \exp[2\Delta(E, B)/C]$. As physical quantities close to E_c are expected to be functions of $(1 - E/E_c)$, we can assume that the superconducting gap follows a simple power law $\Delta(E, B) = \Delta(T_0)[1 - E/E_c(B)]^\mu$. With $\mu = 3/2$ this reproduces the observed form of τ_ε. Regarding the prefactor 3.5 in the experimental result for τ_ε it is tempting to speculate that it follows from $\Delta(T_0) \approx \Delta_0$ at low temperatures and the constant C being around $k_B T_c$. This should, however, be considered as a curiosity, not a substantiated argument.

3.5 Flux-Flow-Resistance Fluctuations

In Sections 3.3 and 3.4 it has been shown that changes in vortex cores have a strong in uence on FF dissipation. We have seen that they can shrink (close to T_c) and expand (at low T), thus one could ask whether their size can uctuate - similarly as the volume of a gas can not only shrink or expand but can also uctuate. This analogy becomes more transparent if one adopts a picture in which the vortex boundary is stabilised by equal pressures of the normal state from inside and the super uid from outside the core.

In measurements of the time-averaged voltage (V), presented so far, these uctuations cannot be detected since the average value of V is recorded by a voltmeter. Therefore, in order to detect possible uctuations of the core size one has to think of measuring the variance $(\delta V)^2$ of V. We see from Eqs.(8)-(10) that FF conductivity depends on two variables: σ_n and the vortex-core area $A_c \sim \xi^2\pi$ linked to B_{c2} via $\xi^2 = \phi_0/2\pi B_{c2}$. If σ_n is constant irrespectively of what happens in the cores, then in a measurement where the applied current is kept constant the uctuation $(\delta V)^2$ is related to the uctuation $(\delta A_c)^2$ of the core size.

Measurements of $(\delta V)^2$ are carried out by recording the voltage noise. In this method one measures a voltage uctuation averaged over some time and performs the Fourier transform to obtain the noise power spectrum $S_V(f)$ in a selected frequency window (f_1, f_2), a function used in theoretical studies of noise. The mathematics of S_V can be found in

Ref.[56], but since the exact formalism is in our case not essential we shall use a simpli ed approach in which $S_V = \langle[\delta V(f)]^2\rangle/\Delta f$, where $\Delta f = (f_2 - f_1) \ll f$ and "$\langle\rangle$" denotes the average.

In many cases S_V can be approximated by a product of two functions, one depending on the average voltage and the other on frequency and some characteristic time ϑ for the uctuations, i.e. $S_V = F(V)K(\vartheta f)$. Voltage noise can originate in a variety of phenomena which lead to different S_V, but three types are observed quite frequently and therefore deserve to be mentioned explicitly.

The rst is the Johnson-Nyquist noise which appears for all resistive samples of a resistance R at a nite T. Up to the quantum limit $k_BT \sim hf$, $F = 4k_BTR$ and $K \approx 1$, so this noise does not depend on V (i.e. appears also if no source is connected to a sample) and is *white* (frequency independent). The second is the shot noise, i.e. a noise produced by moving particles of a *granular* character, such as electrons (quantised electrical charge) or vortices (quantised magnetic- ux charge). In this case, $F = 2\Gamma QV$, where e.g. Q = eR (for electrons) or ϕ_0 (for vortices), and Γ is the so-called Fano factor which depends on the details of a particular particle transport. The related K is again white up to relatively high frequencies. Finally, the third is the $1/f$ noise, named so after the form of $K \propto 1/f$. It is not exaggerated to state that this is the most-frequently-observed but also the least-understood type of noise, see e.g. Ref.[57]. Although its complexity can sometimes extend beyond the reach of our abilities to understand it completely, if $K \propto 1/f$ is found together with $F \propto V^2$ it is likely that the noise corresponds to *resistance □uctuations*. While the exact reason for $K \propto 1/f$ sometimes remains obscure - depending on a particular phenomenon, the $F \propto V^2$ dependence can be explained easily: if a resistance uctuation is δR, at a constant applied current I one measures the voltage uctuation $(\delta V)^2 = I^2(\delta R)^2 = V^2(\delta R)^2/R^2 \propto V^2$.

Noise can be probed by different methods, but we shall concentrate on that used in the experiment presented, i.e. the *cross-correlation technique*. This is a relatively simple and yet powerful method characterised by a high measurement resolution in the standard four-point geometry of transport measurements. It is described in detail in Ref.[58], and here it will be outlined only. The sample is biased by a *dc* source, and the generated voltage, having the average (*dc*) and uctuating components, is ampli ed by two independent low-noise ampli ers kept at room temperature. The internal noise of the ampli ers is canceled by cross-correlating the two ampli er outputs in a spectrum analyser, i.e. only the component of the ampli ed voltage that comes from the sample is correlated and taken as the physical signal. Commercial spectrum analysers perform direct calculation of S_V for a given frequency range. One usually selects a small frequency range $\Delta f \ll f$ and averages all the points within Δf in order to obtain an accurate $S_V(f)$. This procedure is applicable as long as S_V is at within Δf, and a resolution in S_V down to $\sim 10^{-21}$ V^{2}s can be obtained - which is below the internal noise of most ampli ers for at least an order of magnitude. If $S_V(f)$ is too steep for the averaging, then the whole $S_V(f)$ over a range (f_1, f_2) must be taken into account. In order to extract precise vales of the measured noise one must perform a calibration of the measurements setup. Namely, a part of the signal is lost through the capacitances of the leads, which can be determined quantitatively by measuring the Johnson-Nyquist noise in the regime where the product RT can be deduced independently. For well-tuned noise-measurement systems the calibration correction is 10 % at most.

After explaining the experimental method used we return back to the question raised at the beginning of the Section, namely what new information on vortex transport the noise results can bring us, in particular can we learn something additional about vortex cores in motion.

We rst describe the experimental situation of the actual noise measurements. The sample under consideration is a ($L = 50$ μm, $W = 5$ μm) plain a-$Nb_{0.7}Ge_{0.3}$ microbridge with $T_c = 2.91$ K, $S = 2.05$ T/K and $\rho_0 = 2.3$ $\mu\Omega$m. The noise measurements are carried out at a xed temperature $T = 2.4$ K ($t = 0.82$) for B up to $B_{c2} \approx 1.18$ T. The method of averaging S_V in the frequency range 106.5 - 114 kHz is used, which turns out to be satisfactory even when S_V shows a certain f dependence (note that $\Delta f = 7.5$ kHz $\ll f = 110.25$ kHz). The B_{c2} value is determined by tting the magnetoresistivity data to Eq.(10), as explained in Section 3.2, which gives a result similar to that in Fig.7A. More precisely, the plot of Eq.(10) ts the data above ~ 0.65 T all the way up to the B_{c2}, with the $E(J)$ supporting the interpretation of 0.65 T as the B_{irr} for the given temperature.

The measured noise has a complicated structure, depending on the applied magnetic eld and current. For instance, the pinning plays the most important role in the close-to-equilibrium and weak-nonequilibrium regimes for $B < B_{irr}$. When vortices undergo the HVM the noise curves agree reasonably with the vortex shot noise, whereas in the regime $E \propto (J - J_c)$ the noise exhibits a hallmark of an ordering of vortex motion with increasing driving force, see Ref.[32] for details. Furthermore, around the FFI jump the noise has a strong peak [59] - which one would actually expect.

We shall skip these results in order to pay attention to the noise for $B > B_{irr}$, where pinning-independent, general properties of vortex transport are of importance. More precisely, we focus on the noise related to the close-to-equilibrium FF, following Eq.(10) with perfectly linear $E(J)$ over the whole range of measurements, i.e. up to maximum $E \sim 60$ V/m. Although in previous Sections the results have been presented through geometry-independent quantities, i.e. E, J, ρ, for noise measurements it is common to use $V = EL$, $R_f = \rho_f L/Wd$ and S_V, thus this convention will be followed throughout the rest of this Section.

A set of S_V measured as described above for $0.57 \leq b \leq 0.90$ (0.67 T $\leq B \leq$ 1.06 T) is shown in Fig.12A as a function of the average voltage V. By independent measurements at xed $B = 0.67$ T and $V = 1.5$ mV, and over a wide range of f, it was found that this noise is of $1/f$ type, more precisely $K \propto f^{-\upsilon}$ with $\upsilon = 1.5 \pm 0.1$. Note that the values of $S_V(V = 0)$ are nite and increase with increasing b. These intercepts agree with $4k_BTR_f(b)$, which can be subtracted from S_V in order to extract the excess dynamic noise $\Sigma_V = S_V - 4k_BTR_f$. Plotted in Fig.12B, $\Sigma_V(V)$ shows a quadratic dependence represented by the dashed lines. Therefore, the dynamic noise complies with the general requirements for resistance uctuations. Although indicative, this observation alone cannot be taken as suf cient evidence for FF-resistance uctuations, which means that we must relate it to a physical picture.

A step further in this direction is to check whether there is some well-de ned b dependence of Σ_V, as this is the scale in FF dissipation. The existence of such a law is presented in Fig.13, where all the curves from Fig.12 show the same dependence if plotted against $V(1-b)$. The data points for the given f and T follow $\Sigma_V = \gamma(1-b)^2V^2$ (solid line) with $\gamma = 2.1 \times 10^{-13}$ s, and this is the result appropriate for further modelling. Since σ_n

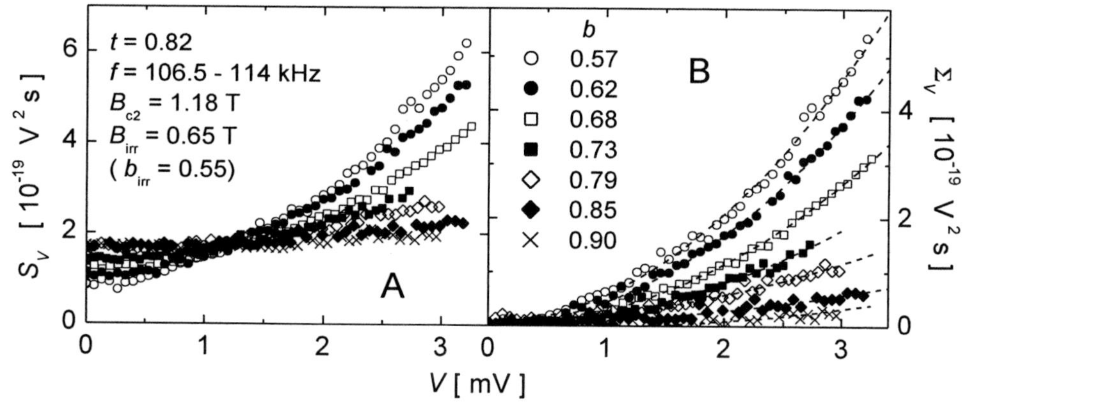

Figure 12: A: $S_V(V)$ for $B > B_{irr}$ at $t = 0.82$, measured in the frequency window 106.5 - 114 kHz as described in the text. By independent measurements it was found that this noise is of $1/f$ type. Note the nite intercepts of S_V at $V = 0$, corresponding to the Johnson-Nyquist noise $4k_BTR_f$. B: The excess noise $\Sigma_V = S_V - 4k_BTR_f$, showing a $\Sigma_V \propto V^2$ dependence as indicated by the dashed lines.

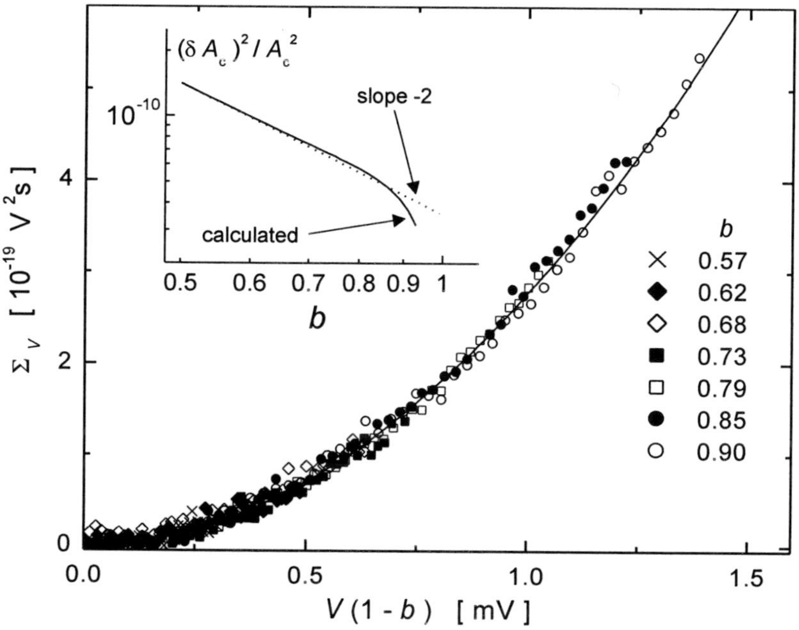

Figure 13: The Σ_V curves from Fig.12B plotted against the scaled variable $V(1-b)$. All the curves display the same quadratic behaviour represented by the solid line. Inset: The relative uctuation $(\delta A_c)^2/A_c^2$ of the vortex-core area (solid line) calculated from the LO FF theory as described in the text. The dashed line indicates $(\delta A_c)^2/A_c^2 \propto b^{-2}$.

of a-Nb$_{1-x}$Ge$_x$ is almost independent of T, the assumed resistance uctuations may re ect uctuations of the vortex-core area A_c. We have already seen that in nonequilibrium vortex cores suffer changes, thus such a hypothesis is not unphysical. Intuitively one expects that a dramatic change of the cores, which is for $t = 0.82$ likely to be the LO shrinking, are preceded by uctuations that do not affect the average values of physical quantities but still change their variances.

The model of FF resistivity uctuations implies $V^2[\langle(\delta\sigma_f)^2\rangle/\sigma_f^2] = \langle(\delta V)^2\rangle = \Sigma_V \Delta f$. We can use Eq.(10) to calculate $(\delta\sigma_f)^2/\sigma_f^2 = Y(b,t)(\delta A_c)^2/A_c^2$, where $Y(b,t) = [dg/db - g/b]^2/[(1-t)^{1/2} + g/b]^2$ is determined from the appropriate interpolation function $g(b)$. This leads to $(\delta A_c)^2/A_c^2 = \gamma(1-b)^2\Delta f/Y(b,t)$, which links the observed form of Σ_V and the uctuation of the vortex-core size. The meaning of this equation is that those uctuations of A_c which have characteristic times corresponding to the frequency range Δf are measured as Σ_V (which hence captures only a portion of the uctuations).

We can plot $(\delta A_c)^2/A_c^2$, calculated by the above procedure, against b in order to try to nd a simple picture of how the core-size uctuates. This plot is given in the inset to Fig.13 by the solid line and compared to a $(\delta A_c)^2/A_c^2 \propto b^{-2}$ dependence (dashed line). One can easily see that over the entire b range of the measurements the two lines agree very well, which means that these two results are experimentally indistinguishable. If we assume that the uctuations occur according to $(\delta A_c)^2/A_c^2 \propto b^{-2}$ we obtain a particularly simple picture of their nature. Namely, we can consider the cores as either compressible or incompressible objects. If the cores were incompressible then we would have $(\delta A_c)^2/A_c^2 \propto (1/b - 1)^2$, because δA_c is proportional to the space available for the uctuations, which is given by the size of the super uid area $\sim (\phi_0/B - \xi^2\pi)$. On the other hand, if the cores can be compressed, δA_c is proportional to the area per one vortex, i.e. ϕ_0/B, which leads to the observed b^{-2} dependence.

The above explanation of the measured Σ_V is qualitative rather than strict, but it relates this interesting experimental result - observed for the rst time quite recently [32], to a picture that does not contradict our current understanding of peculiarities in FF. Its soundness can be, of course, fully tested by subsequent experiments in combination with a detailed theoretical work.

4 Conclusion

In this Chapter a number of transport properties of a-Nb$_{1-x}$Ge$_x$ thin lms in the mixed state have been presented. The motivation for addressing this topic is the suitability of these samples for investigations of fundamental properties of vortices in motion. They are classical superconductors in the dirty limit, with $T_c \sim 2.7-3$ K, $B_{c2}(T=0) \sim 4-5$ T and $\kappa \sim 80-100$, and their overall superconducting properties can be described well within the framework of existing theoretical models. The weak pinning, characteristic of a-Nb$_{1-x}$Ge$_x$ over a large portion of the (B,T) phase diagram of the mixed state, is perhaps the most important property of a-Nb$_{1-x}$Ge$_x$ in the sense of dealing with fundamental properties of ux ow. To understand ux ow in its pure form - when pinning is irrelevant, is important for every extension of vortex-motion dissipation in terms of the pinning force. Moreover, since a-Nb$_{1-x}$Ge$_x$ can be structured to submicron dimensions and/or combined with other materials on mesoscopic scale, we can start with inventing nanoscale devices in which vortex motion plays an important role. Obviously, in this case we again need to know how vortices move under certain conditions.

The Chapter has been structured according to the above guidelines. Namely, there have already been a number of reports on using a-$Nb_{1-x}Ge_x$ as an integral part of mesoscopic devices, but one has to be rather familiar with the general properties of vortex dynamics in the mixed state of a-$Nb_{1-x}Ge_x$ in order to understand such advanced topics. Furthermore, even these general properties could occupy several lengthy articles if considered in every detail or from another point of view, for example if the complementary research of J. Aarts and co-workers (Leiden, The Netherlands) would be presented. I have therefore restricted myself to showing the most basic vortex-motion phenomena, manifested in a simple kind of these samples - a-$Nb_{0.7}Ge_{0.3}$ plain microbridges, and in experimental situations one can frequently encounter. Questions, uncertainties, and even some points of a technical character in dealing with real experimental situations have been identi ed rather than hidden, since the reader may face similar problems in his own experiments.

The basic concepts of vortex motion, presented in Section 2 and introductory part to Section 3, have been used later to discuss the selected experiments. It has been demonstrated how to nd the superconducting parameters of a-$Nb_{1-x}Ge_x$ thin lms directly from transport measurements, how to understand the in uence of an applied magnetic eld and temperature on an equilibrium vortex state, and how this state changes when vortices move. The author has found the vortex-motion-dissipation theory of Larkin and Ovchinnikov applicable in a variety of situations ranging from the close-to-equilibrium conditions to strongly nonequilibrium dynamic regimes in which vortex cores are no longer entities described by B and T only.

In Sections 3.1 and 3.2 the overall properties of the dissipation in the mixed state of a-$Nb_{1-x}Ge_x$ have been presented, which may serve not only as an introduction to the experiments presented in later Sections but also as a guide for a rough understanding of similar experimental results in general terms. Already at this level it has become rather clear that the physics of vortex motion is governed by more than the pinning force, that is, the changes of vortex cores by the electric eld generated therein may have dramatic effects on the dissipation. This issue has been addressed quantitatively in Sections 3.3-3.5.

Section 3.3 deals with a peculiar effect on vortex cores close to T_c - theoretically predicted by Larkin and Ovchinnikov, manifested by a nonlinear increase of the dissipation at applied currents which are relatively large but still much smaller than those corresponding to the depairing-induced destruction of superconductivity. The related microscopic mechanism is a shrinking of vortex cores and a consequent decrease of the vortex-motion viscosity, leading to a dissipation not much smaller than in the normal state. A remarkable quantitative agreement between the measured voltage-current characteristics and the Larkin-Ovchinnikov model has been found up to $\sim$ 75 % of the upper critical magnetic eld, i.e. over a range larger than expected theoretically.

The Larkin-Ovchinnikov mechanism, however, fails to explain the data well below T_c, and at temperatures where the number of thermal phonons is small due to the low absolute temperature. This has been discussed in Section 3.4. The nonlinear voltage-current curves in this case can be modelled by electron heating to above the phonon temperature, resulting in an expansion of vortex cores. The model reproduces both the dramatic nonlinearities of the voltage-current characteristics and the destruction of superconductivity by the overheating. The latter phenomenon implies an extension of the phase diagram of the mixed state by adding the induced electric eld as the third axis.

Another experimental approach for clarifying the dynamic properties of vortex cores has been presented in Section 3.5, namely measurements of the voltage noise. By this method one obtains insight into uctuations expressed through the variance of the measured voltage. In the presented study it has been found that the noise in the close-to-equilibrium conditions above the irreversibility magnetic eld corresponds to ux- ow-resistivity uc-tuations. These uctuations have been interpreted as a consequence of the uctuations of the vortex-core area around its mean value that sets the average voltage. Thus, dynamic vortex cores can either shrink, expand or uctuate, depending on the experimental conditions.

Acknowledgements

I have to express my thanks to those whose involvement has been crucial in carrying out and presenting the experiments discussed in this Chapter. My deepest gratitude goes to my family - Marija and little Ana, whose patience and daily support have been invaluable. Christoph Strunk from Regensburg (Germany) has contributed to every stage of this work, and it is my privilege to continue collaborating with him on the topic. Christoph Sürgers from Karlsruhe (Germany) has been responsible for magnetron sputtering, so without his high level of expertise and precision in fabricating the samples this investigation would not have been possible. The same applies to Johannes Bentner (Regensburg), who has put much effort in performing the electron-beam lithography and taking part in the measurements. I am grateful to Christian Schönenberger from Basle (Switzerland), who acquainted me with the problem of vortex-motion noise and participated to the related research, as well as to Thomas Nussbaumer, also from Basle, for his experimental help. Boran Leontić from Zagreb (Croatia) has read the Chapter very carefully, helping me in my effort to nd a balance of the readability and scienti c rigour.

References

[1] A. A. Abrikosov, *Zh. Eksp. Teor. Fiz.* **32**, 1442 (1957); [*Sov. Phys. JETP* **5**, 1174 (1957)].

[2] U. Essmann and H. Träuble, *Phys. Letters* **24A**, 526 (1967).

[3] J. Bardeen and M. J. Stephen, *Phys. Rev.* **140**, A1197 (1965).

[4] D. Babić, J. R. Cooper, J. W. Hodby and Chen Changkang, *Phys. Rev. B* **60**, 698 (1999).

[5] J. R. Cooper, D. Babić, J. W. Loram, Wai Lo and D. A. Cardwell, *Fizika A* **8**, 333 (1999).

[6] H. Raffy, S. Labdi, O. Laborde and P. Monceau, *Physica C* **184**, 159 (1991).

[7] D. Babić, J. Ivkov, N. Heinrich and B. Leontić, *Phys. Rev. B* **49**, 15965 (1994).

[8] D. Babić, J. Ivkov and B. Leontić, *Physica C* **277**, 203 (1997).

[9] N. Kokubo, R. Besseling, V. M. Vinokur and P. H. Kes, *Phys. Rev. Lett.* **88**, 247004 (2002).

[10] J. Bentner, D. Babić, C. Sürgers and C. Strunk, *Phys. Rev. B* **70**, 184516 (2004).

[11] C. N. Lau, N. Marković, M. Bockrath, A. Bezryadin and M. Tinkham, *Phys. Rev. Lett.* **87**, 217003 (2001).

[12] L. P. Gor'kov, *Zh. Eksp. Teor. Fiz.* **36**, 1918 (1959); [*Sov. Phys. JETP* **9**, 1364 (1959)].

[13] P. H. Kes and C. C. Tsuei, *Phys. Rev. B* **28**, 5126 (1983).

[14] N. Werthamer, E. Helfand and P. Hohenberg, *Phys. Rev.* **147**, 295 (1966).

[15] F. Lindemann, *Phys. Z.* **11**, 69 (1910).

[16] B. A. Hubermann and S. Doniach, *Phys. Rev. Lett.* **43**, 950 (1979).

[17] D. S. Fisher, M. P. A. Fisher and D. A. Huse, *Phys. Rev. B* **43**, 130 (1991).

[18] J. R. Cooper, J. W. Loram, J. D. Johnson, J. W. Hodby and Chen Changkang, *Phys. Rev. Lett.* **79**, 1730 (1997).

[19] D. Majer. E. Zeldov and M. Konczykowski, *Phys. Rev. Lett.* **75**, 1166 (1995).

[20] A. Junod, M. Roulin, J.-Y. Genoud, B. Revaz, A. Erb and E. Walker, *Physica C* **275**, 245 (1997).

[21] D. T. Fuchs, R. A. Doyle, E. Zeldov, S. F. W. R. Rycroft, T. Tamegai, S. Ooi, M. L. Rappaport and Y. Myasoedov, *Phys. Rev. Lett.* **81**, 3944 (1998).

[22] A. I. Larkin and Yu. Ovchinnikov, *J. Low. Temp. Phys.* **34**, 409 (1979).

[23] M. H. Theunissen, PhD Thesis, University of Leiden, 1997.

[24] G. Blatter, B. Ivlev, Y. Kagan, M. Theunissen, Y. Volokitin and P. H. Kes *Phys. Rev. B* **50**, 13013 (1994).

[25] M. Tinkham, *Introduction to Superconductivity*, 2nd Edition (McGraw-Hill, New York, 1996).

[26] E. M. Chudnovsky and A. B. Kuklov, *Phys. Rev. Lett.* **91**, 067004 (2003).

[27] A. I. Larkin and Yu. N. Ovchinnikov, in *Nonequilibrium Superconductivity*, edited by D. N. Lengenberg and A. I. Larkin (North Holland, Amsterdam, 1986).

[28] M. Tinkham, *Phys. Rev. Lett.* **13**, 804 (1964).

[29] A. L. Fetter and P. C. Hohenberg, in *Superconductivity*, edited by R. D. Parks (Marcel Dekker, New York, 1969).

[30] H. R. Kerchner, J. Narayan, D. K. Christen and S. T. Sekula, *Phys. Rev. Lett.* **44**, 1146 (1980).

[31] D. J. van Ooijen and G. J. van Gurp, *Phys. Letters* **17**, 239 (1965); G. J. van Gurp, *Phys. Rev.* **166**, 436 (1968).

[32] D. Babić, T. Nussbaumer, C. Strunk, C. Schönenberger and C. Sürgers, *Phys. Rev. B* **66**, 014537 (2002).

[33] A. Schmid and G. Schön, *J. Low Temp. Phys.* **20**, 207 (1975).

[34] A. I. Bezuglyj and V. A. Shklovskij, *Physica C* **202**, 234 (1992).

[35] M. Kunchur, *Phys. Rev. Lett.* **89**, 137005 (2002).

[36] D. Babić, J. Bentner, C. Sürgers and C. Strunk, *Phys. Rev. B* **69**, 092510 (2004).

[37] J. Bentner, C. Sürgers, D. Babić and C. Strunk, unpublished.

[38] J. Ivkov, N. Radić and D. Babić, unpublished.

[39] M. Ohring, *The Materials Science of Thin Films* (Academis Press, San Diego, 1992).

[40] J. I. Martin, M. Veléz, J. Nogués and I. K. Schuller, *Phys. Rev. Lett.* **79**, 1929 (1997).

[41] J. M. E. Geers, C. Attanasio, M. S. B. Hesselberth, J. Aarts and P. H. Kes, *Phys. Rev. B* **63**, 094511 (2001).

[42] J. R. Cooper and D. Babić, unpublished.

[43] S. Ullah and A. T. Dorsey, *Phys. Rev. Lett.* **65**, 2066 (1990); *Phys. Rev. B* **44**, 262 (1991).

[44] Z. Tesanović and L. Xing, *Phys. Rev. Lett.* **67**, 2729 (1991); Z. Tesanović, L. Xing, L. Bulaevski, Qiang Li and M. Suenaga, *Phys. Rev. Lett.* **69**, 3563 (1992).

[45] S. G. Doettinger, R. P. Huebner, R. Gerdemann, A. Kühle, S. Anders, T. G. Traäuble and J. C. Villégier, *Phys. Rev. Lett.* **73**, 1691 (1994).

[46] S. G. Doettinger, R. P. Huebner and A. Kühle, *Physica C* **251**, 285 (1995).

[47] J. A. Fendrich, U. Welp, W. K. Kwok, A. E. Koshelev, G. W. Crabtree and B. W. Veal, *Phys. Rev. Lett.* **77**, 2073 (1996).

[48] Z. L. Xiao and P. Ziemann. Phys. *Rev. B* **53**, 15265 (1996).

[49] Z. L. Xiao, E. Y. Andrei and P. Ziemann, *Phys. Rev. B* **58**, 11185 (1998).

[50] A. V. Samoilov, M. Konczykowski, N.-C. Yeh and C. C. Tsuei, *Phys. Rev. Lett.* **75**, 4118 (1995).

[51] W. Henderson, E. Y. Andrei, M. J. Higgins and S. Bhattacharya, *Phys. Rev. Lett.* **77**, 2077 (1996).

[52] B. J. Ruck, J. C. Abele, H. J. Trodahl, S. A. Brown and P. Lynam, *Phys. Rev. Lett.* **78**, 3378 (1997).

[53] F. Le och, C. Hoffmann and O. Demolliens, *Physica C* **319**, 258 (1999).

[54] A. E. Koshelev and V. M. Vinokur, *Phys. Rev. Lett.* **73**, 3580 (1994).

[55] A. Rothwarf and B. N. Taylor, *Phys. Rev. Lett.* **19**, 27 (1967).

[56] S. Kogan, *Electronic Noise and Fluctuations in Solids* (Cambridge University Press, Cambridge, 1996).

[57] M. B. Weissman, *Rev. Mod. Phys.* **60**, 537 (1988).

[58] M. Henny, S. Oberholzer, C. Strunk and C. Schönenberger, *Phys. Rev. B* **59**, 2871 (1999).

[59] T. Nussbaumer and D. Babić, unpublished.

In: New Frontiers in Superconductivity Research
Editor: Barry P. Martins, pp. 145-159
ISBN: 1-59454-850-1

Chapter 5

ON THE QUANTUM NATURE OF VORTEX

Rongchao Ma
Northwest Institute for Nonferrous Metal Research, P. O. Box 51, Xi'an, Shaanxi 710016, P. R. China

Yuefei Ma
School of Metallurgical Science and Engineering, Central South University, Changsha, Hunan 410083, P. R. China

Abstract

Vortices in type-II superconductors are usually considered as macroscopic entities and treated with classical mechanics because they have large size. However, a number of experimental results show that vortices can undergo quantum effects, which imply that, in some sense, the nature of vortices should be described by quantum mechanics. In this paper, the quantum effects of vortex in type-II superconductors is studied philosophically by reviewing the ideas in elementary quantum mechanics, in particular, the origin of the quantum nature of vortex is discussed and is concluded to be of that vortex has small mass per unit length. Other vortex matter phenomenology in type II superconductors, such as quantum creep, quantum depinning and quantum melting are also reviewed. It is believed that quantum effects will manifest themselves at sufficiently low temperature because in high temperature (large quantum numbers) the energy differences become arbitrarily small and the energies form almost a continuum. The quantum fluctuations of vortices drive various types of phase transitions and define much the superconducting properties of real materials at low temperature. The extension of the quantum description to the fields beyond the micro world, such as flux pinning classification and economics, is also given.

1 Introduction

Vortices are the consequence of the quantum nature of a physical system and occur in a wide range of gases and fluids, such as in Bose-Einstein condensates of dilute atomic gases [1], superfluid helium [2] and superconductors [3, 4]. In 1934, Onsager [5] firstly conceived the idea of vortex by thinking in terms of vortex sheets in superfluid hydrodynamics. In 1950, London [6] introduced the concept of vortex line and suggested that the flux trapped by a superconducting ring is quantized. This is an important prediction because it foretells a situation in which the quantum effect occurs on a macroscopic scale. In 1955, Feymann [7] independently introduced the concept of quantised vortices in He II and developed its consequences in some detail. However, the concept of vortex did not attract much attention until 1957, Abrikosov published his remarkable paper [8] (the paper was written in 1952 [9]), which predicted that, for applied fields H such that $H_{c1} < H < H_{c2}$, the magnetic field penetrates into type-II superconductors (with GL $\kappa > 1/\sqrt{2}$) in the form of quantized flux tubes which, in the absence of disorder, form a regular lattice of parallel defect lines. Kleiner et al. [10] further developed the theory of vortex lattice in type II superconductors, which was latterly confirmed by a number of experimental results [11-15].

Vortices in type-II superconductors are usually considered as macroscopic entities and treated with classical mechanics [16-17]. The theory of elasticity of a vortex lattice in an isotropic superconductor was worked out by Brandt [18] using the Ginsburg-Landau (GL) description at high inductions, $B \geq 0.6H_{c1}$, and the London theory for small fields. In these theory three elastic moduli, C_{11} for uniaxial compression, C_{44} for tilt and C_{66} for shear, are required for a proper description of the vortex lattice. Later, Larkin and Ovchinnikov [19] extended the validity of the results to the entire temperature range of $0<T<T_c$ based their calculation on the BCS-Gor'kov theory.

The classical picture (elasticity theory) of vortices has already been well studied, and the classical model of flux creep in type-II superconductors predicts a linearly vanishing magnetic relaxation rate at low temperature. However, the low-temperature magnetic relaxation rate in a number of experiments [20-25] has been found not extrapolate to zero, which suggests the existence of vortex motion by quantum tunneling (quantum creep). It has also been shown that the vortices in artificially fabricated Josephson arrays (periodic lattice potential) in a superconductor can be described as Bloch oscillation [26], which implies that vortices have wave particle duality, and their wave functions overlap and energy bands are formed when they move in periodic potential created by the Josephson array.

Up to now, perhaps the most direct proof of the quantum effects of vortex has been obtained from experiments by Wallraff et al. studying the quantum dynamics of an individual vortex in a superconducting Josephson junction [27]. They demonstrate the existence of quantized levels of the vortex energy within the trapping potential well and quantum tunneling of the vortex through the pinning barrier by measuring the statistics of the vortex escapes from a controllable pinning potential. These results show that vortex tunneling occurs at low temperature not only for large collections of vortices, but also for individual vortex.

In this paper, we will present further investigations on the philosophical meaning of quantum effects of vortices in superconductors.

2 Why is it Quantized?

As stated in the introduction, a great number of experimental evidences have shown that quantum effect occurs with vortex; however, vortex is of such large size and can be thought as macroscopic entity, how can it undergo quantum behavior that is so radically different from our "classic feeling"? To answer this question, we now must examine the origin of quantum nature of vortex and make clear which elements determine the quantum effects to occur.

2.1 Quantization of the Nature of Vortex

From statistical mechanics, we known that for a particle system, when the particle density n and the *mean thermal wavelength* $\lambda = h/\sqrt{2\pi mkT}$ satisfy the relationship $n\lambda^3 << 1$, that is, $(n\lambda^3) \to 0$, the physical properties of the system go over smoothly to their classical count part; on the contrary, when $(n\lambda^3) \to 1$, the system is characterized by typical quantum effects. However, this criterion doesn't give any exact quantum information of a single particle. For a deeper discussion of the origin of quantum nature of vortex and its application, we have to go back to the elementary quantum physics, where there are a number of physical quantities, such as energy and momentum, which is quantized under certain bound conditions, that is, can take on only a discrete set of values. In old quantum mechanics the quantized physical quantities is includes in the Sommerfield quantum condition [28-30]

$$\oint p dq = nh \quad \text{(n=1, 2, 3, ...)} \tag{1}$$

where h is the Planck constant. These quantized quantities can also be obtained by solving the Schrödinger equation of the corresponding physical system. However, in most case, it is not convenient to tell which factors determine the quantum effects by referring to the discontinuous energy level only, sometimes even impossible.

As a matter of fact, in quantum mechanics the most general character of a microscopic particle is the wave property, not the discontinuity, which occurs only in the bound conditions. For the sake of convenience, let us now turn back to the old quantum mechanics. In 1911, De Broglie [31-32] suggested that, if waves can have particle properties, then it is reasonable to expect that particles have wave properties. According to this hypothesis, any particle with momentum and wave behavior would be a universal property of moving objects. Thus, if wave particle duality is a general property of matter, all particles will have their own wave. The wavelength for the corresponding ("matter") wave is

$$\lambda = \frac{h}{p} = \frac{h}{mv} \tag{2}$$

where λ is the wavelength, v is the velocity, and m is the mass of the particle. De Broglie has also shown that the Sommerfield quantum condition (1) and its similarity could be deduced from the wave particle duality under certain bound condition by referring to his wave theory. In other words, the Sommerfield quantum condition is a consequence of the wave property of the particle in the case of bound state.

Because in microscopic world, the wave property exists in any case, but the discontinuity exists in the bound condition only, we may conclude that the wave property is more "basic" than the discontinuity.

Equation (2) shows that the wavelength of a particle is inversely proportional to its mass. If we know the mass of a vortex per unit length, we can then calculate the vortex's wavelength under certain velocity. In 1964, Stephon and Suhl [33] proposed a time dependent form of the GL theory, which gives rise to a very small inertial mass of unit length of flux-line. Later, Suhl [34] extended it to the case of a moving vortex, which gives rise to a "core" mass $m_{core} = \frac{3\xi^2 H_c^{\ 2}}{4v_F^{\ 2}}$ and an "electromagnetic" mass $m_{em} = \frac{1}{4}\frac{\xi^2 H_c^{\ 2}}{c^2}(\frac{\lambda}{\lambda_d})^2$ per unit length of the vortex, where ξ is the coherence length, H_c is the bulk critical field, v_F is Fermi velocity, λ is the penetration length and λ_d is a shielding length. For typical arrays, the calculated vortex mass is 500 times smaller than the electron mass, this small value of the expected mass of the vortex suggests that quantum effects are likely to occur with vortices at low temperature [26]. By equation (2) we know that the wavelength of a vortex should be 500 times large than that of electron when they have the same velocity. Thus, we guess, correctly, that quantum effects are very important for vortices.

It is not sufficient to represent the wave property of a particle by its wavelength only because the observability of the wave property of the particle is also related to its size. In other words, the wavelength represents the wave property of a particle, but it does not assure the wave property can be observed. Let us now define an observability parameter, ε, as

$$\varepsilon = \frac{\lambda}{a} \tag{3}$$

where λ is the wavelength of the matter wave, and a is the minimum size of the particle. By this quantity ε we find it is reasonable to tell the wave property of a particle. The size of an electron is very small and the observability of its wave property would be very much significant, however, the size of a vortex is about two times of the penetration depth, that is, 100-4000Å [16], which is thousands times of the wavelength, and the wave property of the vortex would be very difficult to observe. We conclude that the quantum effects of vortex should be "measured" by the discontinued energy and the quantum tunneling, not by the wave property. That is mean, one can not expects to, even with the help of advanced equipments [35], watch a vortex disappears from a place and then re-appear in another place like a ghost though vortex has quantum effects.

Our next step is to analysis Heisenberg's uncertainty principle [36], which reads $\Delta x \cdot \Delta p \geq \hbar / 2$, where Δx is the position uncertainty, Δp is the momentum uncertainty and $\hbar = h / 2\pi$ (h is the Planck constant). For the convenience of discussion, we now re-write the uncertainty principle in the form

$$\Delta x \Delta v \geq h / 4\pi m \tag{4}$$

where Δv is the velocity uncertainty and m is the mass of the particle.

Equation (4) states that the product of the position uncertainty and the velocity uncertainty is at least as large as the Planck constant $h = 6.626 \times 10^{-34}$ divided by 4π times the mass of the particle being considered. The Planck constant h is a very small number and it gives a limit for the transition of quantum mechanics to classical mechanics, where the Planck constant $h \to 0$, or the quantum effects can be neglected when h is small relative to other quantities with the same dimensions [37]. As for the vortex, the uncertainty in position cannot be larger than its size, which is about 4000 Å for high-temperature superconductors [16], and 500 Å for typical classic pure superconductors [38]. We can now see why the wave property is hard to see in vortex and the classical mechanics is such a good approximation to the vortex matter because the uncertainty principle does not put any significant constraint on position and velocity determinations of vortex. The quantum effects of vortex should be detected by the discontinuity quality.

Indeed, when studying the vortex lattice, we are dealing with an enormous number of vortices, and, although the behavior of every single vortex is governed by the laws of quantum physics, the quantization of energy, momentum, and other physical quantities does not manifest itself on the whole due to the chaotic thermal motion of the vortices, the thermal motion masks quantum regularities.

2.2 Quantization of Flux

The preceding arguments have shown that the small mass of a vortex per unit length results in the quantum effects of the vortex. We now turn to another question: why is the flux quantized into vortices? The answers to this question have already been well established by many scientists, we now present a short review.

In 1950, F. London [6] made the suggestion that the flux trapped by a superconducting ring is quantized in analyzing the state of a multiply connected superconductor in the presence of a magnetic field. He introduced the concept of fluxoid ϕ' associated with each normal region passing through the superconductor. By applying the Sommerfield quantum condition (equation (1)), he argued that the values of ϕ' should be restricted to a discrete set, integral multiples of a fluxoid quantum hc/e^*. It must be emphasized that, not anticipating the pairing theory of superconductivity, London predicted that e^* was e. We now known that $e^* = 2e$, so the fluxoid quantum has the value

$$\phi_0 = \frac{hc}{2e} = 2.07 \times 10^{-7}\ \text{Gauss-cm}^2$$

This has been shown by the measurements of the flux trapped in a ring shaped layer [39-40], thus the existence of Cooper pairs is proved experimentally.

As the magnetic field trapped in a superconducting ring can only take a set of discrete values, and the electrical current is related to the magnetic field according to the Biot-Savart law, it turns out that the electrical current in a superconducting ring cannot be of an arbitrary magnitude and change continuously, but can only take on a discrete set of values. This behavior is analogous to that of an electron in an atomic orbit, which can only take on discrete values.

It should be kept in mind that the concept of fluxion quantization is not limited to the superconducting ring, with its semi classical Sommerfeld quantum and the London equation. It also occurs in type-II superconductors as predicted by Abrisokov's remarkable paper [8], in which for applied fields H such that $H_{c1} < H < H_{c2}$, the magnetic field penetrates into type-II superconductors forming a two-dimensional lattice in the plane of cross-section of the cylinder. Although his solution was obtained by solving the GL equation under the limit condition of $H \approx H_{c2}$, it is now clear that the flux quantization is a much more general phenomenon, and occurs in the whole range of $H_{c1} < H < H_{c2}$. The occurrence of the fluxion quantization in type-II superconductors can be easily proved by the viewpoint of Ginzburg-Landau theory [Ref. 9, p44], which is simply based on the existence of a single–valued complex superconducting order parameter ψ and the phase ϕ for a superconductor. By integral the second Ginzburg-Landau equation (under the assumption that $H_{c1} < H < H_{c2}$), one gets

$$\Phi = \int_{\Sigma} \mathbf{h} \cdot d\boldsymbol{\sigma} = n\Phi_0 \tag{5}$$

It is seen that the flux trapped in a superconductor is quantized.

The flux quantization is the macroscopic analog of the quantization of angular momentum in an atomic system. Accordingly, it is not surprising that it provides a powerful tool for dealing with many problems of superconductors penetrated by magnetic flux.

From equation (5) we known that flux quantization can be observed only in a multiply connected geometry, that is, the Abrikosov lattice in type II superconductors. However, in a single connected type I superconductor, the Meissner effect is perfect and n is then equal to zero. The flux quantum can be measured experimentally.

The phenomenology of superconductivity is based on the Ginzburg-Landau theory [41], which provides the starting variational functional (free energy) for the charged superconductor coupled to electromagnetism. In quantum field theory, the relativistic version of the Ginzburg-Landau free energy (Lagrangian) is [42]

$$-\mathcal{L} = m^2\phi^*\phi + \lambda(\phi^*\phi)^2 - (\partial_\mu + ieA_\mu)\phi(\partial^\mu - ieA^\mu)\phi^* + \frac{1}{4}F_{\mu\upsilon}F^{\mu\upsilon} \tag{6}$$

where ϕ is the macroscopic many-particle wave function. This is in fact the Lagrangian of a complex scalar fields (ϕ) coupled to a gauge field (4-vector A_μ), that is, the scalar electrodynamics with spontaneous symmetry breaking (Higgs model). A little more systematical investigating equation (6) indicates that the soliton solution is possible in principle then one can go to the equations of motion to find them. It means that the effect of adding the gauge field is to give the soliton magnetic flux. Superconductivity provides a nice illustration of the Abelian Higgs model. We see that the existence of vortex is a general phenomenon that occurs in a wide range of physics system, and they may be described in a unified way as developed in quantum field theory.

3 Dynamics Model of Vortex

3.1 Dissipative Dynamics Model

In a conventional view, the motion of a vortex involves the motion of magnetic field lines, or the motion of normal electrons. This is essentially based on the belief that vortices are macroscopic object, which couple to the environment and experience dissipation. In a simple approximation to the classical dynamic model of vortex one can assume that a moving vortex experience a viscous drag force characterized by a viscous coefficient η as suggested by Bardeen and Stephen [43].

For a dissipative dynamics system, the application of the Schröedinger equation may be impossible, thus one has to deal the dynamics with other methods [44]. Caldeira and Leggett [45] have developed a theory of the quantum tunneling process that takes the dissipation into account. They found that the interaction with the environment could be accounted for by adding a term

$$\frac{\eta}{4\pi}\int dt\int dt'\int dz\left[\frac{q(t)-q(t')}{t-t'}\right]^2$$

to the Euclidean action which results in the so-called effective action S_E^{eff} of the vortex, and the effect of (linear) friction is to multiply the tunneling probability calculated in its absence by a factor $\exp[-A\eta(\Delta q)^2/\hbar]$, where Δq is the "distance under the barrier" and A is numerical factor. Thus, the linear friction always reduces the tunneling rate. Blatter et al. [46] has developed this theory and applied it to the phenomenon of collective quantum creep in bulk superconductors within the framework of weak-collective-pinning theory.

Since the mass of a vortex per unit length is very small, the dynamics (classical) of a vortex in such a dissipative picture would be always damping dominated [47-48]. Shul et al. [34] has calculated that the relaxation time is less than 10^{-12} s. This indicates that a collective dispersion model of vortices lattice would be very difficult to observe.

3.2 Vanishing Dissipative Dynamics Model

The dissipative dynamics model of vortex is widely accepted. However, on considering the quantum effects of vortex, the application of the dissipative dynamic model may be disputable. As discussed in the preceding section, vortex can undergo quantum tunneling, which indicates that there are may be a series of eigenstates ψ_n for a vortex, on which there will be a probability $P \propto e^{-E_n/k_BT}$ for the vortex to be found [49], where E_n is the eigenvalue of the corresponding eigenstate ψ_n. This is analogous to an atomic system in which there are electrons revolving around a nucleus. In a classical view, the electrons will collapse on to the nucleus by emitting out electromagnetic wave; however, the atomic system is in fact steady because the electrons can stay on a series of discontinued steady states without losing energy due to the quantum effect. Thus, for a vortex confined to pinning centers, we may expect that the vortex will also stay on a series of steady states without losing energy, and then the vortex system is kept steady. As an analog, the vortices should also be described as vanishing dissipative. However, the dissipative dynamics model of vortex may turn out to be describing the quantum phenomena with the classical mechanics.

Whether or not the above interpretation is correct, the dynamics of vanishing dissipation is at least valid in the case of vortices without the normal core, i.e., the special type of vortex in long Josephson junction. These vortices have the characteristics of solitons-nonlinear waves that preserve their shape with time and propagate as ballistic particles [50]. They are distinct from Abrikosov vortices in type-II superconductors, as they have no normal core and thus move with very low damping. Under this belief, we may expect that the dynamics of vortices are not dissipative, thus, it is desirable to treat the vortex dynamics of vanishing dissipation with quantum mechanics.

A simplified model has been employed to discuss the quantum effects of vortex in type-II superconductors under the assumption of vanishing dissipation both in lower magnetic field and higher magnetic field in our paper [51]. Within weak collective pinning theory, we adopt a continuum elastic description of the flux-line lattice, with a potential energy given by the combination of the elastic energy, the pinning energy and the action of the Lorentz force, so we study the single-vortex energy and obtained the Hamiltonian of a decoupling pancake vortex in a strong-layered superconductor as:

$$H = \left(-\frac{\hbar^2}{2m_x}\frac{\partial^2}{\partial x^2} - \frac{\hbar^2}{2m_y}\frac{\partial^2}{\partial y^2}\right) + \frac{\varepsilon L}{2}(x^2 + y^2) - (\gamma\xi^2 L)^{1/2} + j\frac{\Phi_0}{c}Lx \qquad (7)$$

where m_x and m_y denote the effective masses of the vortex along the corresponding axes of the crystal. After much calculation, we get the corresponding eigenvalue of the energy of the vortex:

$$E_{n_x n_y} = (n_x + \frac{1}{2})\hbar\omega_x + (n_y + \frac{1}{2})\hbar\omega_y - (\frac{j\Phi_0}{\varepsilon c})^2 - (\gamma\xi^2 L)^{1/2} \quad (n_x, n_y = 0,1,2,3...) \qquad (8)$$

Eq. (8) shows that the action of Lorentz force reduces the eigenvalue of the energy of the vortex of magnitude $(\frac{j\Phi_0}{\varepsilon c})^2$. This predicts the discontinuous energy level, which can qualitatively explain the experimental results obtained by Wallraff et al. [27].

On the other hand, as pointed out by London [6], magnetic flux trapped in a superconducting ring must be quantized. In this case, the interaction between vortices is the Lorentz force only, there is no motion of normal electron at all and the dynamics should be vanishing dissipative.

Consider a very large and long superconducting cylinder with an applied field *H*, such that $H < H_c$, parallel to *z-axis*, the normal direction of the ring plane. *H* is then a function that only depends on the variables of *x* and *y*, and the influence on the vortex by the boundary conditions can be negligible. The interaction between vortices is the Lorentz forces only, which add up to drive the vortices to their equilibrium position. We can now restrict ourselves to the two-dimensional case. This is in fact similar to the model as developed in our previous paper to describe the vanishing dissipative dynamics of vortex in higher magnetic field [51], where the vortex system is shown could be described as a collection of harmonic oscillators. In these vanishing dissipation dynamics model, a collective dispersion model of vortex lattice should be observed.

The direct results of the quantum nature of vortex are that fluctuations need not be based on a thermal source [16] but can be of quantum origin as well. Since the mass of vortex per unit length is 500 times lighter than that of electron [26], the quantum fluctuations of vortex would be quite large (as usual, the frequency is inversely proportional to the square root of the particle mass). Indeed, many experiments suggest that vortex lattice melting, both in high-Tc materials [52-53] and in low-Tc films and multiplayer [54], is strongly influenced by quantum fluctuations.

The formal applicability of classical mechanics to quantum theory as far as transitions between such states are concerned is the subject of the correspondence principle [55], the second foundational principle of the older quantum theory, which was based on the assumption that quantum theory or at least its formalism contains classical mechanics as a limiting case. In virtue of the quantum conditions, quantum theory approaches classical physics whenever the differences in the quantum numbers are small compared with their absolute values. Accordingly, the quantum fluctuation of vortex can only manifests itself at lower temperature because in high temperature (large quantum numbers) the energy differences become arbitrarily small and the energies form almost a continuum. On the other hand, it is usually thermal fluctuation dominating in higher temperature. At sufficiently low temperature, vortex lines are then subjected to quantum fluctuation. The observations of a strong magnetic relaxation at very low temperatures [22, 56-57] have demonstrated the importance of quantum effects in the phenomenology of high-temperature superconductors.

Equation (8) shows that at absolute zero, vortices are not at rest, but perform so-called "zero-point" vibrations. The physical origin of these zero-point vibrations lies in the Heisenberg uncertainty relations [36], which are of foundational important in the quantum theory and have the form $\Delta p_x \Delta x \geq \hbar/2$, $\Delta p_y \Delta y \geq \hbar/2$, $\Delta p_z \Delta z \geq \hbar/2$, where Δx, Δy, Δz and Δp_x, Δp_y, Δp_z are the uncertainties in the values of vortex coordinates and momentums, respectively. It is impossible to simultaneously specify the position and the momentum of a vortex because there cannot be a state in which $\Delta x = 0$ and $\Delta p_x = 0$, and the same for y and z component. Thus, even at absolute zero the vortices in a lattice are not at rest, but are performing zero-point vibrations. If the energy of the zero-point motion is sufficiently high, that is, if the vibrations are intense enough, the quantum fluctuation should manifest itself in the zero point motion of vortex lines and the flux lattice can melt even at temperature *T=0* [58-60].

4 The Possible Application of Quantum Theory

4.1 Application of Quantum Field Theory to Magnetic Field

In 2.1 and 2.2 we have discussed two kinds of quantization: the quantization of flux and the quantization of the nature of vortex. Compare a vortex with an electron, we find there is a beautiful analogy between them as shown in figure 1.

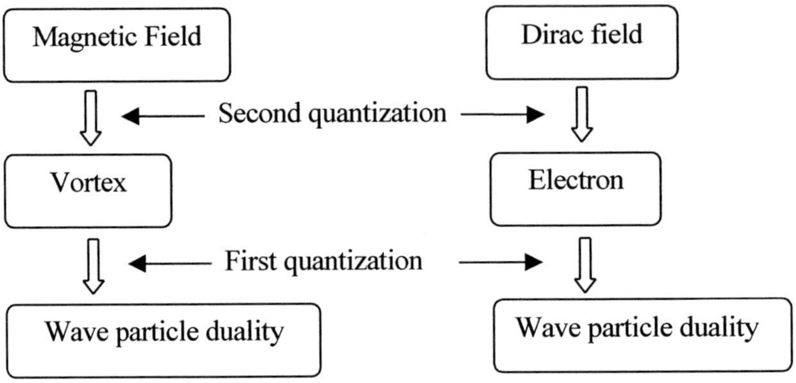

Fig. 1 The parallel between the vortex and electron

Figure 1 show that the description of the quantum nature of vortex is the so-called first quantization as stated in the elementary quantum mechanics, and the quantization of flux to vortices is more or less similar to the second quantization in quantum field theory.

In quantum mechanics it is convenient to introduce the creation operator a_k^+ to represent the addition of one particle to a system and annihilation operator a_k to represent the subtraction of one particle from the system. As discussed in previous section, vortex is a consequence of the quantum nature of the physical system because of the restriction of the single value of the order parameter, furthermore, vortex itself can undergo quantum effects, naturally, we may wonder if we can introduce the creation operator to represent the addition of one vortex to the system and the annihilation operator to represent the subtraction of one vortex from the system?

As can be shown by referring to the relativistic version of the Ginzburg-Landau free energy in quantum field theory, the existence of vortex is a general phenomenon that occurs in a wide range of physics system, which indicates that vortex can be described in a unified model, Abelian Higgs model. The Higgs model has been well developed in quantum field theory and particles physics, and the superconductivity provides a nice illustration to this model. That is mean vortex could be treated as microscopic particle with anisotropy, and it seemed logical that vortex could be described by operators.

The two fields determining the superconductivity are the superconducting order parameter Ψ and the vector potential A. The electromagnetic field as a classical field has been well established for many years and the theory is summed up in the Maxwell equation. In quantum field theory, the quantization of electromagnetic field is usually described as photos; here we intend to quantize the magnetic field as vortices.

To develop this quantization, let us note some similarities between vortex and photos. As is shown before, vortices can be described as harmonic oscillations, so vortices have some similarities with photos because they are all Bosons. At the same time they are also the quanta of electromagnetic field. Recall that for a discrete system of one or more Bosons the commutation relations are

$$[a_i^+, a_k^+] = 0; \ [a_i, a_k] = 0; \ [a_i, a_k^+] = i\delta_{ik} \tag{9}$$

where a_i is the creation operator and a_i^+ is the annihilation operator of vortex respectively. Let $|n_k - 1\rangle$, $|n_k\rangle$ and $|n_k + 1\rangle$ represent the state on which there are n_{k-1}, n_k, n_{k+1} vortices respectively, we have

$$a_k^+|n_k\rangle = \sqrt{n_k + 1}|n_k + 1\rangle; \ a_k|n_k\rangle = \sqrt{n_k}|n_k - 1\rangle \tag{10}$$

To quantized the magnetic field, we now consider expanding the vector potential $\mathbf{A}(x)$ in terms of creation and annihilation operators [61]

$$\mathbf{A}(\mathbf{x}) = \sum_k \left(a_k e^{-i\mathbf{k}\cdot\mathbf{x}} + a_k^+ e^{i\mathbf{k}\cdot\mathbf{x}}\right) \tag{11}$$

In quantum electrodynamics, quite generally the expression for A can be formally obtained from the old wave mechanical expression with wave function replaced by a field operator, that is the second quantization, a satisfactory means of describing systems with a variable number of particles. As a matter of fact, no further quantization has been made; only the extension to multiparticle systems has been involved. However, vortex in three dimensions is with cylindrical symmetry, and the anisotropy make it very difficult to write out the exact form of a_k^+ and a_k, thus, the detailed discussion will be presented in further papers.

4.2 Application of Quantum Language to the Fields that beyond the Microscopic

The quantization of flux as vortices was discussed briefly in the previous section It is also concluded that vortex could be treated as microscopic particle with anisotropy that could be described by operators. Here, the problem we most concerned are not how to write out the exact form of a_k^+ and a_k, but whether this work is effective, that is, whether it is reasonable when we try to describe the vortex with operators? To answer this question, let us now check the meaning of mathematics itself and its application in physics.

Quantum physics deal with a mystical world where there are profound philosophical meanings and involves radical departures from our normal ways of thinking about reality. Solving problems involving microscopic systems requires the mathematical methods of description, so it is very hard to gain insight into the nature of reality as understood in modern physics without the help of mathematics. In fact, it is not too exaggerated to say that half of quantum mechanics is mathematics. That is why there are such a big collection of abstract laws and equations in quantum mechanics. The methods used in this field look like so strange and we may wonder if these methods could be found any usage in other field.

As is known, the laws in mathematics are essentially obtained by logical reasoning, and logic describes the rules of thinking, or the laws of mind, their applicability usually are universal and are not confined to any special real matter condition. If a mathematical law could be used in micro world, naturally, we also expect it could be used in macro world or any other case where it can be fit. Now that the mathematical methods used in quantum mechanics are universally valid, it should be reasonable to describe the vortex with creation operator and the annihilation operator because vortex is the quantization of flux.

It is very interesting that Elinski [62] has tried to synthesis the physics with economy, where he successfully used the creation and annihilation operators to describe the production and the consumption of goods, and deal the commodity flow and funds as a special kind of quantum field. This provides a nice illustration to the extension of the quantum language.

Another example of the application of quantum language to the fields that beyond the microscopic is the superposition principle [63], one of the main principles in quantum mechanics, which states that any particle state can be extended as the addition of a serial of eginstates (the base of the linear space). However, this is nothing new but a method, Hilbert space, borrowed from mathematics. Certain formal correspondences between the theory of vector spaces and functional space enable the formalism of quantum mechanics to be given an abstract geometrical interpretation. The completeness of "base" is highlighted in functional analysis and it provides an effective tool for describing a new state by a serial of known state. This idea has been extended to the case of flux pinning described with the classical mechanics [64]. Furthermore, one can also find there are some obvious analogies between the formulas in quantum mechanics and those in communication science (Fourier analysis).

It could be seen that, no matter how nova a physical insight look like at first, once it is described with mathematics, one can always find some applications of the physical insight into other fields because of the versatility of mathematics. In some sense, the micro world and macro world can be partially united by mathematics.

5 Conclusion

Quantum nature occurs with vortex because vortex has very small mass per unit length. The magnetic field that penetrates into the superconductors must be quantized because the order parameter of superconductivity must be single value.

Quantum effects will manifest themselves at sufficiently low temperature because in high temperature (large quantum numbers) the energy differences become arbitrarily small and the energies form almost a continuum. It is suggested that the quantum tunneling process of vortex is a direct consequence of the quantum nature of the vortex, which results in much vortex matter phenomenology of type-II superconductors in mixed state, such as vortex creep, vortex depinning and vortex melting. The quantum fluctuation of vortices drives various types of phase transitions and defines much the superconducting properties of real materials at low temperature.

It goes without saying that a consistent picture of the quantum nature of vortex is hard to attain currently because of the complexity of vortex itself, and the main pieces of vortex matter phenomenology are only interpreted as what the authors believe to be. Some ideas used in quantum mechanics are also suggested could be used in macroscopic world, even in social science because the microscopic world and the macroscopic world can be unified in many aspects by mathematics.

Acknowledgement

The authors would like to thank Gaoshan Li for careful reading of the manuscript.

References

[1] Abo-Shaeer, J. R.; Raman, C.; Vogels J. M.; Ketterle, W. *Science* 2001, 292, 476.

[2] Blaauwgeers, R. et al. *Nature* **200**, 404, 471.

[3] Little, W. A.; Parks, R. D. *Phys. Rev. Lett.* 1962, 9, 9. Little, W. A.; Parks, R. D. *Phys. Rev.* 1964,133, A97.

[4] Kawasaki, T.; Matsuda, T.; Endo, J. et al. Jpn. *J. Appl. Phys*. 1990, 29, L508.

[5] Onsager, L. *Nuovo Cimento* 1949, 6, suppl. 2 249.

[6] London, F. *Superfluid*; Wiley: New York, 1950; vol. I, p.152.

[7] Feynman R. P., *Progress in Low Temperature Physics*; Ed. Gorter, C. J.; Amsterdam: North-Holland, 1955; vol. I, ch2.

[8] Abrikosov, A. A. *Soviet Phys. JETP* 1957, 5, 1174.

[9] Saint-James, D.; Sarma, G.; Thomas, E. J. Type II superconductivity; *Pergamon Press Ltd*., 1969, p41.

[10] Kleiner, W. H.; Roth, L. M.; Autler, S. H. *Phys. Rev*. 1964, 133, A1226.

[11] Gribier, D.; Jacrot, B.; Madhav Rao, L.; Farnoux, *B. Phys. Lett*. 1964, 9, 106.

[12] Sarma, G. *Compt. Rend. Acnd. Sci.* Paris 1964, 258, 1461.

[13] Essmann, U.; Trauble, H. *Sci. Am.* 1971, 224, 75.

[14] Gammel, P. L.; Bishop, D. J.; Dolan, G. J.; Kwo, J. R.; Murray, C. A.; Schneemeyer, L. F.; Waszczak, J. V. *Phys. Rev. Lett*. 1987, 59, 2592.

[15] Murray, C. A.; Gammel, P. L.; Bishop, D. J. *Phys. Rev. Lett*. 1990, 64, 2312.

[16] Blatter, G.; Feigel'man, M. V.; Geshkenbein, V. B.; Larkin A. I.; Vinokur, V. M. *Rev. Mod. Phys.* 1994, 66, 1125.

[17] Brandt, E. H. Rep. *Prog. Phys*. 1995, 58, 1465.

[18] Brandt, E. H. J. *Low Temp. Phys*. 1977, 26, 709; Brandt, E. H. J. *Low Temp. Phys*. 1977, 26, 735; Brandt, E. H. J. *Low Temp. Phys*. 1977, 28, 263; Brandt, E. H. J. *Low Temp. Phys*. 1977, 28, 291.

[19] Larkin, A. I.; Ovchinnikov, Yu. N. J. *low Temp. Phys*. 1979, 34, 409.

[20] Luo, Sheng; Li, Chunguang; He, Yusheng; Zhang, Guohua; Jin, Duo; Chen, Zhaojia; Xiong, Guangcheng; Lian, Guijun; Yin, Daole; Gough, C. E. *Supercond. Sci. Technol.* 2002, 15, 1300.

[21] Liu, Y.; Haviland, D. B.; Glazman, L. I.; Goldman, A. M. *Phys. Rev. Lett.* 1992, 68, 2224.

[22] Fruchter, L.; Malozemoff, A. P.; Campbell, I. A.; Sanchez, J.; Konczykowski, M.; Griessen, R.; Holtzberg, F. *Phys. Rev. B* 1991, 43, 8709.

[23] Mota, A. C.; Pollini, A.; Visani, P.; Muller, K. A.; Bednorz, J. G. *Phys. Scr.* 1988, 37, 823.

[24] Mota, A. V.; Visani, P.; Pollini, P. A. *Phys. Rev*. B 1988, 37, 9830.

[25] Mitin, A. V. *Sov. Phys. JETP* 1987, 66, 335.

[26] Rosario Fazio; Herre van der Zant Physics Reports 2001, 355, 235.

[27] Wallraff, A.; Lukashenko, A.; Lisenfeld, J.; Kemp, A.; Fistul, M. V.; Koval, Y.; Ustinov, A. V. *Nature* 2003, 425, 155.

[28] Wilson, W. Phil. *Mag*. 1915, 29, 795.

[29] Sommerfeld, A. *Ann. der Physik* 1916, 5, 5.

[30] Planck, M. *Ann. der Physik* 1916, 50, 385.

[31] L. de Broglie, *Comptes Rendus* 1923, 177, 507 548 630.

[32] L. de Broglie, *Phil. Mag*. 1924, 47, 446.

[33] Stephen, M. J.; Suhl, H. *Phys. Rev. Lett*. 1964, 13, 797.

[34] Suhl, H. *Phys. Rev. Lett*. 1965, 14, 226.

[35] Kawasaki, T.; Matsuda, T.; Endo, J. et al. *Jpn. J. Appl. Phys*. 1990, 29, L508.

[36] Heisenberg, W. *Z. Physik* 1927, 43, 172.

[37] Planck, M. Barth, Theory of Heat Radiation, Dover, New York, 1959, p143.

[38] Michael Tinkham, Introduction to Superconductivity; second edition; McGraw-Hill Inc.: New York; 1996; p.11.

[39] Deaver, B. S.; Fairbank, W. M. *Phys. Rev. Lett*. 1961, 7, 43.

[40] Doll, R.; Näbauer, M. *Phys. Rev. Lett*. 1961, 7, 51.

[41] Ginzburg, V. L.; Landau, L. D. *Zh. Eksp. Teor. Fiz*. 1950, 20, 1064.

[42] Lewis H. Ryder. Quantum field theory; Cambridge University Press, 1986; p406.

[43] Bardeen, John; Stephen, M. J. *Phys. Rev.* 1965, 140, A1197.

[44] Callan, C. G.; Coleman, S. *Phys. Rev. D* 1977, 16, 1762.

[45] Caldeira, A. O.; Leggett, A. J. *Phys. Rev. Lett.* 1981, 46, 211.

[46] Gianni Blatter; Vadim B. Geshkenbein; Valeri M. Vinokur *Phys. Rev. Lett.* 1991, 66, 3297.

[47] Stephen, M. J.; Bardeen, J. *Phys. Rev. Lett.* 1965, 14, 112.

[48] Strnad, A. R.; Hempstead, C. F.; Kim, Y. B. *Phys. Rev. Lett.* 1964, 13, 794.

[49] Landau, L. D.; Lifshitz, E. M. Statistical Physics; 3rd edition Part I; Pergamon: Oxford, New York, Toronto, Sydney, Paris, Frankfurt, 1980; Course of Theoretical Physics Vol. 5, p80.

[50] Ustinov, A. V. *Physica D* 1998,123, 315.

[51] Ma, Rongchao; Ma, Yuefei; Sun, Yuping. *Physica C* 2005, 422, 57.

[52] de Andrade, R.; de Lima, O. F. *Phys. Rev. B* 1995, 51, 9383.

[53] Fukuzumi, P. Y. et al. *Phys. Rev. Lett.* 1996, 76, 684.

[54] Attanasio, C. et al. *Phys. Rev. B* 1996, 53, 1087.

[55] Bohr, N. Z. *Physik* 1920, 2, 423.

[56] Mota, A. C.; Juri, G.; Visani, P.; Pollini, A.; Teruzzi, T.; Aupke K.; Hilti, B. *Physica C* 1991, 185, 343.

[57] Griessen, R.; Lensink, J. G.; Schnack, H. G. *Physica C* 1991, 185, 337.

[58] Chudnovsky, E. M. *Phys. Rev. B* 1995, 51, 15351.

[59] Onogi, H.; Doniach, S. *Solid State Commun.* 1996, 98, 1.

[60] Rozhkov, A.; Stroud, D. *Phys. Rev. B* 1996, 54, R12697.

[61] Landau, L. D.; Lifshitz, E. M. Quantum Electrodynamics; Second edition; Pergamon: Exeter, 1982; *Course of Theoretical Physics* Vol. 4, p5.

[62] Kirill Ilinski, Physics of Finance: Gauge Modeling in Non-equilibrium Pricing; John Wiley & Sons Ltd.: Chichester, England, 2001; p249.

[63] Landau, L. D.; Lifshitz, E. M. Quantum Mechanics; Third edition; Pergamon: Surry, 1977; *Course of Theoretical Physics* Vol. 3, p6.

[64] Ma, Rongchao; Ma, Yuefei; Song, Wenhai; Zhu, Xuebin; Liu, Shengman; Du, Jiaju; Sun, Yuping; Li, Chengshan; Ji, Ping; Feng, Yong; Zhang, Pingxiang. *Physica C* 2004, 411, 77.

In: New Frontiers in Superconductivity Research
Editor: Barry P. Martins, pp. 161-197
ISBN 1-59454-850-1

Chapter 6

Electric Noise and Local Photon-Induced Nonequilibrium States in a Current-Carrying Nanostructured Superconductor

Andreas Engel
Physics Institute of the University of Zurich,
Winterthurerstr. 190, 8057 Zurich, Switzerland
Alexei Semenov and Heinz-Wilhelm Hübers
DLR, Institute of Planetary Research,
Rutherfordstr. 2, 12489 Berlin, Germany
Konstantin Il'in and Michael Siegel
Institute for Micro- and Nanoelectronic Systems,
University of Karlsruhe, Hertzstr. 16,
76187 Karlsruhe, Germany

Abstract

Superconducting thin films and micro- and nanostructured superconducting elements have enormous potential for applications in electronics and as highly sensitive detectors. We present recent studies of the superconductivity of nanostructured Nb and NbN thin films as well as their potential use as very fast, sensitive and energy-dispersive photon-counters. Using photo- and electron-beam-lithography 5 to 20 nm thick films were patterned into single bridges and meander structures with widths of the conduction paths as narrow as 50 nm. Dependencies of the critical temperature and critical current density on variations in bridge width and thickness can be modelled rather well using existing theories applicable to dirty superconductors with very short coherence lengths.

The meander structures have already been shown to be very fast and sensitive single photon detectors in the visible and near-infrared range. Their ultimate sensitivity is hampered however, by electric noise caused at least partly by fluctuations in the superconducting state. The detection mechanism exploits the creation of a localized hot-spot by the absorption of a photon with sufficient energy. Albeit being a relatively small perturbation of the superconducting state, it can be detected as a voltage transient given the right experimental conditions. These conditions require cross-sectional areas of the conduction paths comparable with the cross-section of the hot-spot and an applied bias current only slightly less than the critical current at the operating temperature. Under these conditions fluctuations, like e.g. phase-slip centers, fluctuations in the number density of Cooper-pairs or fluctuations within the vortex system, may also cause voltage pulses very similar in magnitude and shape to those caused by absorbed photons and may be a significant contribution to the dark count rate. Measured dark count rates have been analyzed with respect to these various noise sources. Besides the noise characteristics or noise equivalent power the performance of the single-photon detectors was investigated concerning the quantum efficiencies as a function of the incident photon energy and material parameters of the superconducting meander line.

1 Introduction

Nonequilibrium states produced in a superconducting film by absorbed photons have been a subject of intense investigations for a long time. These states are interesting themselves from a physical point of view since they shine light on nonequilibrium properties of superconductors. The understanding of these phenomena are also of direct practical interest, because when superconductors are used in various applications they are often in a state deviating essentially from equilibrium. The literature on non-equilibrium superconductivity has grown enormously during the last three decades. Reviews including experimental as well as theoretical aspects can be found in Refs. [1–3].

Below we concentrate on the so called quantum regime in that the nonequilibrium areas or spots induced in a superconductor by single photons are not overlapping. In this case nonequilibrium spots can be treated independently while the overall effect of irradiation can be described as a statistical combination of such spots. Of particular interest is the regime when the nonequilibrium state induced by only one photon or ionizing particle can be detected. Intuitively, the size of the corresponding nonequilibrium area is not very large. However, in order to generate a measurable effect on the resistance of a superconducting film, it should be at least as large as the cross-section of the superconductor. Thus, the single-particle detection regime is likely to be realized in relatively narrow superconducting lines. N. K. Sherman was the first [4] who practically demonstrated this detection regime. The possibility to detect in such a way single X-ray quanta was first shown by A. Gabutti *et al.* [5].

The possibility to resistively detect a single photon in a relatively large superconducting film was first discussed by Kadin and Johnson [6] in the framework of the hot-electron approximation, which was introduced for superconductors near the transition by Shklovskii [7] and further developed by Perrin and Vanneste [8] and Semenov *et al.* [9]. Kadin and Johnson suggested that a photon absorbed somewhere in the resistive film initiates a growing hot spot. The resistance inside the hot spot is larger than in the surrounding area. Even if the size of the hot spot is much smaller than the size of the film, the voltage drop across the current-biased film *feels* the presence of the hot spot. This detection scheme requires a fast read-out because on time scales longer than the electron-phonon interaction time the deposited energy will be shared between electrons and phonons making the resistance change practically undetectable. The usable time scale enormously increases at millikelvin temperatures because the electron-phonon interaction weakens at low temperatures.

Almost uniform electron heating can be realized at these temperatures at a millisecond time scale even in relatively large superconducting films. The approach was first developed by Cabrera [10] to demonstrate transition edge sensors (TES) capable of detecting single infrared photons. Earlier Irwin showed [11] that a negative electro-thermal feedback may substantially shorten the relaxation time of the TES and increase its sensitivity. A photon absorbed in the sensor heats the electron system above its equilibrium temperature, leading to an increase of the sensor's resistance and, consequently, to the decrease of the bias current and dissipated Joule power. The integral of the drop in current (read out by an array of dc SQUID's) gives the energy absorbed by the sensor with no free parameters.

Semenov *et al.* [12] proposed to use a bias current close to the critical current in order to relax requirements on the width of a superconductor and to extend single-photon sensitivity to the infrared range. In this current-assisted detection approach, a superconducting strip that is operated well below its transition temperature carries a bias current only slightly smaller than the critical value at the operating temperature. Generation of a hot spot at the position where the photon has been absorbed creates a local region with suppressed superconductivity. The supercurrent is forced to flow around this normal (resistive) spot, through the parts of the film that remain superconducting. If the diameter of the resistive spot is such that the current density in the superconducting portion of the film reaches the critical value, a resistive barrier is formed across the entire width of the strip, giving rise to a voltage pulse with the magnitude proportional to the bias current. The detection of near-infrared photons using this approach was first demonstrated by Gol'tsman *et al.* [13]. The physical difference of this quantum detection mechanism is that the resistive state and, thus, the response appears due to the collaborative effect of the bias current and the hot-spot formation.

The use of a tunnel junction in order to collect nonequilibrium quasiparticles produced by a photon or an ionising particle was suggested by Wood and White [14] and demonstrated by Kurakado and Mazaki [15]. In this approach the superconductor is not supposed to undergo the superconducting transition. Typically, a superconducting tunnel-junction (STJ) detector consists of a relatively thick absorber film with an underlying thinner trapping layer, which forms one junction electrode. A photon captured in the absorber generates a high-energy photoelectron that relaxes via hot-electron multiplication into the energy gap of the absorber. Nonequilibrium quasiparticles excited during the cascade diffuse to the adjacent trapping layer, which has a smaller energy gap. There, quasiparticles scatter inelastically, reaching an energy level corresponding to the trapping-layer energy gap. The latter process is called *trapping* because it confines the charge to the region close to the tunnel barrier. The tunnel junction is externally biased in such a way that trapped quasiparticles can tunnel directly to the electrode characterized by the lower-energy gap. The same potential barrier prevents them from returning. They can, however, break Cooper pairs in the low-gap electrode and then form new pairs with unpaired electrons in their own electrode. Thus, the process returns unpaired electrons to the initial electrode, increasing the number of tunneling events per quasiparticle and providing intrinsic charge amplification. The time integral of the current transient gives, with no free parameters, the charge that has been transferred through the tunnel junction. This latter value is proportional to the number of quasiparticles created in the cascade and, consequently, to the x-ray quantum energy.

The first thorough model describing the formation of the nonequilibrium state following absorption of a single X-ray photon in an infinitely large superconductor was developed by Twerenbold [16] in order to interpret his own experimental data with STJ detectors. A more general approach, including time evolution of nonequilibrium distribution functions of quasiparticles and phonons, was developed by Nussbaumer *et al.* [17]. The authors solved the Chang-Scalapino equations numerically for the quasiparticle and phonon distribution functions in a spatially homogeneous situation and supplemented the solution by one-dimensional diffusion. The full theory includes the parameters that are important for a real detector, such as back tunneling and losses of quasiparticles at the edges of the device, resulting in good agreement between the calculated transient response signals and the experimentally measured pulse shapes.

In this report we concentrate on the current-assisted photon detection in narrow superconducting strips. We present recent developments in the preparation of devices made from Nb and NbN thin films and the characterization of their superconducting properties [18, 19] as well as progress in the detection model and detector characteristics [20,21]. The report is organized as follows: In Sec. 2 we discuss the thin film and nanostructuring technology used to prepare the devices. It is followed by a thorough characterization of the superconducting parameters of thin films and bridges (Sec. 3). Then we introduce a refined photon detection model in thin and narrow superconducting strips in Sec. 4 before we present results for meander lines used as single photon detectors in the visible and near-infrared spectral range in Sec. 5. We conclude in Sec. 6 with a summary and a short outlook on future developments.

2 Thin Film and Nanostructuring Technology

2.1 Deposition of Nb and NbN Thin Films

Nb and NbN films with a thickness ranging from 5 nm to 200 nm were deposited by dc magnetron sputtering of a Nb target in either pure argon atmosphere or in Ar+N_2 gas admixture, respectively, at a total pressure of about 10^{-2} mbar. Residual gas pressure was less than 10^{-6} mbar. The films were deposited on different substrates, like Si, MgO, and sapphire, with a nominal size of 10×10 mm^2. The thickness of the deposited film was inferred from the sputtering time and a pre-defined deposition rate. In order to grow films required by detector applications [20], i.e. films with a reduced density of electronic states, substrates were kept at room temperature during deposition. Deposition rates were 2 and 1.2 nm/sec for Nb and NbN films, respectively.

2.2 Nanopatterning of Nb and NbN Thin Films

Deposited films with thickness thinner than 20 nm (interesting from application point of view) were patterned by means of conventional photo- and electron-beam lithography, ion milling (IBE) and lift-off technique. Preparatively, a set of global and local alignment marks for the following e-beam lithography and a set of photolithography alignment marks were made by photolithography, a gold layer of thickness about 30 nm was deposited, followed by lift-off process of metallization in acetone. Then in the center part of a 100 μm writing field defined by local alignment marks we made the e-beam lithography to form a set of bridges (16 samples per chip) with different sizes or a set of meander line structures (6 samples per chip). The e-beam lithography was made using a scanning electron microscope (SEM) *LEO 1530* with a *GEMINI* column[1], which has been refitted to an electron-beam writer with an *ELPHY Plus* system[2]. We worked with a low energy electron-beam (5 kV) in order to reduce the effect of backscattered electrons in the relatively thick e-beam resist PMMA with molecular weight 950 and thickness about 150 nm. Using the resist as a protective mask against ion milling we transferred an image created in the e-beam resist into the superconducting film by ion-beam etching (IBE). Contact pads of size about 0.75×1 mm^2 were formed by photolithography and a final IBE etching step.

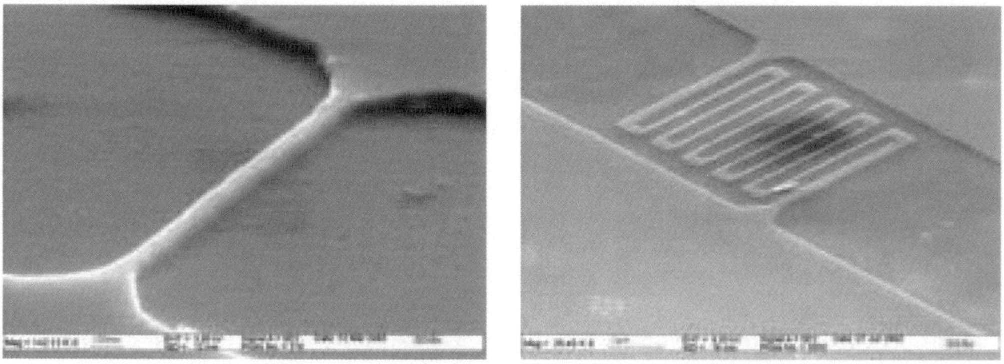

(a) (b)

Figure 1: (a) SEM image of a 85 nm wide bridge made from 10 nm thick Nb film on Si substrate. In (b) an image is shown of a 90 nm wide meander structure made from the same Nb film as the bridge in (a). The bright spot near the lower end of the meander is a remnant of a PMMA resist.

Within a set of bridges, the width was varied from 50 nm to 10 μm. Bridges narrower than 2 μm were made 1.5 μm long. Wider bridges had the length equal to the width. After patterning the actual sizes of the structures were controlled with a scanning electron microscope. A representative SEM picture of a 85 nm wide bridge is shown in Fig. 1(a). The SEM study proved that all studied bridges had a rather precise rectangular form. The roughness of bridge-edges did not exceed 5 nm for all studied specimens.

In order to increase the active area of the detector and to improve the coupling of visible and near-infrared radiation the active sensor element was made as a meander line covering an area of approximately 4×4 μm. The width of the meander lines was in the range from 80 to 120 nm with a separation between lines of about 300 nm and a total length approximately 37 μm. In Fig. 1(b) the SEM image shows a typical meander line structure limited by contact pads made from the same superconducting film as the bridge in Fig. 1(a).

---

[1]http://www.smt.zeiss.com/nts

[2]http://www.raith.de/nanolithography/nano_elphyplus.html/

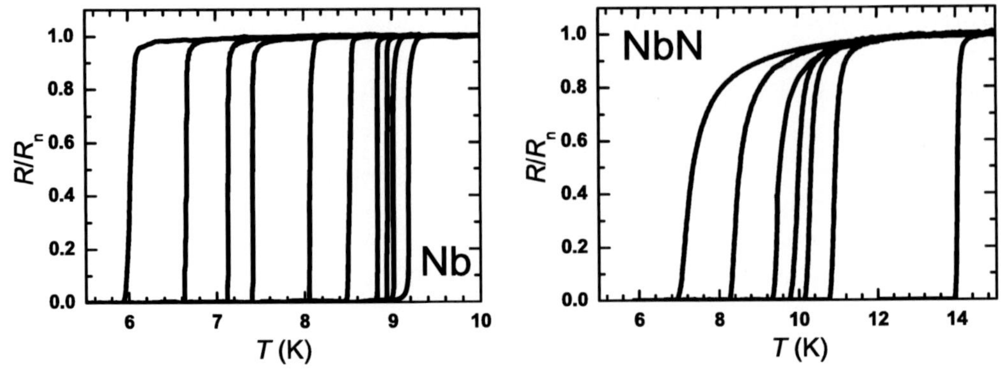

Figure 2: Temperature dependence of the normalized resistance of Nb (a) and NbN (b) films of different thickness: from left to right Nb: 8 - 200 nm; NbN: 5 - 100 nm

3 Characterization of Superconducting Films and Nanostructures

3.1 Superconducting Transition Temperature of Films

The resistive superconducting transition of as-deposited films was measured by the standard four-probe method. We have not found any systematic difference between T_c of superconducting films deposited on different substrates. The transition temperature was defined as the midpoint between the temperatures corresponding to 10% and 90% of the normal resistance of the sample. The width of the superconducting transition of Nb films varied from a few hundredth of a Kelvin for the thickest films up to 0.1 K for the thinnest films. Typical transition widths of NbN films were a few tenths of a Kelvin for 100 nm thick films and about half a Kelvin for 5 nm thick films. In Fig. 2 there are series of resistive transitions measured for Nb (a) and NbN (b) films of different thickness. The resistance is normalized to the normal state resistance R_n for better comparison. Both materials demonstrate similar phenomena, i.e. the decrease of the film thickness results in a shift of the $R(T)$ curve to lower temperature.

In Fig. 3 we plotted the transition temperature T_c versus the film thickness d for Nb and NbN films to show the systematic variation. For Nb films, T_c remains almost constant for $d > 50$ nm and approaches a value of about 9.35 K, which is close to T_c of bulk Nb. The transition temperature dropped continuously for $d < 30$ nm reaching 6 K for 8 nm thick films. In the range of film thicknesses from 5 to 30 nm, the transition temperatures of NbN films demonstrate similar behavior. The thinnest 5 nm NbN films had a typical T_c of 7 K. Contrary to Nb films, the T_c of NbN films continued to grow with d up to the largest film thickness. This effect may be due to variation of the film composition with film thickness. The transition temperature of NbN strongly depends on the nitrogen content. The nitrogen content in the initial, thinnest continuous layer of the film strongly depends on both the lattice mismatch between the film and substrate materials and on the sputtering regime, i.e. substrate temperature, Ar and N_2 partial pressure, sputtering rate, etc. The film growth process above the initial layer is less influenced by the substrate with increasing film thickness. Therefore, the nitrogen content in a thicker NbN film differs from the nitrogen content in the initial layer. In the following we will discuss results obtained for the NbN films thinner than 10 nm, which are believed to have the same composition.

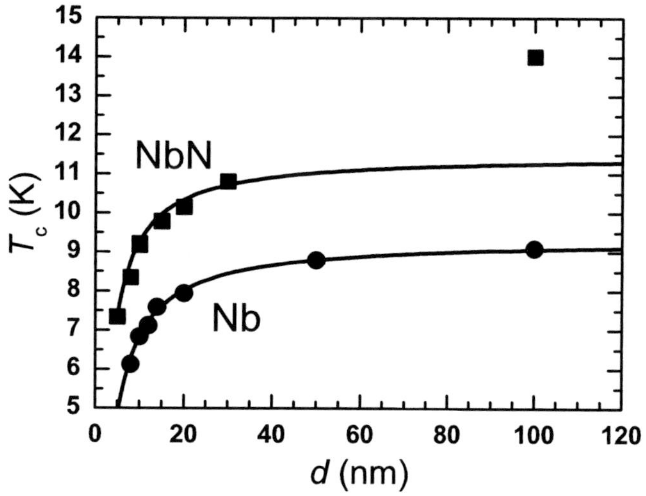

Figure 3: Dependence of the critical temperature T_c of as deposited Nb (circles) and NbN (squares) films on thickness of the film. The solid lines are best fits according to Eq. (1).

The reduction of the critical temperature of thin superconducting films with decrease of the film thickness has been reported by many authors [22–24]. The proximity effect model, which was first suggested by Cooper [25], is a frequently used approach to describe the reduction of T_c in superconducting thin films. Cooper suggested that superconducting interactions should be destroyed in thin layers adjacent to the film surface and the film-substrate interface. He further demonstrated that in ultra-thin superconducting films with a thickness of about or less than BCS coherence length, ξ_0, the suppression of superconductivity was due to the proximity effect between these normal layers and a remaining, superconducting part of the film. Recently, Fominov and Feigel'man [26] have presented a general theory of superconducting properties of thin superconductor-normal-metal (SN) bi-layers. They have shown that, in the diffusive limit of a dirty material, the proximity effect in a SN bi-layer is governed by the interlayer interface resistance ρ_{int}. In case of a perfect interface $\rho_{\text{int}} \to 0$ (the interface transparency approaches unity) they have obtained Cooper's classical result [25]. The authors of Ref. [26] have also shown that their results can be directly applied to a NSN three-layer system if one assumes equal thickness of both normal layers. According to them, the transition temperature T_c of a NSN system in the Cooper's limit can be found from the following equation:

$$\ln\frac{T_c^0}{T_c} = \frac{\tau_N}{\tau_S+\tau_N}\left[\Psi\left(\frac{1}{2}+\frac{(\tau_S+\tau_N)\,\hbar}{k_B T_c \tau_S \tau_N}\right) - \Psi\left(\frac{1}{2}\right) - \ln\sqrt{1+\left(\frac{\tau_S+\tau_N}{\tau_S\tau_N\omega_D}\right)^2}\,\right], \quad (1)$$

where $\Psi(x)$ denotes the digamma function, T_c^0 is the critical temperature of an isolated superconductor, ω_D is the Debye frequency of the S material, k_B and $\hbar$ are the Boltzmann and Planck constants. The $\tau_{\rm S}$ and $\tau_{\rm N}$ quantities for the NSN three-layer film are

$$\tau_{\rm S} = \pi \frac{d_{\rm S}}{v_{F_{\rm S}}} \rho_{\rm int}, \; \tau_{\rm N} = 2\pi \frac{v_{F_{\rm N}} d_{\rm N}}{v_{F_{\rm S}}^2} \rho_{\rm int}, \tag{2}$$

with $d_{\rm S,N}$ and $v_{F_{\rm S,N}}$ being the thickness and the Fermi velocity of the S and N layers, correspondingly.

Before we apply the above proximity effect model, we have to assure it is an appropriate model and discuss the origin of the non-superconducting layers in our superconducting Nb and NbN thin films. It is well known that Nb has a relatively high affinity to oxygen. Therefore, at least an oxidized top layer is expected. In order to clarify the situation we studied our Nb films using secondary ion mass spectroscopy (SIMS)[3]. For the SIMS measurement an especially layered 600 nm thick Nb film structure has been deposited. At first, 400 nm of Nb was deposited on silicon substrate. After deposition, the film has been exposed to normal air for 24 hours and followed by another deposition of 200 nm of Nb on top of the first layer. The SIMS study of our Nb films (see Fig. 4) has detected the presence of two elements besides Nb: oxygen, as it was expected, and carbon. Both were found in the surface layer and in the layer adjacent to the film-substrate interface. Niobium mono-oxide is a superconductor too but with a lower transition temperature of about 1.4 K [27]. Carbon atoms on the other hand act as impurities and destroy superconductivity. Therefore, we believe that our Nb films are most probably NSN rather than a structure with three superconducting layers S'SS', where S' is a superconductor with a lower $T_{\rm c}$ value. Unfortunately, the spatial resolution of SIMS does not allow for a direct measurement of the thickness of the normal layers in our niobium films.

Due to the chemical bonding between Nb and N atoms, NbN has a reduced oxygen affinity compared to pure Nb. Nevertheless, we also expect to find O and C impurities at the surface and film substrate interface of NbN films, not least due to the fact that all films were made in the same magnetron sputtering system. Even though it is likely that the normal layers are thinner for the NbN films, we consider both, Nb and NbN films as NSN three-layer structures. Furthermore, absence of a real mechanical interface between N and S layers suggests the transparency of the NS interfaces to be close to unity. This ensures that $\rho_{\rm int} \rightarrow 0$ and that the NSN structures are in the Cooper's limit and Eq. (1) is applicable.

Results of the fitting of the experimental data with Eq. 1 are shown in Fig. 3 by the solid lines. In case of the Nb films we used $d_{\rm N}$, the effective thickness of the normal conducting top and bottom layers, as the only fitting parameter. Accordingly, we define $d_{\rm S}$ as the difference between the nominal film thickness d and the total thickness $2d_{\rm N}$ of the normal layers. For the critical temperature of the pure superconductor, we used the experimental value of the thickest films $T_{\rm c}^0 = 9.35$ K. As it has been mentioned above, this value is close to the transition temperature of bulk Nb. The Debye frequency has been calculated from the well known value of the Debye temperature for niobium $\Theta_D = 275$ K. From the best fit of the $T_{\rm c}(d)$ dependence we obtained a value for the effective thickness of the Nb films $d_{\rm N} = 0.5$ nm.

The situation for the NbN films is more complicated, because we do not know the exact value of $T_{\rm c}^0$ for our NbN films. Therefore, we fitted the experimental results for the

[3]SIMS study has been performed by Dr. U. Zastrow, Forschungszentrum Juelich, Germany.

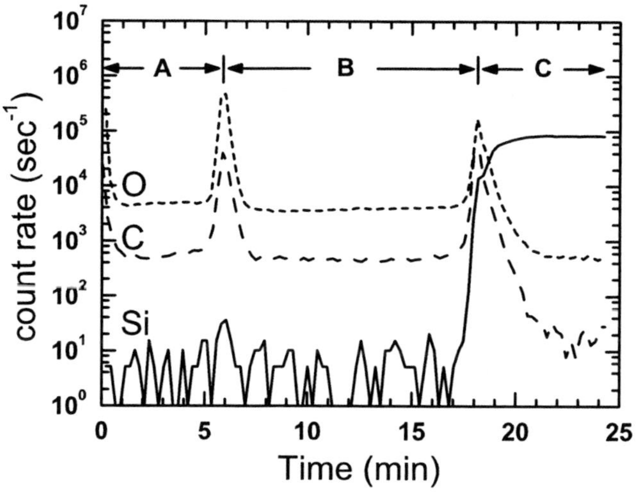

Figure 4: Result of the SIMS study of a 600 nm Nb film. The solid, dashed, and dotted lines are count rates of silicon (substrate), carbon and oxygen, respectively. Areas A and B correspond to 200 nm and 400 nm thick Nb layers; area C marks the Si substrate. From left to right there are three pronounced O and C peaks: at the film surface, at the first and second Nb layer interface and at the film substrate interface.

NbN films (excluding the 100 nm thick film) using two fitting parameters: T_c^0 and d_N. From the best fit we found a transition temperature $T_c^0 = 11.8$ K for an infinitely thick NbN film with a composition similar to the composition of our thin films in the range from 5 to 30 nm thickness. This value of T_c^0 is lower than the transition temperature of bulk NbN with perfect stoichiometry. The difference is most likely due to the particular sputtering regime, which we used growing our ultra-thin films and resulted in films with a high concentration of defects and a somewhat reduced N content. The effective thickness of the normal conducting layers obtained from the fitting procedure is about 0.4 nm, which is smaller than d_N for the pure Nb films just as expected.

3.2 Second Critical Magnetic Field of Films

The temperature dependence of the second critical magnetic field H_{c2} has been measured for a series of Nb and NbN films of different thickness. The measurements were done in an *Oxford Instruments Heliox VL* insert[4] equipped with a superconducting solenoid capable of reaching magnetic fields up to 2 T. During measurements films were mounted on a brass holder along with the magnetic field independent temperature sensor. Magnetic fields were oriented perpendicular to the film surface. At each operating temperature the external magnetic field was increased up to a value at which the resistance of the studied film reached the value of $R_n/2$, where R_n is the resistance of the film in the normal state. This particular value of the applied magnetic field was taken as the value of the second critical magnetic field at the given temperature, $H_{c2}(T)$. The temperature was changed step by step in the

[4]Operational temperature range of the insert is from 300 mK up to 300 K with a magnetic field up to 2 T. Thermal stability is ±3 mK below 1.5 K.

range from 4.2 K up to the critical temperature of the film. The accessible temperature range in our measurements was limited especially in case of the NbN films, which are generally characterized by a much stronger temperature dependence of the second critical magnetic field near T_c in comparison with Nb films. Nevertheless, an automatic temperature stabilization system allowed us to change the temperature in minimum steps of less than 40 mK. The temporary temperature stability of about 5 mK in the actual temperature range was long enough to perform measurements of H_{c2}.

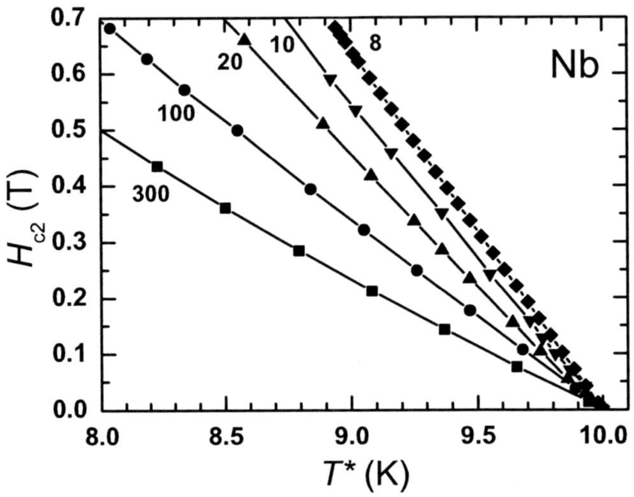

Figure 5: Temperature dependence of the second critical magnetic field H_{c2} measured on Nb films of different thickness. For better comparison all curves are shifted in temperature to an equal value of the critical temperature T_c (10 K). Numbers next to the data curves indicate the nominal film thickness in nanometers.

The results of the measurements of the temperature dependence of the second critical magnetic field in Nb films are shown on Fig. 5. All experimental curves are shifted to the same value of the critical temperature for better comparison. It is seen that a decrease of the Nb film thickness results in an increase of the slope of the experimental $H_{c2}(T)$ curve. Also important to note that all curves are linear near T_c as required by Ginzburg-Landau (GL) theory.

Results we have obtained for NbN films were qualitatively different. In Fig. 6(a) H_{c2} is plotted versus the reduced temperature T/T_c. Obviously, the $H_{c2}(T)$ behavior is non-linear near T_c and all experimental curves are in line for $T/T_c > 0.98$. At lower temperatures the curves start to deviate from the common line. At first the deviation is observed for the thickest 10 nm NbN film, then for the 8 nm thick film. The two curves corresponding to the thinnest 5 nm thick films deposited on silicone and sapphire substrates coincide with each other at least down to $T/T_c \approx 0.95$. After deviation from the common line all experimental $H_{c2}(T)$ curves become linear at lower temperatures. It is intriguing that the slope of the linear part of all curves is almost equal which is best seen in Fig. 6(b) where all curves are shifted to the same T_c. Thus in contrast to the Nb films there is no noticeable influence of the film thickness on the slope of $H_{c2}(T)$ in NbN films in the range from 5 to 10 nm of film thickness. It is also of interest that in Fig. 6(b) the four curves separated into two groups of

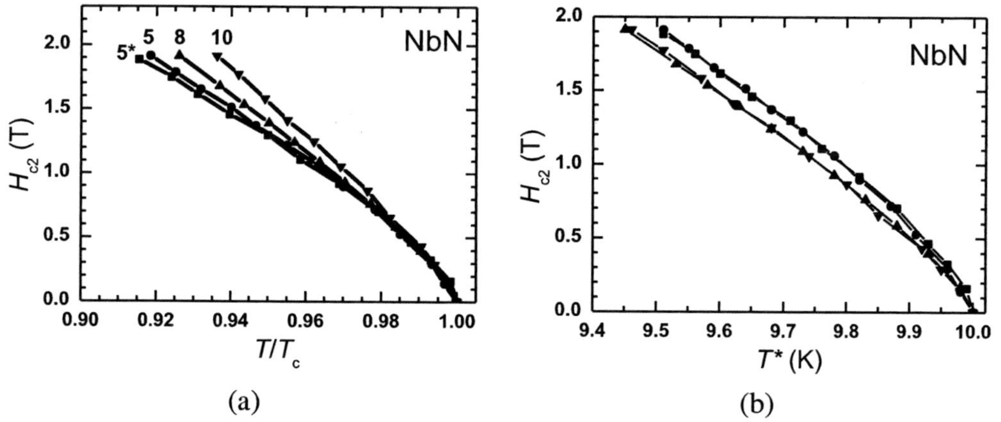

Figure 6: (a) Dependence of the second critical magnetic field of NbN films on reduced temperature T/T_c. Numbers indicate the film thickness in nanometers. A 5 nm film was also sputtered onto a sapphire substrate 5^*, all others onto silicone substrates. (b) Dependence of the second critical magnetic field on temperature for the same NbN films as in (a). Similar to Fig. 5 all curves are shifted to an equal value of the critical temperature T_c (10K). The data separate into two groups: both 5 nm thick films (upper data) and the 8 and 10 nm thick films (lower data).

almost coinciding curves: both 5 nm thick films are in one group and the two thicker films are in the other one.

The value of the temperature derivative of the second critical magnetic field $\mathrm{d}H_{c2}/\mathrm{d}T$, versus thickness of Nb and NbN films is shown in Fig. 7. The magnitude of $\mathrm{d}H_{c2}/\mathrm{d}T$ for the NbN films was determined from the linear part of the $H_{c2}(T)$ curves. As already pointed out, the value of $\mathrm{d}H_{c2}/\mathrm{d}T$ of the NbN films of thickness between 5 and 10 nm is almost constant with a value of about 3 T/K. This is in contrast to the Nb films where a reduction of the film thickness from 300 down to 25 nm resulted in a gradual increase of $\mathrm{d}H_{c2}/\mathrm{d}T$ starting from about 0.25 T/K up. This increase of $\mathrm{d}H_{c2}/\mathrm{d}T$ became stronger for films thinner than 25 nm approaching about 0.6 T/K for the 8 nm thick Nb film. In the dirty limit with $\xi_0 \gg l$ (here ξ_0 is the BCS coherence length and l is the electron mean free path) the value of $\mathrm{d}H_{c2}/\mathrm{d}T$ is related to the electron diffusion coefficient D as [5]

$$D = -\frac{0.407\pi k_B}{e}\left(\frac{\mathrm{d}H_{c2}}{\mathrm{d}T}\right)^{-1}. \tag{3}$$

From Eq. (3) it is obvious that even the thinnest of the Nb film is characterized by an about four times larger value of the electron diffusivity than the NbN films (also compare Table 1).

Linearly extrapolating the experimental second critical magnetic field to zero temperature gives the value of the upper critical field $H_{c2}(0)$. Thus derived values of $H_{c2}(0)$ as a function of Nb and NbN film thicknesses are shown on Fig. 8. Niobium films with a thickness less than 25 nm have an almost constant upper critical magnetic field $H_{c2}(0) \approx 3.5$ T

[5]The relation may be derived from equations found in Ref. [28], for example.

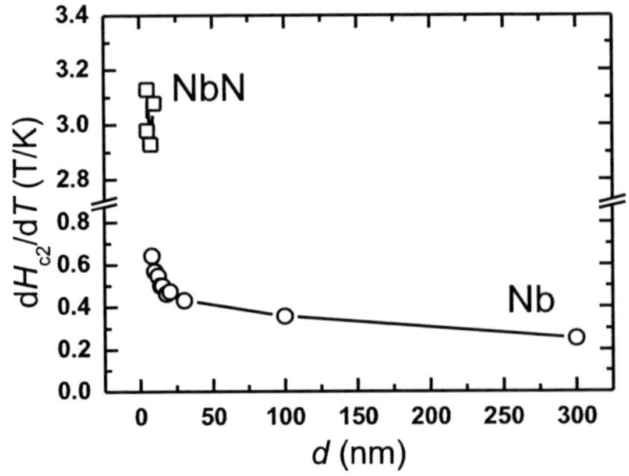

Figure 7: The temperature derivative of the second critical magnetic field of Nb (circles) and NbN (squares) films versus the film thickness.

which decreases slightly for thicker films down to 2.3 T. Unlike the NbN films of thickness less than 10 nm which show a decreasing value of $H_{c2}(0)$ with decreasing film thickness. Using the extrapolated value of $H_{c2}(0)$ we can calculate the GL coherence length $\xi_{\mathrm{GL}}(0)$ at zero temperature as

$$\xi_{\mathrm{GL}}(0) = \sqrt{\frac{\Phi_0}{2\pi\mu_0 H_{c2}(0)}}, \tag{4}$$

with $\Phi_0 = h/2e$ the flux quantum. This parameter characterizes the radius of the normal core of a magnetic vortex penetrating into the thin film of a type-II superconductor under an external magnetic field. In the dirty limit the GL coherence length $\xi_{\mathrm{GL}}(0)$ is proportional to $(\xi_0 l)^{0.5}$, where $\xi_0 \propto T_c^{-1}$. In NbN films the decrease of the film thickness leads to a reduction of the critical temperature (see Fig. 3) and thus to an increase of the BCS coherence length. In general niobium nitride is characterized by a very short electron mean free path of the order of the lattice constant, especially in case of highly disordered film fabricated by deposition on an unheated substrate. If we assume a constant Fermi velocity v_F the electron mean free path becomes independent of the film thickness, because of the nearly constant electron diffusion coefficient D (compare Table 1) and the general relation $D \propto v_F l$. Therefore the GL coherence length of the NbN films increases for thinner films. A different situation is realized in the Nb films. The critical temperature of the Nb films also drops with a decrease of the film thickness (Fig. 3). But opposite to NbN films the temperature derivative of the second critical magnetic field of Nb films becomes larger for the same range of the film thickness. Thus both changes in ξ_0 and l almost compensate each other and result in an almost constant value for $\xi_{\mathrm{GL}}(0)$.

The next point we want to address is the negative curvature of the $H_{c2}(T)$ curves near T_c measured for the NbN films (see Fig. 6). This behavior can be understood if one considers a granular structure of the NbN films. Such a granular structure of NbN films was confirmed experimentally by different methods [29–32]. In a granular film NbN grains are surrounded by material with excess nitrogen [33] making it a non-superconducting intergrain barrier. The deposition conditions are decisive in determining the grain size. Typical

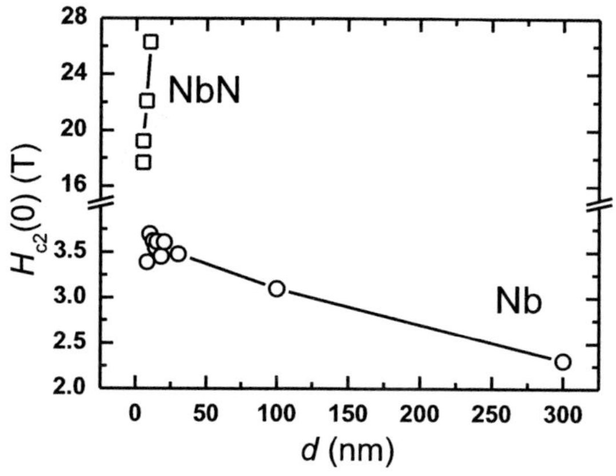

Figure 8: The upper critical magnetic field of Nb (circles) and NbN (squares) films as a function of film thickness.

grain diameters were found in a range from 20 up to 120 nm depending on film thickness [30] and Ar/N_2 partial pressure during sputtering [34]. Other parameters that influence the grain size are the substrate temperature and deposition rate. High substrate temperatures and low deposition rates allow for a continuous growth of initial islands resulting in large grain sizes in the final film. Contrarily, low substrate temperature and high deposition rate lead to a reduced mobility of NbN clusters formed on the surface and the final grain size approaches the size of the initial islands. The latter deposition regime is realized for our NbN films. The low substrate temperature also supports a constant stoichiometry within one grain as the grain height or film thickness increases. We should note that this is valid for relatively small film thicknesses. For thicker films the composition may change with film growth as it was already discussed in Section 3.1 concerning the critical temperature of NbN films of relatively large thickness.

In granular superconducting films the superconducting properties may be influenced by the relative size of the grains and the GL coherence length ξ_{GL}. At low temperatures when $\xi_{GL}(T)$ is much smaller than the size of the grains, magnetic vortices form a normal

Table 1: Important parameters of various Nb and NbN films.

	d (nm)	T_c (K)	D (cm^2/s)	$\xi_{GL}(0)$ (nm)	d_{grain} (nm)
Nb	20	8.2	2.3	9.55	--
	12	7.3	2	9.53	--
	8	6.1	1.7	9.85	--
NbN	10	9.2	0.36	3.54	46
	8	8.3	0.38	3.86	45
	5	7.3	0.35	4.14	37
(on sapphire)	5	6.8	0.37	4.32	39

Abrikosov lattice inside the grains and the value of the second critical magnetic field reflects the properties of the pure NbN material. The granular morphology of the film only changes the pinning potential landscape for the vortex system. At temperatures close to T_c the coherence length diverges and becomes comparable to or even larger than the size of the grains. This leads to a restructuring of the vortex lattice and a deformation of individual vortices. Effectively, the second critical magnetic field is reduced. This in turn allows one to estimate the mean diameter of the NbN grain for our films by assuming that the $H_{c2}(T)$ curve deviates from the linear behavior at a characteristic temperature T_{gr}, at which ξ_{GL} becomes equal to the grain radius. Thus obtained values of the grain diameters of our NbN films are listed in Table 1. We have found almost equal values of about 38 nm for both 5 nm thick films and somewhat larger values $\approx$ 45 nm for the 8 and 10 nm thick films.

3.3 Superconducting Transition Temperature of Bridges

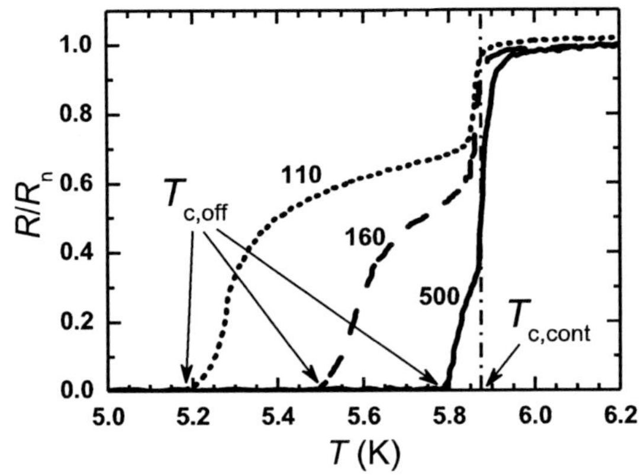

Figure 9: Normalized resistance versus temperature of 10 nm thick Nb bridges of different width (indicated in nanometers). The arrows show the position of the temperature which was taken as the critical temperature of the bridge $T_{c,off}$. The vertical dashed-dotted line indicates the critical temperature of the contact pads $T_{c,cont}$.

Representative examples of resistive transitions of structures containing nano-bridges with different width w are shown in Fig. 9 for the case of a 10 nm thick Nb film. The two-step transition is typical for structures with $w \leq 1$ μm. Structures containing broader bridges do not demonstrate this behavior. We attribute the first drop in resistance to the superconducting transition of the contact pads and denote the corresponding transition temperature as $T_{c,cont}$. The value of $T_{c,cont}$ depends on the thickness of the film and is independent of the width of the bridge. Further decrease of the resistance is believed to be the superconducting transition of the bridge itself. This transition occurs at a lower temperature than the transition of the contact pads and can be attributed solely to the bridge. For smaller bridges, the transition broadens and shifts to lower temperatures. We define the superconducting transition temperature of a bridge $T_{c,off}$ as the temperature corresponding to the offset of the resistance, i.e. the temperature that corresponds to the resistance less than a factor 10^{-3} of the normal state resistance (indicated by arrows in Fig. 9). The measuring

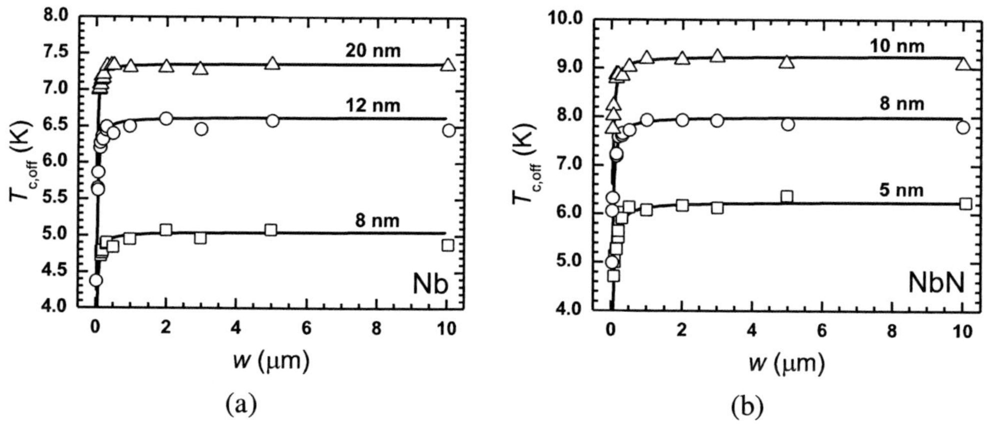

Figure 10: Dependence of the critical temperature $T_{\text{c,off}}$ on the width of the bridge made from (a) Nb and (b) NbN films of different thickness shown on the graphs. The solid lines are best fits according to Eq. (1).

current $\approx 0.15\ \mu$A did not influence the $T_{\text{c,off}}$ value even for the bridges with the smallest cross-section area.

In Fig. 10 we plot experimental values of the critical temperature $T_{\text{c,off}}$ as function of the width of the bridge for Nb (Fig. 10(a)) and NbN (Fig. 10(b)) bridges and varying thickness. For both materials and for all thicknesses, we obtained similar results. In a range of w from about 2 μm up to the largest value of 10 μm, $T_{\text{c,off}}$ is almost independent of width and equal to the transition temperature of the contact pads. The small variation of $T_{\text{c,off}}$ in this range of w might be due to spatial non-homogeneities of the film. For narrower bridges, the transition temperature strongly reduces and reaches values $1 - 2$ K lower than $T_{\text{c,off}}$ of the micrometer-wide bridges. Comparing the dependencies of the transition temperature on film thickness and width of the bridge they appear to be very similar: a plateau at large d and w, respectively, and a significant reduction of T_{c} once the lateral dimension is reduced below a certain value. Therefore, it appears to be reasonable to assume the proximity effect to be the responsible mechanism for the suppression of superconductivity in both cases.

To explain the origin of the normal conducting edges of the bridges, we suggest that they were mechanically damaged by Ar ions during patterning of the film. Subsequently the damaged areas oxidized once the bridges had been exposed to air and became normal conducting. Furthermore, we consider the bridges as planar NSN structures and apply to our experimental data on $T_{\text{c,off}}$ the same approach as we used in order to describe the dependence of T_{c} on the film thickness. To fit the experimental data we used only one fitting parameter w_{N} (instead of d_{N} in Eq. (2)), which is the effective width of normal conducting strips running along both edges of the bridge. The value of T_{c}^{0} used for evaluating $T_{\text{c,off}}(w)$ using Eq. (1) was set equal to the value of $T_{\text{c,off}}$ for the widest bridges of the given thickness. The resulting least-square fits are shown in Fig. 10 by the solid lines. From the fitting procedure we found for the Nb films $w_{\text{N}} \approx 1$ nm with thicker films tending to have thinner normal conducting edges. The same trend was found in NbN films, but with two to three times wider normal edges than in Nb structures with comparable film thicknesses.

3.4 Critical Current Density of Bridges

Current-voltage characteristics (CVC) were recorded by means of the four-probe method simultaneously with the transition temperature measurements. CVC were acquired in the current-bias mode using a slow-sweep current-generator. In Fig. 11 representative CVC are shown for superconducting bridges with different widths. Hysteretic CVC similar to the ones shown in Fig. 11 were obtained for all bridges independent of the material, film thickness and width of a bridge. Such hysteretic behavior is a distinctive feature of the formation and development of a normal domain in the superconducting strip. The critical current I_c was defined as the current value at which the jump from the superconducting to the normal state occurs. As expected, the critical current increased with the increase of both the width and the thickness of the bridges. Anticipating a uniform critical current density, one would expect a linear dependence of I_c on the bridge width. The raw data of $I_c(4.2\ \mathrm{K})$ versus the geometrical strip width are shown in Figs. 12 (a) and (b) for a number of Nb and NbN bridges, respectively. At first sight the linear relationship seems to be confirmed except for very narrow bridges.

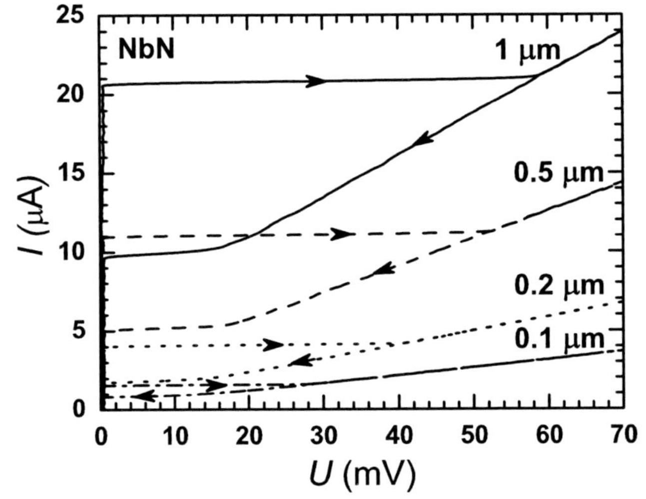

Figure 11: I-V characteristics recorded at 4.2 K for NbN bridges with the same thickness (5 nm), but different widths. The nominal width is indicated near each curve. (Data are taken from Ref. [19].)

A very different situation emerges if one plots the nominal critical current density using the geometrical cross-sectional area $j_c^* = I_c/wd$ versus the strip width w. This is shown in Figs. 12 (c) and (d) using I_c-values recorded at 4.2 K. Especially the Nb bridges exhibit a distinctive variation of the critical current density with the strip width. In the range of the bridge width from 10 to 3 μm the nominal critical current density changed very weakly with a tendency to increase. Further decrease of w led to a stronger rise of j_c and the formation of a well-pronounced maximum for about 200–300 nm wide Nb bridges made from 12 and 20 nm thick films. Independently of the film thickness, the nominal critical current density at 4.2 K dropped for bridges narrower than 200 nm. For narrow NbN bridges $j_c(4.2\ \mathrm{K})$ was also dependent on the width. In contrast to the Nb bridges, a well-pronounced maximum

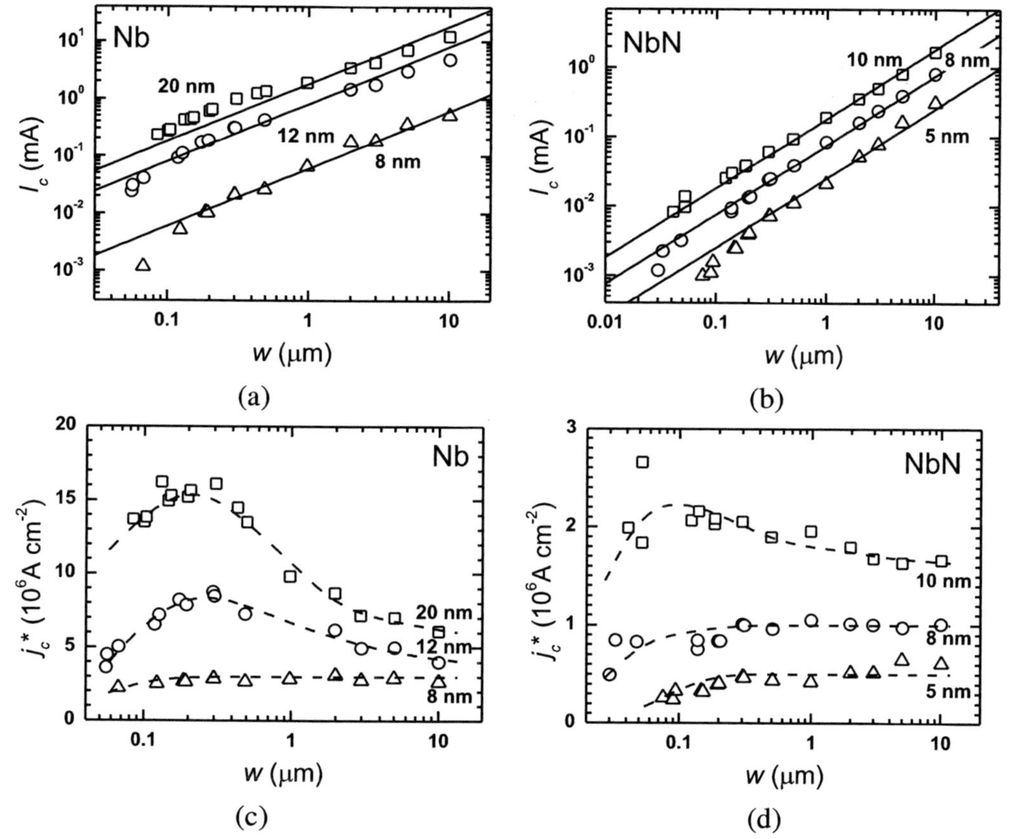

Figure 12: Critical current measured at 4.2 K versus the strip width w on a double logarithmic scale for bridges made from Nb (a) and NbN (b) films of different thickness as indicated. Solid lines represent a linear relation ship between critical current and width. The graphs (c) and (d) show the nominal critical current density $j_c(4.2\text{ K})$ calculated using the geometrical cross-section as a function of the strip width w for the Nb and NbN bridges, respectively. As detailed in the text, some bridges have a maximum in the critical current density for $w \approx 200$ nm. The dashed lines are guides to the eye, only.

has only been found for the NbN bridges made from 10 nm thick films. Bridges with smaller thicknesses demonstrated a gradual decrease of the nominal critical current density with a decrease of the bridge width.

To discuss our experimental results consistently we first compare the size of our bridges with the effective penetration depth λ_{eff} and the Ginzburg-Landau coherence length at zero temperature, $\xi_{\text{GL}}(0)$. In superconducting strip lines with a width larger than λ_{eff} and ξ_{GL}, the critical current is usually determined by depinning of magnetic vortices. In other words, a voltage appears when pinned magnetic vortices start to move across the strip under the influence of the Lorentz force. The value of the depinning critical current density j_{pin} is determined by the concentration, type and size of the pinning sites. Thus j_{pin} should be strongly influenced by the film deposition technique. The depairing current, i.e. the current leading to Cooper-pair breaking, is the ultimate limit of the current-carrying capacity of a superconductor. From an experimental point of view the depairing current can be observed in a superconducting structure having a width less than both the effective superconducting penetration depth λ_{eff} and the Ginzburg-Landau superconducting coherence length ξ_{GL}. The first condition ensures uniform distribution of the current over the cross-section of the strip. The second condition eliminates vortex nucleation and flow, which would give rise to dissipation before the value of the depairing current is reached. Likharev [35] predicted that the vortex flow will occur at about one-half of the depairing current. He also suggested that no vortices can exist/nucleate in a superconducting strip line if it has a width smaller than $4.4\,\xi_{\text{GL}}$. In the following we will call this condition Likharev's criteria.

Using the value of the magnetic field penetration depth λ reported in Ref. [18], we estimated for our Nb films the values of $\lambda_{\text{eff}} = \lambda^2/d$ ($d \ll \lambda$) equal to 6.6, 2.1 and 0.845 μm for film thicknesses 8, 12 and 20 nm, respectively. NbN films are usually characterised by a much larger penetration depth. Values of about 400 nm have been reported [30, 36, 37] for NbN films with a thickness between 0.5 and 1.5 μm. Using the smallest reported value $\lambda \approx 300$ nm for NbN, we obtain even for the 10 nm thick film an effective penetration depth approximately equal to the width of the largest bridge. Therefore, for all NbN bridges and the bridges made from the 8 nm thick Nb film the effective penetration depth at $T = 0$ K is larger than the strip width, $\lambda_{\text{eff}} \geq w$. The condition becomes even stricter at higher temperatures because of the temperature dependence of the penetration depth. Thus variations of the current density over the cross-section have to be taken into account only for the 12 and 20 nm thick Nb bridges with widths larger than 4 and 2 μm, respectively.

As we have already stated above in Sec. 3.2, the coherence length $\xi_{\text{GL}}(0)$ is another important parameter for understanding the nature of the critical state. Using the measured temperature dependence of the second critical magnetic field $H_{c2}(T)$, we have estimated the coherence length at zero temperature $\xi_{\text{GL}}(0)$ to be approximately 9.5 and 4 nm for Nb and NbN films, respectively (also compare Table 1). It appears that even our narrowest 50 nm bridges do not satisfy Likharev's criteria $w < 4.4\,\xi_{\text{GL}}(0)$. Estimations made above show that for almost all our bridges nucleation and flow of magnetic vortices should be the predominant mechanism of the current induced critical state.

For further analysis we take into account the reduction of the transition temperature of our bridges and the temperature dependence of the critical current. To account for the temperature dependence of the critical current we calculated the nominal density of the critical current at zero temperature j_c^0 assuming the mean-field behavior of $I_c(T)$ predicted by the GL-theory for the depairing critical current

$$\frac{I_c(T)}{wd} = j_c^0 \left(1 - t^2\right)^{3/2} \left(1 + t^2\right)^{1/2}, \; t = \frac{T}{T_{\text{c,off}}}. \tag{5}$$

This temperature dependence is appropriate for the two-fluid model at all temperatures [38]. It reduces to the familiar $(1 - t)^{3/2}$ dependence near the transition temperature and saturates for $T \to 0$. We experimentally verified this temperature dependence of the critical current for a number of bridges with different cross-section areas made from both Nb and NbN films. Figure 13 shows the temperature dependence of the normalized critical current for a representative 1 μm wide and 15 nm thick Nb bridge and also for a 80 nm wide and 5 nm thick NbN bridge. One can see that, in spite of different bridge sizes and materials, both data sets show an almost identical temperature dependence that corresponds very well to the theoretical relation (5) suggested by the GL-theory.

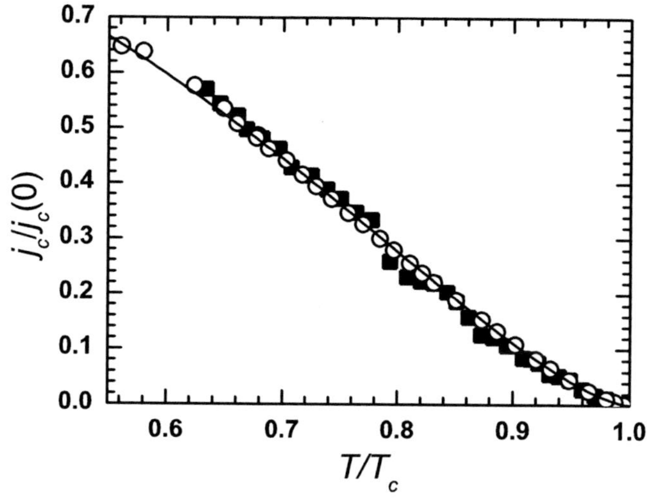

Figure 13: Temperature dependence of the critical current for a 1 μm wide, 15 nm thick Nb bridge (filled squares) and for a 80 nm wide, 5 nm thick NbN bridge (open circles). The solid line represents the temperature dependence given by Eq. (5).

Values of j_c^0 obtained using Eq. 5 are plotted in Fig. 14 as function of the width for both Nb and NbN bridges. Whereas the data for the thicker bridges look similar to the measured j_c data at 4.2 K they qualitatively differ for the 5 nm thick NbN and 8 nm thick Nb bridges. Looking at the 5 nm NbN bridges j_c^0 gradually decreases in the range of w from 5 to approximately 1 μm, reaches a minimum and then increases even exceeding the j_c^0 values of 10 nm thick NbN bridges of the same width. The 8 nm thick Nb bridges show a similar behavior: j_c^0 noticeably increases for bridges narrower than 500 nm.

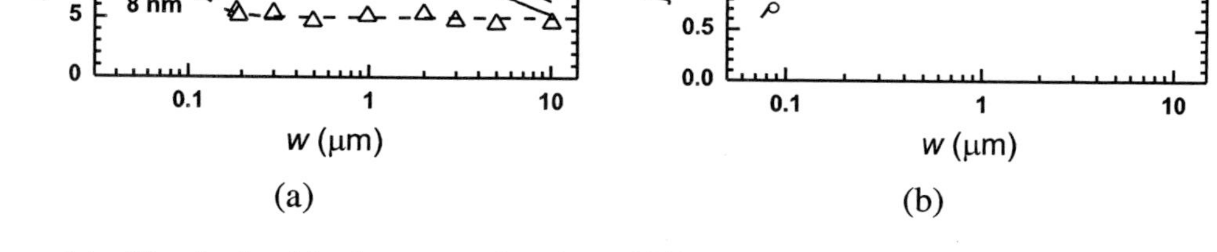

Figure 14: Nominal critical current density of Nb (a) and NbN (b) bridges extrapolated to $T = 0$ K. The thickness of bridges is indicated in the graph. Solid lines in Fig. 14(a) represent best fits to the experimental data using Eq. (8). The dashed lines in (a) and (b) are guides to the eye.

Before we come back to the thin bridges we will try to understand the variation of j_c^0 as a function of the strip width for bridges made from thicker films. The frequently used approach for the description of enhancement of the critical current density in narrow strips relies on an uneven distribution of the super-current over the cross-section of a strip. Such a distribution arises if the bridge width is comparable or larger than the effective penetration depth, $w \geq \lambda_{\text{eff}}$. The distribution of the local current density across the strip width is then given by [39,40]

$$j(x) = \frac{I_{\text{T}}}{\pi d \sqrt{\left(\frac{w}{2}\right)^2 - x^2}}, \tag{6}$$

where the coordinate x varies from zero to $\pm w/2$. According to Eq. (6), the local current density diverges near the strip edges when x approaches $w/2$. For any value of I_{T} there are portions near the edges of the strip where the local current density exceeds the depairing current density. In order to eliminate this inconsistency, we adopt common arguments [39–41] and limit the super-current density at the edges to a value specific to a particular critical state. In Ref. [40] it was shown that the current I_{T} distributed across the strip according to the Eq. (6) generates at the edges the magnetic field $H_{\text{edge}} = \mu_0 I_{\text{T}}/(\pi\sqrt{2wd})$. When H_{edge} reaches the first critical magnetic field H_{c1} vortices start to enter the strip and the nominal critical current density becomes

$$j_{\text{vc}} = \frac{\pi H_{\text{c1}}\sqrt{2}}{\mu_0\sqrt{wd}}. \tag{7}$$

The nominal critical current density may exceed this value if vortices are pinned near the strip edges. We introduce an effective penetration distance of vortices w_{v}, which will be used as a fitting parameter. The smallest reasonable value of w_{v} is about one half of Likharev's criteria. The critical state is achieved when the local current density at a distance w_{v} from the edges reaches the density of the depinning critical current j_{pin}. We suppose that at this stage the local current density inside w_{v}-regions adjacent to the edges also equals j_{pin}. The corresponding nominal density of the critical current is then given by [41]

$$j_{\text{c}} = \frac{j_{\text{pin}}}{w}\left\{2w_{\text{v}} + \sqrt{ww_{\text{v}} - w_{\text{v}}^2}\left[\arcsin\left(\frac{w - 2w_{\text{v}}}{w}\right) - \arcsin\left(\frac{-w + 2w_{\text{v}}}{w}\right)\right]\right\}. \tag{8}$$

Equation (8) is valid for bridges wider then $2w_{\text{v}}$ and describes an increase of j_{c} when the bridge width decreases. This happens mostly due to *optimization* of the current distribution across the bridge. In smaller bridges $2w_{\text{v}} > w > 4.4\,\xi_{\text{GL}}$ the critical current is uniformly distributed resulting in $j_{\text{c}} = j_{\text{pin}}$ for any value of w within this interval.

As we have already pointed out in Sec. 3.1 the reduction of the superconducting cross-section of a bridge with respect to its geometrical cross-section causes a decrease of the transition temperature of narrow bridges. We will now analyze the influence of the reduction of the cross-section on the critical current density. On the right hand side of Eq. (8) we have to replace w by $(w - 2w_{\text{n}})$, where w_{n} is the width of the portion of the bridge, which has become normal due to patterning. For $4.4\,\xi_{\text{GL}} < w < 2w_{\text{v}}$ the nominal critical current density becomes $j_{\text{c}} = j_{\text{pin}}(w - w_{\text{n}})/w$ and decreases with the decrease of the bridge width. Using a thus modified Eq. (8) we computed the nominal critical current density and fitted it to the experimental data. The width of the normal portion of the bridge and the vortex penetration distance were used as fitting parameters. The solid lines in Fig. 14 (a) show the best results for 20 and 12 nm thick Nb films with fitting parameters $w_n = 8$ and 12 nm and $w_v = 65$ and 100 nm, respectively. The increase of w_v with decreasing d correlates with the increase of the magnetic field penetration depth. The thin Nb bridges as well as all of the NbN bridges do not seem to by gathered by this scenario.

The above analysis is valid for bridges having a width larger than the effective penetration depth of the magnetic field. Because $\lambda_{\text{eff}} = \lambda^2/d$ and $d \ll \lambda$, the effective penetration depth remains larger than w in ultra-thin films even for micrometer-wide bridges. In this case the current is uniformly distributed over the width of the bridge. The local current density is a constant value that equals the nominal current density. As long as $w \gg 4.4\,\xi_{\text{GL}}$ the nominal critical current density equals j_{pin} and is independent of w. Once the Likharev's criteria is met and vortices can no longer form within the bridge, the critical current will be determined by the breaking of Cooper pairs. According to Ref. [35], the corresponding depairing current density should be twice the density of the depinning current. Therefore, when the width decreases and approaches $4.4\,\xi_{\text{GL}}$ the critical current density should increase due to a crossover from depinning to depairing. This rise will be partly counterbalanced by the decrease of j_c caused by the normal edges of the bridge. There is another scenario that may result in an apparent rise of j_c in narrow bridges. According to Eq. (7) the nominal density of the supper-current, which corresponds to the nucleation of vortices, increases when the width decreases and may eventually exceed the density of the depinning critical current. In this case right after nucleation the vortices will start to move resulting in the critical state. The measured critical current, however, will be determined by the nucleation of vortices.

To summarize, the dependence of the nominal critical current density on the bridge width may comprise all discussed features: a maximum corresponding to the *optimal* distribution of the super-current in a bridge with normal edges or a plateau corresponding to the uniform current density. Such a maximum is clearly the case for thicker Nb bridges while all NbN bridges demonstrate a rather uniform current distribution. When the width further decreases, either shape can display a minimum due to the crossover from the depinning to the depairing or to the nucleation regime. We think that the nucleation scenario is more likely for the 5 nm thick NbN bridges while depairing may be more relevant for the 8 nm thick Nb bridges.

4 Current-Assisted Hot-Spot Photon Detection

A recently proposed hot-spot scenario [12] of the resistive state formation in a thin, narrow, current-carrying superconducting strip due to absorption of a single infrared photon was subsequently suggested as a mechanism of single-photon detection [13] in NbN microbridges. Although some experimental evidences of the envisaged mechanism have been soon reported [42] other experimental data [13,43,44] are hard to interpret in the framework of the hot-spot model.

The model [12] describes the formation and evolution of a normal spot around the site where a single photon has been absorbed in an infinite homogeneous superconducting film. The energy of the captured photon is quickly shared among electrons via electron-electron and electron-phonon interaction leading to an avalanche-like multiplication [45] of the number of non-equilibrium quasiparticles (QP). The normal state appears when the excess QP concentration becomes large enough in order to locally destroy superconductivity. At some stage the normal hot-spot reaches the maximum size that is a trade-off between quasiparticle multiplication and their diffusion out of the absorption site. In a narrow strip carrying a small sub-critical current, no resistive state is formed unless the maximum size of the normal spot, a_n, exceeds the strip width. If the spot-size is less than the strip-width, the resistive state may appear due to the supplementary action of the bias current that has a density j close to the critical current density j_c. After the normal spot has grown larger than the coherence length, the supercurrent is expelled from the spot. Since the effective cross-section available to the current flow decreases, the current density may eventually exceed the critical current density. If this happens, a normal domain spans the entire cross-section of the strip. Once formed, the resistive state causes a voltage pulse developing between the strip ends. The number of photons, which result in a voltage response, related to the number of photons crossing the geometrical area of the strip measures the quantum efficiency of the resistive state formation.

If a_n is not sufficiently large in order to enhance the current density to its critical value, there is still a finite probability of the resistive state formation due to multi-photon or fluctuation assisted processes [13,43]. Because a_n increases with the photon energy there should be a current dependent photon energy ε_0 that demarcates these two regimes of the resistive state formation. The threshold energy ε_0 is defined by the simple geometric criteria $a_n(\varepsilon_0)/w = 1 - (j/j_c)$ where w is the width of the strip. For photon energies $\varepsilon > \varepsilon_0$ but smaller than $h\nu_0$ (ν_0 is the plasma frequency), the quantum efficiency should equal the probability of photon absorption and therefore should not depend on the photon energy. At $\varepsilon < \varepsilon_0$ the quantum efficiency is dominated by multi-photon and fluctuation assisted events, which generally have a low probability. So far the cut-off of the quantum efficiency at $\varepsilon \approx \varepsilon_0$ has been experimentally observed [20] only in disordered NbN films, which have a low density of electronic states and a reduced transition temperature. In polycrystalline NbN films with the B1 structure, a monotonous decrease of the quantum efficiency with the photon wavelength was found [43, 44]. Knee-like features have been observed in the overall exponential dependence of the quantum efficiency on the bias current. Although they were attributed to the above discussed cut-off, the concluded maximum spot size was far too large in comparison to model estimates [46].

The photon absorption in a superconductor obeys Poisson statistics, in that the probability of absorbing simultaneously n photons from a weak photon flux is proportional to the flux intensity to the power of n. Variation of the quantum efficiency corresponding to $n = 1$, 2 and 3 have indeed been found [13] in NbN microbridges. However, this observation would generally manifest quantum nature of any detector and is not uniquely specific to any particular response mechanism. A time delay of the resistive state formation in NbN meander lines has been recently reported [42]. The delay was attributed to the supplementary action of the bias current in the formation of the normal state but the quantitative description of the delay time implied the spot size to be much larger than the diffusive hot-spot model would predict.

In what follows we refine the hot-spot model and show that the resistive state formation in the absence of fluctuations may be triggered by a single photon even though no normal spot appears. This approach eliminates the contradictions within the earlier published experimental data and provides a quantitative explanation of the way fluctuations in the current-carrying superconductor enhance the quantum efficiency beyond the cut-off.

4.1 Model Refinement

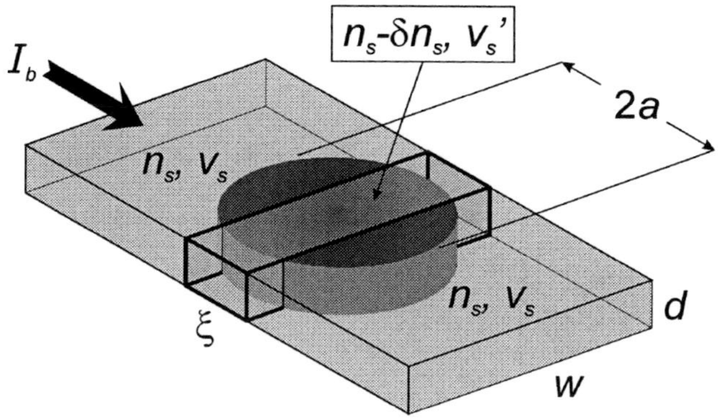

Figure 15: Schematics of the superconducting strip-line carrying a current I_b. Bold lines demarcate the smallest volume where a reduction of the superconducting electron pairs causes the change of their mean velocity. Grey cylinder depicts the cloud of nonequilibrium quasiparticles with diameter $2a$.

Assume a strip made of a superconducting film carrying a supercurrent as it is shown in Fig. 15. The film thickness is chosen smaller than both the line width and the coherence length ξ assuring the applicability of a two-dimensional diffusion model for nonequilibrium electrons. The small film thickness results in a large magnetic penetration length that exceeds the line width even at temperatures well below the superconducting transition temperature. In this geometry, the magnetic field of the supercurrent may be neglected and, in the absence of magnetic vortices, the local supercurrent density should be constant throughout the cross-section of the strip. However, since our strips are wider than the coherence length, one cannot exclude the presence of vortices. If there are vortices they should obey a Berezinskii-Kosterlitz-Thouless (BKT) topological transition. Reported square resistances of NbN films are always smaller than the resistance quantum h/e^2 resulting in a BKT transition temperature [47] only slightly below T_c. At typical operation temperatures $T \approx T_c/2$ all vortices are then bundled into pairs and do not contribute to the resistance unless the bias current is very close to the critical current. Although bundled vortices modulate the supercurrent density, we will neglect this current distribution and associate the local supercurrent density $j = 2e\, n_s\, v_s$ with the mean velocity v_s and the mean density n_s of paired electrons with charge $2e$. The local current state may noticeably change only if n_s changes over a distance ξ_{GL} or larger along the current path; shorter perturbations are tunnelled by electron pairs without energy dissipation. The smallest volume of the strip relevant to a current change, the ξ-slab, is marked in Fig. 15 with bold lines. If n_s in the ξ-slab decreases by an amount δn_s, the mean pair velocity increases and becomes

$$v_s' = \frac{n_s}{n_s - \delta n_s} v_s \tag{9}$$

as required by charge flow conservation. The characteristic conversion time of the pair velocity practically equals the GL relaxation time $\approx \hbar/\Delta$ (Δ is the temperature dependent energy gap) and is small compared to the electron thermalization time τ_{th}. Thus, the mean velocity in the slab instantaneously follows changes of the pair density. The ξ-slab switches into the normal state if the pair velocity exceeds the critical value v_{sc} that corresponds to the critical current density $j_{\mathrm{c}} = 2e\, n_{\mathrm{s}}\, v_{\mathrm{sc}}$ in the absence of excitations.

The concentration of nonequilibrium electrons $C(r,t)$ at the distance r from the photon absorption site evolves in time t due to multiplication of electrons and their diffusion out of the absorption site [12]

$$C(r,t) = \frac{M(t)}{4\pi Ddt} \exp\left(-\frac{r^2}{4Dt}\right), \tag{10}$$

where $M(t)$ describes the increase of the number of nonequilibrium electrons in time [48]. For times $t > \tau_{\mathrm{th}}$ nonequilibrium electrons have thermalized to the energy level Δ effectively becoming quasiparticles. Their number reaches the maximum value $M(\tau_{\mathrm{th}}) = \zeta\varepsilon/\Delta$, where $\zeta \leq 1$ is the efficiency of the QP multiplication. After the thermalization time τ_{th} has elapsed and before subgap phonons appear, the concentration of nonequilibrium quasiparticles locally equals the reduction of superconducting electrons $2\delta n_{\mathrm{s}}$. Assuming that well below the transition temperature $n_{\mathrm{s}} \approx \frac{N_0\Delta}{2}$, one can rewrite Eq. (9) and find the smallest number of nonequilibrium QP that is sufficient for switching the ξ-slab into the normal state

$$\delta N^* = N_0 \Delta \xi_{\mathrm{GL}} wd \left(1 - \frac{I}{I_{\mathrm{c}}}\right). \tag{11}$$

If the spread of the QP cloud is comparable to the coherence length, all quasiparticles contribute to the change of the condensate velocity and $\delta n_{\mathrm{s}} = M(\tau_{\mathrm{th}})/(wd\xi_{\mathrm{GL}})$. If the cloud is larger (see Fig. 15), only those quasiparticles confined within the ξ-slab have to be taken into account. The size of the electron cloud is twice the radius of the disc that confines all nonequilibrium electrons. Equating $M(t)$ to the integral of the electron concentration Eq. (10) over the cylinder with the radius $a(t)$ and the thickness d, we find the time dependent size of the electron cloud $a(t) = 2[Dt \ln(M(t))]^{1/2}$. If $M(\tau_{\mathrm{th}}) \geq 100$ and typical parameters of NbN films are considered, $a(t)$ exceeds the coherence length already at an early thermalization stage $t < \tau_{\mathrm{th}}$. For $a \gg \xi_{\mathrm{GL}}$ the absolute number δN of nonequilibrium electrons confined to the ξ-slab can be evaluated analytically. Integration of the electron concentration, Eq. (10), over the slab results in

$$\delta N = M(t) \frac{\xi_{\mathrm{GL}}}{\sqrt{\pi Dt}}. \tag{12}$$

The number of nonequilibrium electrons in the slab peaks at $t \approx \tau_{\mathrm{th}}$ when nonequilibrium electrons can already be treated as quasiparticles. The cut-off of the quantum efficiency occurs when the maximum number of nonequilibrium QP in the slab drops to δN^*. The criteria becomes $\delta N(\tau_{\mathrm{th}}) \geq \delta N^*$ that allows one to determine the intrinsic efficiency of the quasiparticle multiplication once the cut-off parameters (I, ε_0) are known:

$$\zeta = \frac{N_0 \Delta^2 wd \sqrt{\pi D \tau_{\mathrm{th}}} \left(1 - \frac{I}{I_{\mathrm{c}}}\right)}{\varepsilon_0}. \tag{13}$$

With material parameters of disordered NbN films one arrives at a multiplication efficiency of the order of 0.1 in the visible and near-infrared range. Multiplication efficiencies determined this way for disordered NbN films [20] are shown in Fig. 16.

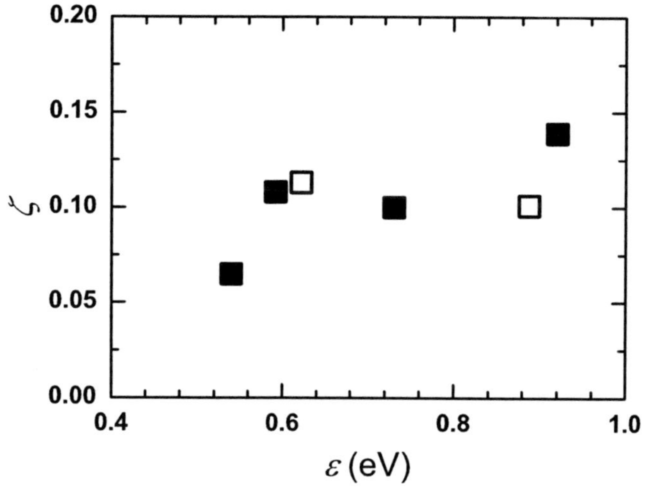

Figure 16: Multiplication efficiency of quasiparticles for different photon energies for disordered NbN films. Data marked with filled symbols were obtained from the spectral dependence of the photon count rate (see Fig. 20 (b)) and open symbols stand for multiplication efficiencies concluded from the current dependence (Fig. 17). (Data are taken from Ref. [20].)

The maximum size of the actual normal spot $a_{\rm n}$ can be estimated using above value for the multiplication efficiency and solving the equation $C(a_{\rm n}, \tau_{\rm th}) = N_0\Delta$. When applied to conventional films [13, 42] with the stoichiometric B1 composition this analysis suggests that no normal spot is formed. Even in disordered films the normal spot would turn out to be smaller than the coherence length $2a_{\rm n} < \xi_{\rm GL}$. According to the simplified model [12] such spots should not trigger the formation of the resistive state. However, as we have shown above, the contradiction can be eliminated if one takes into account the full number of nonequilibrium quasiparticles and their lateral distribution. Finally, it is worth noting that the size of the normal spot defined by the criteria $2a_{\rm n}/w = (1 - I/I_{\rm c})$ has the following meaning: it is the diameter of the cylinder that would confine all nonequilibrium quasiparticles at times $t \geq \tau_{\rm th}$ if their concentration was uniform and equal to $N_0\Delta$.

4.2 Fluctuations and Quantum Efficiency beyond the Cut-off

Besides the fundamental interest in fluctuations, they may result in dark counts in narrow strips or meander lines (see also Sec. 5.2). This is an important issue for sensor applications where small photon fluxes have to be detected. In a one-dimensional superconductor, depending on the cross-section, dark counts may originate from the spontaneous formation of either classical [49] or quantum [50] phase-slips. When the cross-section confining the current path extends over the coherence length, dark counts can appear due to discrete fluctuations of the superconducting order parameter [51, 52] and/or number fluctuations in the gas of bundled magnetic vortices [53]. It has been observed in many experiments that the dark count rate strongly depends on the bias current. Since at temperatures below T_c the density of the superconducting condensate does not vary much with the bias current [54], we suggest that this current dependence is connected only with the velocity of the superconducting electron pairs. In particular, the velocity controls the minimal excitation energy of the pairs. Although variations of the pair density would modify the current dependence of the dark count rate, experimental data show [52] that such modifications are a week factor in comparison to the current dependence itself. A photon counting event beyond the cut-off can be thought of as a dark count event associated with the portion of the detector where the density of electron pairs has been decreased due to photon absorption. The increase of the pair velocity caused by a photon in the ξ-slab is determined by the relative decrease of the mean pair density $\delta n_s/n_s$, Eq. (9). For the current I^* corresponding to the cut-off the relation $\delta n_s/n_s = 1 - I^*/I_c$ holds. Therefore, for a particular photon energy, the pair velocity in the ξ-slab scales down at any bias current with the factor I^*/I_c. Fig. 17 illustrates how these ideas apply to the approximation of photon count rates beyond the cut-off for disordered NbN films.

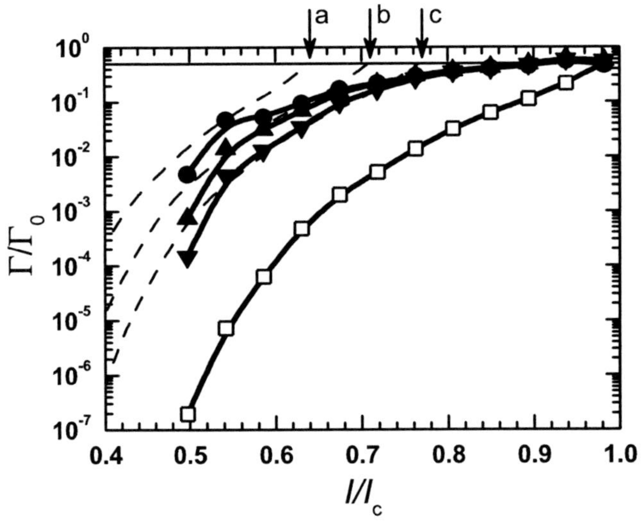

Figure 17: Normalized photon count rates (filled symbols and thick solid lines, from top to bottom $\lambda = 0.8$ μm, 1.4 μm and 2.0 μm, resp.) and dark count rate (open squares) for a meander made from a disordered NbN film (Ref. [20]). Arrows mark the currents corresponding to the cut-off. Thin dashed lines approximate the signal count rates beyond the cut-off.

Each thin dashed line in Fig. 17 (reproduced from Ref. [20]) was obtained from the current dependence of the dark count rate (open squares in Fig. 17) by scaling the bias current. The scaling factor equals the normalized bias current at which the approximating curve reaches the magnitude of the signal count rate corresponding to $I = I_c$. The thin horizontal straight line marks this level in Fig. 17. The scaling factor was varied to achieve the best fit to the measured signal count rate beyond the cut-off. The best fit values of the scaling factor are marked with arrows. These values were used to estimate the multiplication effectiveness presented in Fig. 16 with open symbols. The method holds as long as the number of electron pairs destroyed by the absorbed photon is not too large. Or, equivalently, as long as the number of remaining pairs compares to their equilibrium number. If the pair density changes noticeably, the current dependence of the fluctuation probability will change like it does with temperature. An obvious spin-off of this approach is an understanding of the spectral dependence of the signal count rate beyond the cut-off. The spectral dependence should mimic the temperature dependence of the dark count rate at a fixed bias current. Qualitatively such behavior was found for Nb and polycrystalline B1 NbN films [43, 44, 52], too, thus suggesting the mixed, fluctuation assisted mechanism of the resistive state formation in these materials.

5 Single-Photon Detector Characterization

5.1 Experimental Setup

The experimental setup for the characterization of single photon detectors was built around a ^{4}He-bath cryostat from *IRLabs, Inc.*, a schematic drawing is shown in Fig. 18(a). The devices were mounted on and thermally anchored to a copper holder, which in turn was in good thermal contact with the cold plate of the cryostat. The sample space itself was evacuated. Temperatures below 4.2 K down to $\approx$ 2 K were reached by reducing the vapor pressure above the ^{4}He-bath and were stabilized by varying the pumping speed. A calibrated thermistor was also mounted on the sample holder. However, thermal gradients, whose cause could not be clarified conclusively, prevented an accurate temperature measurement. Alternatively, we used the carefully and independently measured temperature dependence of the critical current as an intrinsic thermometer of the detector. Still, the temperature accuracy may not be as good as it could be expected. Thermal stability on the other hand was better than at least ± 0.05 K even over longer measurement periods.

Biasing of the device was done using a custom-built, battery-operated bias-box in constant voltage mode. To prevent thermal runaway once a normal domain has formed due to photon absorption or fluctuations a shunt capacity and resistor were put in parallel to the device; additionally a passive current/voltage divider was used when better fine control of the biasing current was necessary. The hf voltage signal was fed to a chain of microwave amplifiers. The first low-noise amplifier with a noise temperature of 6 K was installed as close as possible to the detector on the cryostat cold plate. Further room temperature amplifiers were deployed as signal levels required; the maximum total gain was up to 50 dB. The overall bandwidth of the amplifier chain was 0.1 to 1.6 GHz. Thus amplified voltage signals were then fed to the readout electronics, which consisted of either a digital oscilloscope [6] or a voltage-level threshold counter[7]. Figure 18(b) shows a single-photon signal pulse as recorded with the oscilloscope. The bandwidth $0 - 0.25$ GHz of the oscilloscope allows to reproduce without distortion pulses as short as 1.4 ns. Therefore, we suppose that both the time constant of the bias and the lower edge of the amplifier chain band-pass determined the actual shape of the signal.

[6] Agilent Technologies 5481A

[7] *Stanford Research Systems* SR 400 Dual Channel Gated Photon Counter

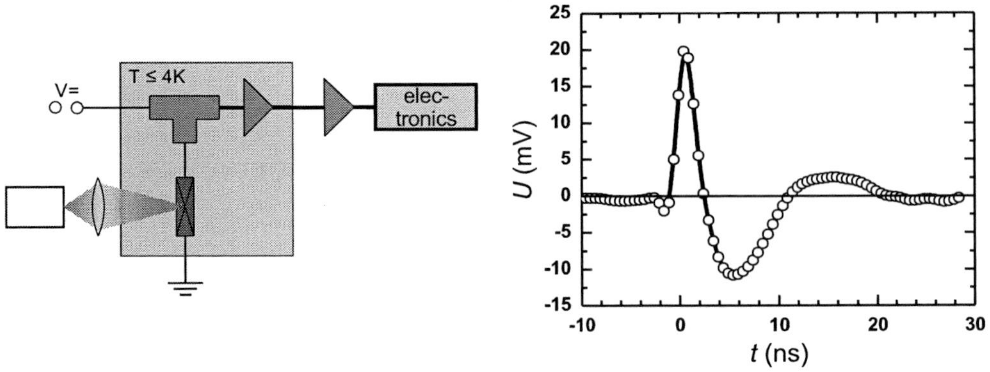

(a) (b)

Figure 18: (a) Schematic drawing of experimental setup. (b) Example of a photon pulse as recorded with the oscilloscope.

For illumination we used an incandescent light source and a prism spectrometer. The light output of the spectrometer with typically $\Delta\lambda/\lambda \approx 0.02$ was focused onto the detector using a parabolic mirror. The mean count rate, that is proportional to the quantum efficiency of the detector, was normalized to the light intensity measured with either a silicon or a lead-selenide photodiode (at longer wavelengths), which alternatively substituted the detector. For dark count measurements the optical entrance of the cryostat was blocked completely.

5.2 Dark Counts and Fluctuations

The importance of fluctuations within the superconducting nanostructures has already been touched in Sec. 4.2. In the following we will discuss experimental measurements of the dark count rate, i.e. the mean number of voltage pulses per second not caused by absorption of an optical or near-infrared photon, and their possible origin. The dark count rate is a measure of the detector's intrinsic noise and therefore it is particularly important for low-intensity applications when a very high sensitivity is required.

Dark count rates Γ_{dark} obtained for a meander line with a geometric cross-section 5 nm thick and 84 nm wide are shown in Fig. 19 as a function of the reduced bias current I/I_{c} for various temperatures below the transition temperature $T_{\text{c}} = 4.8$ K. The meander covered an area 4×4 μm. The apparent increase of Γ_{dark} with decreasing temperatures may be counterintuitive, but for a fixed nominal current the count rate decreases with lower temperature as one would expect. Nevertheless a strong current and temperature dependence remains that needs to be explained and there are certainly numerous possible sources for dark count events. Background radiation, i.e. black-body radiation from the surrounding walls, is insignificant. All surfaces facing the detector are at about 4 K or 77 K, and due to the intrinsic cut-off wavelength of the detector a quick estimate puts the number of detectable photons crossing the detector area at less than 10 s^{-1}. Bearing in mind a maximum quantum efficiency of a few percent, background radiation from the surrounding surfaces can be safely neglected. Cosmic radiation may be another source of dark count events. Because the detection mechanism should allow for a moderate energy discrimination [12, 55]

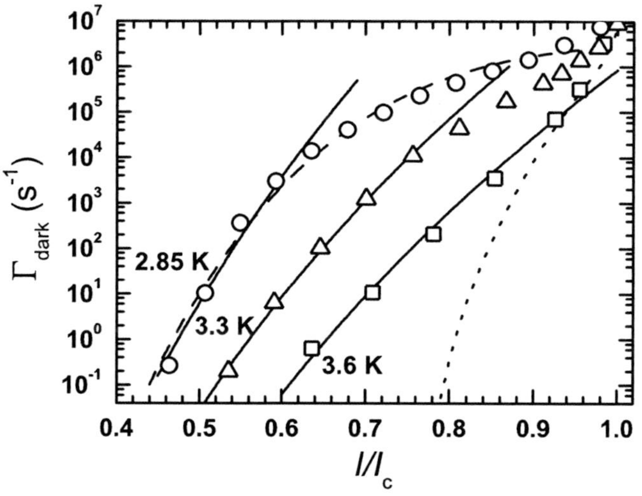

Figure 19: Dark count rates Γ_{dark} versus reduced bias current measured for a 5 nm thick, 84 nm wide NbN meander line. Temperatures are indicated on the graph. Solid lines are best fits according to our model of current-induced unbinding of VAPs. The fit at lower temperatures can be improved substantially by allowing for a current dependence of $\varepsilon(l_j)$ (dashed curve). The dotted line is the possible contribution of fluctuations of the Cooper-pair density fitted to the 3.6 K-data. It may only become important for bias currents extremely close to I_c.

such high energy particles would cause voltage pulses with significantly higher amplitude and longer duration. On the other hand, the absorption probability for high energy quanta in a 5 nm thick NbN film is extremely low. Since we have never seen any signals from high energy quanta, cosmic radiation is also neglected as a significant source of dark count events. Although great efforts have been made to reduce noise superimposed on the bias current, e.g. using a battery-powered custom-made current-source and low-pass filters, contributions from current noise cannot be ruled out so easily. Especially for bias currents extremely close to I_c current noise will add to the dark count rates as well as temperature fluctuations and associated fluctuations in I_c. However, the observed temperature-dependence is not easily explained by such extrinsic fluctuations.

It is a well-known general trend also observed in superconductors [56] that fluctuations become more important if the physical dimensions of the system are reduced. For example, in the limiting case of a one-dimensional superconducting wire, i.e. for a rectangular cross-section both the width and the thickness are approximately equal or less than the GL coherence length ξ_{GL} and the length $l \gg \xi_{\text{GL}}$, so called phase-slip centers (PSCs) can occur [2, 57, 58]. In a PSC the order parameter is temporarily reduced to zero leading to a mean voltage between the ends of the wire in time-averaged measurements or a succession of voltage pulses if the time resolution is fast enough. The lateral size of a PSC is either given by the width and thickness of the wire or the coherence length ξ_{GL}, whatever is smaller; the length is determined by ξ_{GL}. Because the width of our NbN strips $w \gg \xi_{\text{GL}}$ at temperatures of interest, using a simple geometric argument and assuming such fluctuations do still occur in such relatively wide strips one can show that PSCs can only trigger voltage pulses for $I/I_c \gtrsim 0.9$. In addition, at the operating temperatures considerably below T_c the probability of thermally activated PSCs is already very low. The alternative possibility of quantum PSCs [50, 59], which should also occur at temperatures well below T_c, has the same problems with respect to the relative size of ξ_{GL} compared to the strip width. Attempts to fit the experimental data based on PSCs have failed or resulted in unrealistic fitting parameters.

Within our refined detection model we could show that for triggering a voltage transient the order parameter needs not to be reduced to zero within a small volume. Instead, it is sufficient if the number of Cooper-pairs within the ξ-slab is reduced by a minimum amount. The number of Cooper-pairs within a given volume naturally fluctuates around its mean value [60, 61] due to the thermally activated generation and subsequent annihilation of quasiparticles. As we will show below, such fluctuations are also of minor importance and may only be significant if the bias current is extremely close to the critical current.

In two-dimensional superconducting films a very different type of fluctuations is possible: thermally activated vortex-antivortex pairs (VAPs) that are bound together by magnetic flux loops. As already noted in Sec. 4.1 these fluctuations are a manifestation of topological excitations that undergo a Berezinskii-Kosterlitz-Thouless (BKT) transition [62,63]. Below the superconducting transition temperature T_c thermal activation leads to a mean density of VAPs. They themselves may be thermally activated causing a certain fraction to be unbound individual vortices. An applied bias current exerts a Lorentz force on them and unless these unbound vortices are pinned they start to move. This movement in turn gives rise to dissipation, i.e. ohmic resistance even in the limit of a vanishing bias current. Below the BKT transition temperature T_{BKT} all VAPs remain bound together and an extended 2D film becomes a *true* superconductor. The current-voltage characteristic becomes algebraic because of the current exerting a torque on the VAPs and lowering their binding energy. Thus, with increasing current more and more VAPs can thermally unbind even below T_{BKT}.

Such a scenario is likely to happen in our superconducting strips, as well. As outlined in Sec. 3.4, the bridges are wide enough for vortices to nucleate, certainly at the temperatures the dark count rates have been recorded. It has to be stressed here that VAPs are different from the vortices discussed in Sec. 3.4, which enter from the strip edges at high applied currents. VAPs exist in 2D superconducting films even in zero magnetic field and zero applied current. We have also measured the resistance with an emphasis on how the resistance vanishes below the transition temperature T_c. These R-T data are consistent with the model of a BKT-transition of VAPs. For the meander line for which we present dark count measurements, the BKT temperature was determined to be $T_{BKT} \approx 4.15$ K, well above the operating temperatures during dark count measurements. Therefore, we assume bound VAPs to be present in our strips. If a bias current is applied we will have a certain number of current-assisted thermally unbound vortices and antivortices, as well, which will start to move towards opposite strip edges due to the now unbalanced Lorentz force acting on them, where they will disappear. Using the free vortex velocity of the Bardeen-Stephen model and taking the travelling distance to be half the strip width, we estimate sub-nanosecond voltage pulses with an amplitude comparable to those caused by absorbed photons, therefore contributing to the observed dark count rate.

For a thermally activated unbinding process the observed count rate should be proportional to $\exp(-U_b/k_B T)$, with U_b the binding energy. For the binding energy one finds the following expression [64]

$$U_b = 2\mu_c + \frac{A(T)}{\varepsilon(l_j)}(l_j - 1). \tag{14}$$

Here $l_j = \ln(2.6j_c/j)$ is a current scale, j_c the critical current density and j the applied current density, and $\varepsilon(l_j)$ is a renormalization factor close to unity. The temperature dependent energy scale $A(T) = (\pi^2\hbar/2e^2R_s)\Delta(T)\tanh[\Delta(T)/k_BT]$ can be calculated from device parameters, where $\Delta(T)$ is the superconducting energy gap at zero current and R_s the normal state sheet resistance. The core energy of one vortex μ_c is assumed to be approximately proportional to the energy scale $A(T)$, i.e. $\mu_c \approx A(T)/\gamma$, $\gamma = \text{const}$. Then the count rate may be expressed as

$$\Gamma_{\text{BKT}}(I,T) = \Omega_{0,\text{BKT}} \exp\left[\frac{\beta A(T)}{k_BT}\right]\left(\frac{I}{2.6I_c}\right)^{\frac{1}{\varepsilon(l_j)}\frac{A(T)}{k_BT}}, \tag{15}$$

with $\Omega_{0,\text{BKT}}$ the attempt rate and $\beta = [\gamma - 2\varepsilon(l_j)]/[\gamma\varepsilon(l_j)]$ a parameter of order unity.

Assuming $\beta = 1$ Eq. (15) can be used to fit the experimental dark count rate with just two fitting parameters, the attempt rate $\Omega_{0,\text{BKT}}$ and the renormalization factor $\varepsilon(l_j)$. At the highest temperature of $T = 3.6$ K this results in a satisfactory fit with $\varepsilon(l_j) = 0.67$ and $\Omega_{0,\text{BKT}} = 9 \cdot 10^9\ \text{s}^{-1}$ and the lower solid line in Fig. 19. The other two solid lines in the graph are the results for lower temperatures and the substantial deviations at reduced currents approaching unity is obvious. Their fitting parameters are $\varepsilon(l_j) = 0.85$, $\Omega_{0,\text{BKT}} = 2.5 \cdot 10^9\ \text{s}^{-1}$ and $\varepsilon(l_j) = 1.0$, $\Omega_{0,\text{BKT}} = 3.8 \cdot 10^{10}\ \text{s}^{-1}$ at $T = 3.3$ and 2.85 K, respectively. Taking into account that variations in β would have a strong effect on the attempt rate, the parameters for all three temperatures are reasonably compatible. So far, the current dependence of $\varepsilon(l_j)$, indicated by the subscript j, has been neglected. We have made this simplification partly because the BKT-theory makes only vague statements on its current dependence. Yet, if one allows for a current dependence of $\varepsilon(l_j)$, albeit phenomenologically, the fit at low temperatures can be substantially improved. The dashed curve in Fig. 19 was obtained for ε decreasing from 1.0 at low currents to about 0.75 very close to I_c. Although there remain a number of issues to be resolved, e.g. the current dependence of $\varepsilon(l_j)$ and its variation with temperature, the proposed model of current-assisted thermally unbound VAPs seems to be a promising candidate to explain the current and temperature dependence of the observed dark count rate.

A close inspection of the dark count rates reveals an upturn for currents exceeding about $0.9I_c$. This is when fluctuations in the Cooper-pair density may come into play. Within the refined detection model it makes no difference whether the Cooper-pair density is reduced due to photon absorption or due the fluctuations. If we allow for fluctuations in a disk-like volume with diameter $2a \geq \xi_{\text{GL}}$ (see Fig. 15) then the minimum number of fluctuating Cooper-pairs to trigger the transition into the normal-conducting state in the ξ-slab is up to a factor close to unity given as

$$\delta N_s^* \approx awd\, N_0\Delta(I_b,T)\left(1 - \frac{I}{I_c}\right). \tag{16}$$

The thermodynamic probability for such a fluctuation is given by $\exp[-\Delta F/k_BT]$ using the change in free energy ΔF associated with the fluctuation. For the change in free energy we used the minimum excitation energy $E(I, T)$ required to break one Cooper pair times the number of pairs broken:

$$\Delta F = \delta N_s E(I_b, T), \tag{17}$$

where $E(I, T)$ equals the gap energy $\Delta(I, T)$ including the explicit current-dependence of the minimum excitation across the energy-gap in the presence of a transport current. The dominant current-induced effect is a shift of the Fermi sphere in k-space which reduces the minimum excitation energy [49] which we approximate with $E_{\text{min}} \propto (1 - I/I_c)$, whereas for the temperature dependence we use the analytical approximation $E_{\text{min}} \propto [1 - (T/T_c)^4]^{1/2}$. An applied bias current further reduces the energy gap or pairing potential [54, 65]. This contribution is generally of minor importance except at temperatures very close to T_c. We account for this contribution by an additional factor $[1 - \alpha(I/I_c)^2]$ and the phenomenological parameter $\alpha = 0.03(1 - T/T_c)^{-0.5}$. Because fluctuations larger than the minimum fluctuation also trigger dark count pulses, we take the integral over the thermodynamic probability of all possible fluctuations with $\delta N_s \geq \delta N_s^*$ and multiply it with a constant attempt rate to obtain the dark count rate. The integrals have to be solved numerically. In Fig. 19 the result for $T = 3.6$ K is plotted as the dotted curve. The only adjustable parameter is the attempt rate, which was chosen to be 1.6×10^8 s^{-1}.

The simulations have shown that fluctuations of the Cooper-pair density similar to possible PSC-type fluctuations may only play a significant role for high currents exceeding $I/I_c \approx 0.9$. Furthermore, despite the excellent agreement between experimental data and simulations taking the sum of dark counts caused by unbound VAPs and fluctuations in the Cooper-pair density, it has to be noted, that for bias currents sufficiently close to I_c current noise will certainly add to the observed dark count rate and the observed upturn may, at least in part, be caused by this noise contribution.

5.3 Spectral Detection Efficiency

A fundamental characteristic of any photon detector is its spectral detection efficiency, i.e. how many of the incoming photons trigger a detectable signal as a function of the photons' energy. According to our refined detection model photons with an energy that is large enough to produce the minimum number of QP required to trigger the resistive state, Eq. (11), should be registered with a probability that is equal to the absorption probability, which we expect to be relatively flat. On the other hand there should be a well pronounced cut-off wavelength beyond which the detection efficiency is expected to be drastically reduced. Recasting Eq. (13) we find for the cut-off wavelength

$$\lambda_{\text{cut-off}} = \frac{\zeta h c e^2}{\sqrt{\pi \tau_{\text{th}}}} \rho \sqrt{D} \frac{1}{\Delta^2 A \left(1 - \frac{I}{I_c}\right)}, \tag{18}$$

where we have used the normal state resistivity $\rho = (N_0 e^2 D)^{-1}$, with D being the electron diffusivity, and the cross-sectional area $A = wd$. From above equation one can see, that the cut-off wavelength should be strongly dependent on the bias current, but also on material parameters such as the transition temperature T_c via the energy gap Δ and the normal state resistivity.

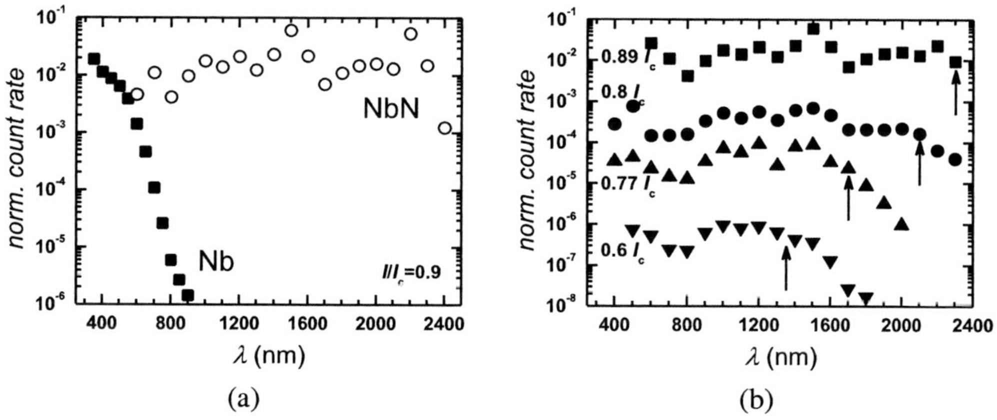

Figure 20: Normalized count rate versus photon wavelength. In the left graph (a) we compare the spectral detection efficiency for detectors made from different materials: Nb and NbN. In both cases $I/I_c = 0.9$ and $T \approx T_c/2$. In (b) the spectral detection efficiency of the NbN detector is shown for different bias currents as indicated in the graph; temperature $T \approx T_c/2$. Data have been shifted along the vertical axis for clarity. Arrows indicate the cut-off wavelength $\lambda_{\text{cut-off}}$ used to calculate the multiplication efficiency ζ (Sec. 4.1, Fig. 16).

In Fig. 20 (a) we show normalized count rates as a function of the photon wavelength for two detectors made from different materials. The NbN detector is the same for which we presented the dark count measurements. The other one was made from a 10 nm thick Nb film. The geometric strip width equaled 98 nm and the meander covered an area of approximately $4 \times 4\ \mu\text{m}^2$. The transition temperature of the Nb meander was slightly higher at $T_c = 5.1$ K. Both meanders were operated at temperatures $T \approx T_c/2$ and bias currents $I = 0.9I_c$. Despite a similar transition temperature and a cross-sectional area varying by only factor of ≈ 2, the spectral response is markedly different. The normalized count rate of the NbN stays approximately constant with a detection efficiency of about 1% exceeding wavelengths of 2 μm. Contrariwise, the detection efficiency of the Nb meander is decreasing continuously in the accessible spectral range. We estimate the cut-off wavelength of the Nb meander to be just below the visible spectrum in the ultraviolet, $\lambda_{\text{cut-off}} \lesssim 400$ nm. This is roughly an order of magnitude below the cut-off wavelength of the NbN meander for comparable operating conditions, $\lambda_{\text{cut-off}} \approx 2.4\ \mu$m. As large as the difference may seem, it is easily understood in view of Eq. (18) and taking into account the normal state resistivities. They have been determined to $\rho = 607\ \mu\Omega\,\text{cm}$ and $6\ \mu\Omega\,\text{cm}$ for the NbN and Nb meander, respectively.

To show the variation of $\lambda_{\text{cut-off}}$ with the bias current we show the spectral detection efficiency of the NbN meander for various bias currents and constant temperature $T \approx T_c/2$ in Fig. 20 (b). Firstly, it has to be noted that the cut-off is not as well pronounced as one might expect. In Sec. 4.2 it has already been mentioned that fluctuations enhance the detectivity beyond the cut-off. Non-uniformities of the meander line smear out the cut-off further. Best estimates of $\lambda_{\text{cut-off}}$ are indicated by arrows in Fig. 18 (b). Despite these uncertainties, it is undoubtedly clear that the cut-off wavelength decreases with decreasing bias current. Verification of the $(1 - I/I_c)^{-1}$-dependence is hampered by the energy dependence of the multiplication efficiency ζ (Sec. 4.2). However, the data are consistent with a current dependence according to Eq. (18).

6 Summary

In this report we outlined our current understanding of superconductivity in nanostructured Nb and NbN bridges. The potential of single-photon detectors of meander lines based on these bridges has been elaborated. The superconducting thin films and nanostructured bridges can be very well described as conventional type-II superconductors. Normal conducting layers at the top and bottom of the films and additional normal conducting edges in the case of bridges significantly influence the superconducting properties via the proximity effect. In very thin and narrow bridges there appears to be a qualitatively understood crossover from unpinning of vortices to pair-braking of Cooper-pairs as the mechanism limiting the critical current density.

A very recently developed extension of the model of single-photon detection in nanostructured superconducting meanders gave us new insight into the detection processes. The key finding of the refined detection model, that a voltage transient may form even without the formation of a normal spot, allowed us a consistent quantitative description of the detection process. Furthermore, we could determine the multiplication efficiency of the QP thermalization cascade.

Comparison of the spectral sensitivity between Nb and NbN meanders demonstrated the importance of material parameters, most notably normal state resistivity and electron diffusivity and the superconducting transition temperature. In light of these results and the equation relating the cut-off wavelength to these parameters, we are optimistic to be able to extent the spectral range of these detectors significantly by using the right materials. Dark counts caused by fluctuations limit the ultimate sensitivity of the detectors. Of the many different fluctuation processes that can occur in superconducting wires and thin films, the current-assisted thermal unbinding of VAPs seems to be the dominant fluctuation mode for the interesting temperature and current range. Although we cannot yet fully explain the observed current and temperature dependence of the dark count rate, our limited model already describes most of the principal observations.

This work is partly supported by the DFG-Center for Functional Nanostructures at the University of Karlsruhe. One of the authors (A. E.) acknowledges support by the NCCR program *Materials with Novel Electronic Properties (MaNEP)* sponsored by the Swiss National Science Foundation.

References

[1] J. A. Pals, K. Weiss, P. M. T. M. van Attekum, R. E. Horstman, and J. Wolter, *Phys. Rep.* **89**, (1982) 323.

[2] R. Tidecks, *Current-Induced Nonequilibrium Phenomena in Quasi-One-Dimensional Superconductors* (Springer Verlag Berlin Heidelberg, 1990).

[3] A. D. Semenov, G. N. Gol'tsman, and R. Sobolewski, *Supercond. Sci. Technol.*, **15**, (2002) R1.

[4] N. K. Sherman, *Phys. Rev. Lett.*, **8**, (1962) 438.

[5] A. Gabutti, R. G. Wagner, K. E. Gray, R. T. Kampwirth, and R. H. Ono, *Nucl. Instr. and Meth. A*, **278**, (1989) 425.

[6] A. M. Kadin and M. W. Johnson, *Appl. Phys. Lett.*, **69**, (1996) 3938.

[7] V. A. Shklovskii, □*z. Tverd. Tela*, **17**, (1975) 3076, [*Sov. Phys. Sol. State*, **17**, (1975) 2040].

[8] N. Perrin and C. Vanneste, *Phys. Rev. B*, **28**, (1983) 5150.

[9] A. D. Semenov, R. S. Nebosis, Y. P. Gousev, M. A. Heusinger, and K. F. Renk, *Phys. Rev. B*, **52**, (1995) 581.

[10] B. Cabrera, R. M. Clarke, P. Colling, A. J. Miller, S. Nam, and R. W. Romani, *Appl. Phys. Lett.*, **73**, (1998) 735.

[11] K. D. Irwin, *Appl. Phys. Lett.*, **66**, (1995) 1998.

[12] A. D. Semenov, G. N. Gol'tsman, and A. A. Korneev, *Physica C*, **351**, (2001) 349.

[13] G. N. Gol'tsman, O. Okunev, G. Chulkova, A. Lipatov, A. Semenov, K. Smirnov, B. Voronov, A. Dzardanov, C. Williams, and R. Sobolewski, *Appl. Phys. Lett.*, **79**, (2001) 705.

[14] G. H. Wood and B. L. White, *Appl. Phys. Lett.*, **15**, (1969) 237.

[15] M. Kurakado and H. Mazaki, *Phys. Rev. B*, **22**, (1980) 168.

[16] D. Twerenbold, *Phys. Rev. B*, **34**, (1986) 7748.

[17] T. Nussbaumer, P. Lerch, E. Kirk, A. Zehnder, R. Füchslin, P. F. Meier, and H. R. Ott, *Phys. Rev. B*, **61**, (2000) 9719.

[18] K. S. Il'in, S. A. Vitusevich, B. B. Jin, A. I. Gubin, N. Klein, and M. Siegel, *Phys. C*, **408**, (2004) 700.

[19] K. Il'in, M. Siegel, A. Semenov, A. Engel, and H.-W. Hübers, *phys. stat. sol. (c)*, **2**, (2005) 1680.

[20] A. Semenov, A. Engel, H.-W. Hübers, K. Ilin, and M. Siegel, in Optical Sensing, *Proceedings of the International Conference SPIE Int. Soc. Opt. Eng.*, Strasbourg, France, 2004, edited by B. Culshaw, A. G. Mignani, and R. Riesenberg, volume 5459, p. 237.

[21] A. Engel, A. Semenov, H.-W. Hübers, K. Il'in, and M. Siegel, *phys. stat. sol. (c)*, **2**, (2005) 1668.

[22] S. A. Wolf, J. J. Kennedy, and M. Nisenoff, *J. Vac. Sci. Technol.*, **13**, (1976) 145.

[23] S. A. Wolf, F. J. Rachford, and M. Nisenoff, *J. Vac. Sci. Technol.*, **15**, (1978) 386.

[24] S. I. Park and T. H. Geballe, *Physica B+C*, **135**, (1985) 108.

[25] L. N. Cooper, *Phys. Rev. Lett.*, **6**, (1961) 689.

[26] Y. V. Fominov and M. V. Feigelman, *Phys. Rev. B*, **63**, (2001) 094518.

[27] M. Gershenzon and V. Koshelets, *Zh. Tekh. Fiz.*, **50**, (1980) 572, (*Sov. Phys. - Tech. Phys.*, **25**, 343, 1980).

[28] R. D. Parks, ed., *Superconductivity*, 2 Vols. (Decker, New York, 1969).

[29] C. C. Chin, D. E. Oates, G. Dresselhaus, and M. S. Dresselhaus, *Phys. Rev. B*, **45**, (1992) 4788.

[30] S. Kubo, M. Asahi, M. Hikita, and M. Igarashi, *Appl. Phys. Lett.*, **44**, (1984) 258.

[31] J. Gavaler, A. Santhanam, A. Braginski, M. Ashkin, and M. Janocko, *IEEE Trans. Mag.*, **17**, (1981) 573.

[32] R. Chicault and J. C. Villegier, *Phys. Rev. B*, **36**, (1987) 5215.

[33] J. Villegier, L. Vieux-Rochaz, M. Goniche, P. Renard, and M. Vabre, *IEEE Trans. Mag.*, **21**, (1985) 498.

[34] K. L. Westra, M. J. Brett, and J. F. Vaneldik, *J. Vac. Sci. Technol.*, **8**, (1990) 1288.

[35] K. K. Likharev, *Rev. Mod. Phys.*, **51**, (1979) 101.

[36] D. E. Oates, A. C. Anderson, C. C. Chin, J. S. Derov, G. Dresselhaus, and M. S. Dresselhaus, *Phys. Rev. B*, **43**, (1991) 7655.

[37] R. Hu, G. Kerber, J. Luine, E. Ladizinsky, and J. Bulman, *IEEE Trans. Appl. Superc.*, **13**, (2003) 3288.

[38] W. J. Skocpol, M. R. Beasley, and M. Tinkham, *J. Appl. Phys.*, **45**, (1974) 4054.

[39] E. H. Brandt and M. Indenbom, *Phys. Rev. B*, **48**, (1993) 12893.

[40] E. Zeldov, J. R. Clem, M. McElfresh, and M. Darwin, *Phys. Rev. B*, **49**, (1994) 9802.

[41] J. M. E. Geers, M. B. S. Hesselberth, J. Aarts, and A. A. Golubov, *Phys. Rev. B*, **64**, (2001) 094506.

[42] J. Zhang, W. Slysz, A. Pearlman, A. Verevkin, R. Sobolewski, O. Okunev, G. Chulkova, and G. N. Gol'tsman, *Phys. Rev. B*, **67**, (2003) 132508.

[43] A. Verevkin, Z. Zhang, R. Sobolewski, A. Lipatov, O. Okunev, G. Chulkova, A. Korneev, K. Smirnov, G. N. Gol'tsman, and A. Semenov, *Appl. Phys. Lett.*, **80**, (2002) 4687.

[44] A. Korneev, P. Kouminov, V. Matvienko, G. Chulkova, K. Smirnov, B. Voronov, G. N. Goltsman, M. Currie, W. Lo, K. Wilsher, J. Zhang, W. Słysz, A. Pearlman, A. Verevkin, and R. Sobolewski, *Appl. Phys. Lett.*, **84**, (2004) 5338.

[45] A. G. Kozorezov, A. F. Volkov, J. K. Wigmore, A. Peacock, A. Poelaert, and R. den Hartog, *Phys. Rev. B*, **61**, (2000) 11807.

[46] A. Semenov, A. Engel, K. Il'in, G. Gol'tsman, M. Siegel, and H.-W. Hübers, *Eur. Phys. J. AP*, **21**, (2003) 171.

[47] M. R. Beasley, J. E. Mooij, and T. P. Orlando, *Phys. Rev. Lett.*, **42**, (1979) 1165.

[48] A. D. Semenov and G. N. Gol'tsman, *J. Appl. Phys.*, **87**, (2000) 502.

[49] M. Tinkham, *Introduction to Superconductivity* (McGraw-Hill, Inc., New York, 1996), second edition.

[50] C. N. Lau, N. Markovic, M. Bockrath, A. Bezryadin, and M. Tinkham, *Phys. Rev. Lett.*, **87**, (2001) 217003.

[51] A. Garzarella and C. J. Martoff, *J. Appl. Phys.*, **79**, (1996) 2426.

[52] A. Engel, A. Semenov, H.-W. Hübers, K. Ilin, and M. Siegel, *Nucl. Instr. and Meth.*, **A520**, (2004) 32.

[53] R. F. Voss, C. M. Knoedler, and P. M. Horn, *Phys. Rev. Lett.*, **45**, (1980) 1523.

[54] A. Anthore, H. Pothier, and D. Esteve, *Phys. Rev. Lett.*, **90**, (2003) 127001.

[55] A. Semenov, A. Engel, H.-W. Hübers, K. Il'in, and M. Siegel, Optical and Infrared Detectors for Astronomy, *proceedings of the International Conference SPIE Int. Soc. Opt. Eng.*, Glasgow, Scotland, 2004, edited by J. D. Garnett and J. W. Beletic, volume 5499, p. 310.

[56] W. J. Skocpol and M. Tinkham, *Rep. Prog. Phys.*, **38**, (1975) 1049.

[57] J. S. Langer and V. Ambegaokar, *Phys. Rev.*, **164**, (1967) 498.

[58] D. E. McCumber and B. I. Halperin, *Phys. Rev. B*, **1**, (1970) 1054.

[59] N. Giordano, *Physica B*, **203**, (1994) 460.

[60] C. M. Wilson, L. Frunzio, and D. E. Prober, *Phys. Rev. Lett.*, **87**, (2001) 067004.

[61] C. M. Wilson and D. E. Prober, *Phys. Rev. B*, **69**, (2004) 094524.

[62] Z. L. Berezinskii, *Zh. Eksp. Teor. Fiz.*, **59**, (1970) 907; *Zh. Eksp. Teor. Fiz.*, **61**, (1971) 1144.

[63] J. M. Kosterlitz and D. J. Thouless, *J. Phys. C*, **6**, (1973) 1181.

[64] J. E. Mooij, in *Percolation, Localization, and Superconductivity* (Plenum Press New York, 1984), edited by A. M. Goldman and S. A. Wolf, p. 325.

[65] J. L. Levine, *Phys. Rev. Lett.*, **15**, (1965) 154.

In: New Frontiers in Superconductivity Research
Editor: Barry P. Martins, pp. 199-213
ISBN: 1-59454-850-1

Chapter 7

Recent Progress on Enhanced Critical Current Density in Air-Processed Gd-Ba-Cu-O Bulk Superconductors

***Caixuan Xu*[1,2], *Anming Hu*[3], *Naomichi Sakai*[2], *Izumi Hirabayashi*[2], *Mitsuru Izumi*[1]**

[1]Department of Marine Electronics and Mechanical Engineering, Tokyo University of Marine Science and Technology, 2-1-6, Etchu-jima, Koto-ku, Tokyo 135-8533, Japan

[2] Superconductivity Research Laboratory, International Superconductivity Technology Center, 1-10-13 Shinonome, Koto-ku, Tokyo 135-0062, Japan

[3] Department of Physics, University of Waterloo 200 Univ. Ave. West, Waterloo, ON, N2L 3G1, Canada

Abstract

Melt-processed Gd-Ba-Cu-O bulk superconductors have drawn worldwide interest due to their superior superconducting performance. In particular, air processing is attractive from the economical viewpoints. In this paper, we report on the superconducting performance of $GdBa_2Cu_3O_{7-\delta}$ (Gd123) fabricated in air with focus placed on the effects of the addition of BaO_2, refined Gd_2BaCuO_5 (Gd211), and a new second phase $Gd_2Ba_4CuMoO_y$ (Gd2411). First, we discussed the solubility and the stability of $Gd_{1+x}Ba_{2-x}Cu_3O_{7-\delta}$ solid solutions and the relation with the properties of single domain Gd-Ba-Cu-O bulks prepared in air, in that the doping of BaO_2 was found to be crucial for the suppression of Gd substitution on Ba sites. We further discussed the relation between the sizes of the second phase Gd211 particles and the critical current (J_c) of Gd123 bulks. The addition of Gd2411 powders sintered in air was also performed and its effect on the superconducting properties and microstructure of Gd-Ba-Cu-O was studied. With the proper amount additions of BaO_2, Gd2411 and refined Gd211, the J_c of Gd-Ba-Cu-O bulks grown in air reached 90,000 A/cm^2 at 77 K in self-field. We also propose potential applications of air-processed Gd-Ba-Cu-O to rotating machines used for motors and generators.

1 Introduction

In the last decade, enormous efforts have been made on the melt-textured large grain Gd-Ba-Cu-O superconductors that exhibit strong ability on trapped magnetic field compared with Y-Ba-Cu-O bulks, which led to the development of studies on wide engineering applications [1-5]. The trapped magnetic field up to 3.05 Tesla was obtained from the Gd123 bulk grown in oxygen-controlled-melt-growth (OCMG) process [6]. It is well known that the trapped magnetic field is approximately proportional to critical current density (J_c) and the size of single domain according to the critical state model. The key to the enhanced trapped magnetic field is the achievement of a high J_c in large single domain. On the other hand, for commercialization, it is essential to lower the cost for fabrication while keeping the Gd123 bulk performance high.

As we know, Gd123 superconductors exhibit $Gd_{1+x}Ba_{2-x}Cu_3O_{7-\delta}$ type solid solutions similar to other LRE-Ba-Cu-O (LRE= La, Nd, …Eu) systems and that LRE/Ba substitution deteriorates superconducting properties [7 - 12]. Different methods [13-15], such as high crystallization rate, post-annealing in Ar and using Ba-rich precursors have been developed to suppress this substitution and hence to improve the superconducting performance of the LRE-Ba-Cu-O bulks melt-processed in ambient atmosphere. It is worth noting that doping of Ba compounds such as $BaCuO_{2-\delta}$ and BaO_2 was reported to suppress the $RE_{1+x}Ba_{2-x}Cu_3O_{7-\delta}$ solid solution and enhance the superconducting properties effectively.[16-17]

The J_c of bulk RE-Ba-Cu-O superconductor is strongly dependent on the microstructure of the sample [16-18]. The introduction of second phase $REBaCuO_5$ (RE211) contributes to the enhancement of the J_c [11]. Influence of the size of RE211 starting powder on J_c of bulks has been studied by several groups [5, 18,19]. Finer RE211 particles in RE-Ba-Cu-O bulks led to higher J_c. However, fined Gd211 starting particles are easy to coarsen themselves because of the push-out effect and Ostwald ripening during the melt growth process [20]. Although Pt and/or CeO_2 were introduced to hold back the coarsening process [21], it is still difficult to get sub-micrometer size Gd211 inclusions in bulk samples [6, 11]. This indicates that further investigation is necessary on the enhancement of J_c in Gd123 system with Gd211 as second phase.

Recently $Y_2Ba_4CuMO_z$ (M = Nb, Ta, W, Mo, Zr, Hf) was reported as another second phase candidate in Y123 system [22]. Hari Babu *et al.* have studied Y2411 addition in bulk Y123 system, in which very fine inclusions with a size of a few hundred nanometers were stably embedded in the Y123 matrix [23]. This indicates that Y2411 is a promising candidate instead of Y211 for flux pinning sites. Thus, it is interesting to introduce Gd2411 into the Gd123 system for improving the bulk superconductivity. Furthermore, a band structure with inhomogeneous distribution of Y211/Y2411 was revealed in Y123 composition [24, 25]. Although there has been no report yet, the band structure may bring a drawback due to the spatial modulation of superconducting properties. Very recently, both nano-size ZrO_2 [26] and $BaZrO_3$ addition [27] are also reported to induce an inhomogeneous distribution of Y211 in Y123. Obviously, an experimental understanding of the relationship between RE2411 addition and microstructure evolution is necessary for homogeneous engineering RE211/RE2411 inclusion with enhanced superconducting properties.

In this paper, we study the superconducting properties and microstructures of bulk Gd123 superconductors with several additions of BaO_2, fined Gd211, and Gd2411. The recent progress on the development of engineering application of Gd123 bulks is reviewed.

2 Experiment

Commercial powders $GdBa_2Cu_3O_{7-\delta}$ (3 N, Gd123), Gd_2BaCuO_5 (3 N, Gd211), Gd_2O_3 (3N), MoO_3 (3N), BaO_2 (3N) and CuO (3N) were used as starting powders. In order to prevent possible chemical contamination from CO_2 and moisture, the Gd_2O_3 powder was calcined at 900 ℃ for 10 hours before weighing. The experimental process can be divided into four series as follows.

1. To study the solid solutions in Gd-Ba-Cu-O system, the nominal composition of $Gd_{1+x}Ba_{2-x}Cu_3O_{7-\delta}$ with x = - 0.1, 0, 0.05, 0.1, 0.15, 0.2 (denoted as Gd (1+x) hereafter) were synthesized in air by annealing at 900 ℃ with duration of 10 hours followed by furnace cooling. The powder samples were pressed into pellets with diameter as 3 – 5 mm. Argon post annealing (ArPA) was carried out in Ar flow at 940 ℃ for 10 hours and oxygen post annealing in O_2 flow at 400 ℃ for 100 hours. Powder X-ray diffraction (XRD) pattern was employed to determine the single-phase range of $Gd_{1+x}Ba_{2-x}Cu_3O_{7-\delta}$. The continuous scan with the rate 1.5 ° /min. was used for phase identification while the $2\theta-\theta$ step-scan mode with $\Delta\theta = 0.02°$ was used for the accumulation time 10 s to refine the crystal structure by using Rietveld analysis. Rietan2000 was employed to refine the crystal structure for the obtained powder diffraction data [28]. Differential thermal analysis (DTA) and thermogravimetry (TG) were used to determine the phase transition temperature and simulate the crystallization process [29]. Commercial superconducting quantum interference device (SQUID, Quantum Design, MPMS2) was employed to detect the superconducting transition.
2. BaO_2 was added into Gd123 bulks to suppress the solid solutions and enhance the superconducting properties. The precursor was fabricated by mixing Gd123, Gd211 and BaO_2 according to a nominal composition of Gd123 + 0.4Gd211 + $yBaO_2$ in a molar ratio with y = 0, 0.1, 0.2 and 0.3 (labeled as $yBaO_2$ hereafter).

During the fabrication process of bulks, the top-seeded melt-textured growth process (TSMTG) had been employed to overcome the problem of multi-nucleation process. In order to avoid the containment from seed melt, we used hot top-seeding method [5]. For all melt-processed bulks, both 10 wt.% Ag_2O and 0.5 wt.% Pt were added into the Gd123 bulks to improve the mechanical property and reduce the coarsening of Gd211 precipitate particles, respectively. The thoroughly mixed powder was first pressed into pellets of 30 mm in diameter and 10 mm in thickness. And then it was further solidified via cold iso-static pressing (CIP) in a rubber bag. A melt-textured Nd123 (001) seed crystal was placed on the surface of the pellet to control the crystallization. The grown samples were further annealed at 940 ℃ for 10 hours in flowing Ar (ArPA). Finally, the ArPA samples were annealed at 400 ℃ for 300 hours, 350 ℃ for 50 hours and 300 ℃ for 50 hours in flowing O_2.

3. To study the relation between Gd211 particle sizes and performance of bulks, Gd211 was synthesized at 900 ℃ in air and further fined by ball-milling machine which can control the particle size by ball-milling time. 1 h, 2 h and 4 h have been selected as ball-milling time for study. Commercial Gd211 powder has also been employed only for comparison. Starting materials were Gd123 + 0.4Gd211 + $0.1BaO_2$ in a molar ratio with different sizes of Gd211 particles.
4. Gd2411 single phase was synthesized in air with the oxide starting powders. The mixing oxide powder was pre-reacted at 950 ℃ for 10 hours in air twice and the single phase was obtained by calcining at 1460℃ twice in air, each time with the duration of 10 hours. From X-ray diffraction pattern, the single phase has been determined.[30] The Gd2411 single phase was introduced into Gd123 bulks with nominal composition following as Gd123 + (0.4-z) Gd211 + zGd2411 + 0.1 BaO_2 with z = 0.2, 0.12, 0.08, 0.04, 0.02, 0.004 and 0.002 (labeled as zGd2411), in a molar ratio.

After ArPA and O_2 annealing, small slabs along a-direction in grow-section of every bulk with dimension as 2 mm x 2 mm x 1 mm (a x b x c) were cut from the edge of single domain for measurement. SQUID was employed for superconducting transition. The microstructure was observed by scanning electron microscopy (SEM) and local composition was qualitatively identified with electron probe micro-area analysis (EPMA).

3 Results and Discussion

3.1 $Gd_{1+x}Ba_{2-x}Cu_3O_{7-\delta}$ Solid Solutions and Gd123 Single Domain Growth

From XRD results, single-phase $Gd_{1+x}Ba_{2-x}Cu_3O_{7-\delta}$ solid solutions are obtained for composition ranging from x = 0 to x = 0.1 for the as grown samples in air. But after ArPA, Gd211 can be detected by XRD in all the samples [11]. With the increase of x, the intensity of XRD peak of $Gd_{1+x}Ba_{2-x}Cu_3O_{7-\delta}$ decreases and the intensity for Gd_2BaCuO_5 increases. We obtained the solubility dependent lattice parameters *a*, *b* and *c* by Rietan 2000. The results are shown in Fig. 1. Both *b* and *c* decrease while *a* slightly increasing with an increasing of x. It indicates that the orthorhombicity decreases with increasing x. This is in consistent with reports for $LRE_{1+x}Ba_{2-x}Cu_3O_{7-\delta}$ in other light rare earths [9, 31-34].

Fig. 2 shows the critical onset transition temperature (T_c) of $Gd_{1+x}Ba_{2-x}Cu_3O_{7-\delta}$ solid solutions samples with and without ArPA prior to O_2 annealing process. The results exhibit that the onset T_c of the samples without ArPA dramatically decreases with increasing x. But after ArPA, the onset T_c kept nearly constant. Obviously, ArPA narrows the onset T_c distribution of $Gd_{1+x}Ba_{2-x}Cu_3O_{7-\delta}$ superconductors. On the basis of XRD study, we conclude that ArPA improves the onset T_c of $Gd_{1+x}Ba_{2-x}Cu_3O_{7-\delta}$ by decomposing such a solid solution with higher x values [11].

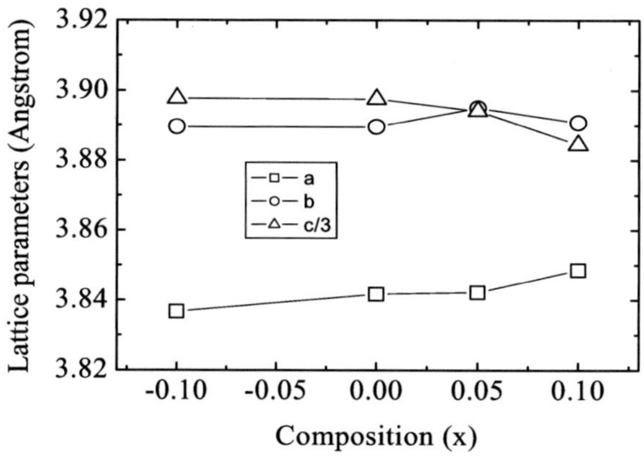

Fig. 1 Xu *et al.:* Tendency of lattice parameters a, b and c / 3 for $Gd_{1+x}Ba_{2-x}Cu_3O_{7-\delta}$ compositions with x = -0.1, 0, 0.05 and 0.1.

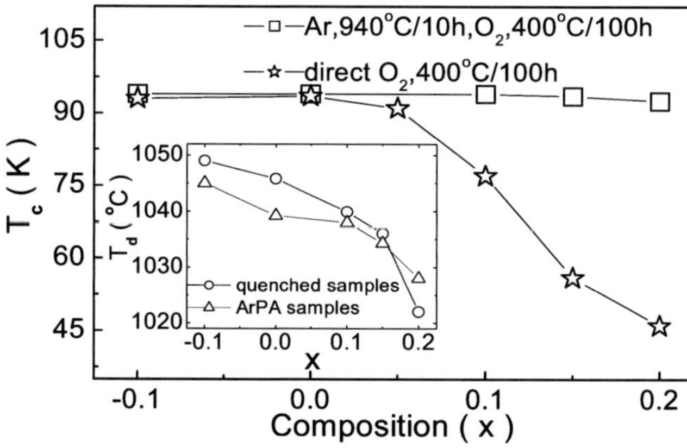

Fig. 2 Xu *et al.:* Onset T_c of $Gd_{1+x}Ba_{2-x}Cu_3O_{7-\delta}$ solid solutions (x = -0.1, 0, 0.05, 0.1 and 0.2) samples both with and without ArPA. Inset shows the decomposition temperature (T_d) plotted against the composition x for both quenched samples and ArPA samples.

Inset of Fig. 2 displays the decomposition temperature (T_d) plotted against the composition x for both quenched samples and ArPA samples. T_d was determined from the DTA data by a tangent method. From this inset figure, the decomposition temperature decreases with increasing x. This indicates a decrease of thermal stability in $Gd_{1+x}Ba_{2-x}Cu_3O_{7-\delta}$ compositions.

Fig. 3 shows a typical surface image of melt-process bulk sample grown by air - process. We used the thermal gradient method as heat profile to control the single single-domain growth [4]. There are no appreciable cracks visible on the surface.

Fig. 3 Xu *et al.*: Surface image of typical melt-processed bulk sample grown in air.

3.2 Role of BaO_2 Addition

Fig. 4 shows the J_c versus applied field B for various BaO_2 addition amounts under 77 K, $B // c$. Because all the slabs were cut at positions near the edge of the pellets, the J_c-B plots do not show obvious peak effects, probably due to an accumulation of Gd211 toward the edge of samples [35]. However, the J_c at low magnetic fields (< 1 T) is enhanced dramatically to reach 63 000 A/cm^2 (B = 0) for 0.1 BaO_2. Such improvement in J_c is attributed to the effective suppression of the overall LRE-Ba substitution and the optimization of the locally composition modulation in the superconducting matrix [16]. It is worth noting that 0.2 BaO_2 and 0.3 BaO_2 showed low J_c values. This is consistent with the results of T_c measurements shown in the inset of Fig. 4. Although T_c of 0.1 BaO_2 had almost the same onset T_c as 0 BaO_2 addition, the transition width was much narrower for the former composition. 0.2 BaO_2 and 0.3 BaO_2 have showed a lower onset T_c, which indicates an excess amount addition. An increase in T_c and narrower superconducting transition width indicated the suppression of Gd-Ba substitution by adding suitable amount of BaO_2. It has been proved that a Ba-rich composition dramatically narrows the range of LRE-Ba substitution [14-16]. Moreover, it is worth noting that excess BaO_2 leads to a decrease in T_c. It may come from the deteriorated superconducting junction with the formation cracks along the c-axis [15]. However, with an appropriate amount addition, superconducting properties has been remarkably enhanced. In this case, it is possible that a minor amount of BaO_2 is almost consumed during the peritectic reaction.

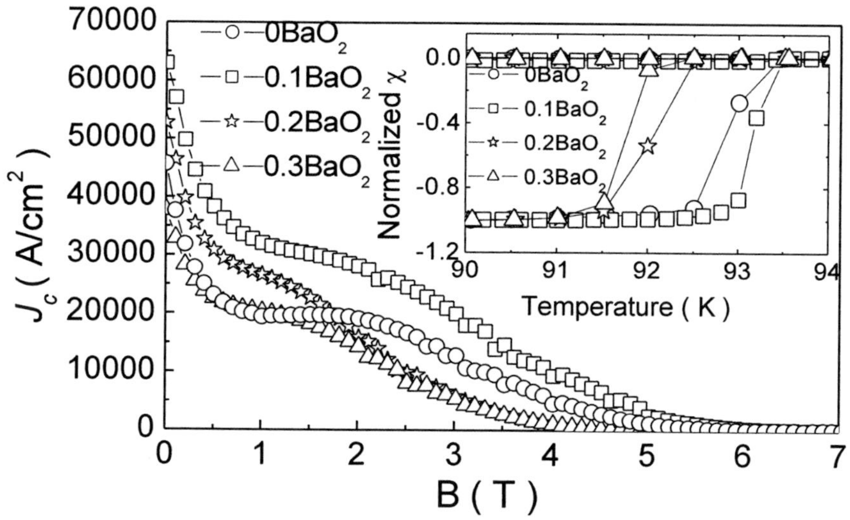

Fig. 4 Xu *et al.*: Critical current J_c versus applied field *B* for various BaO_2 addition amounts (y = 0, 1, 2 and 3) under 77 K, B // *c*. Inset showed the according onset T_c.

3.3 Influence of Particle Sizes of Gd211 Fined by Ball-Milling Process

Fig. 5 presents plots of the J_c - B for bulks with various sizes of Gd211 particles addition. It is significantly evident that J_c can be enhanced remarkably by fining Gd211 particles. For the samples with the composition of self-made Gd211 particle after 2 h ball-milling, the J_c at zero applied field reaches 90 000 A/cm^2, the record high J_c value of Gd-Ba-Cu-O superconductors melt-processed in air [21, 36, 37]. However, for the 4-hour ball-milling Gd211 particles, J_c decreases to 64,000 A/cm^2 at 77 K and under self field. This kind of suppression for J_c has also been observed in YBCO system.[38] The reason may come from push-out and coarsening effects of fined Gd211 particles.[38]

SEM micrographs for three kinds of samples with different sizes Gd211 particle additions are shown in Fig. 6. With the commercial powder in Fig. 6 (a) the majority of Gd211 particles remarkably coarse with the averaged size around 10 μm. Fig. 6 (b) shows a sample with self-made Gd211 annealed at 900 ℃. The average size of Gd211 inclusion is about 1 μm. The sample with the 1 hour ball-milling Gd211 particles is exhibited in 6 (c). It is found that the most particles are rod-like Gd211 particles with a size pronouncedly less than 1 μm. Obviously, adding ball-milling Gd211 particles directly results in refining the Gd211 inclusion. This may owe to that either the main resource of Gd211 inclusion obtained in melt-processed samples is the initial addition of Gd211, or superfined Gd211 acts as the nucleated site and rich nucleated sites may hinder the further coarsening of Gd211 particles. However, further work is needed to elucidate this issue.

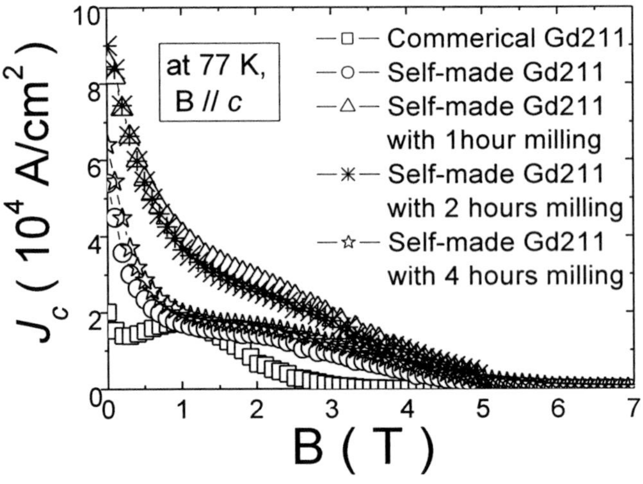

Fig. 5 Xu *et al.*: The plot of the J_c - B for bulk superconductor with various sizes of Gd211 particles addition.

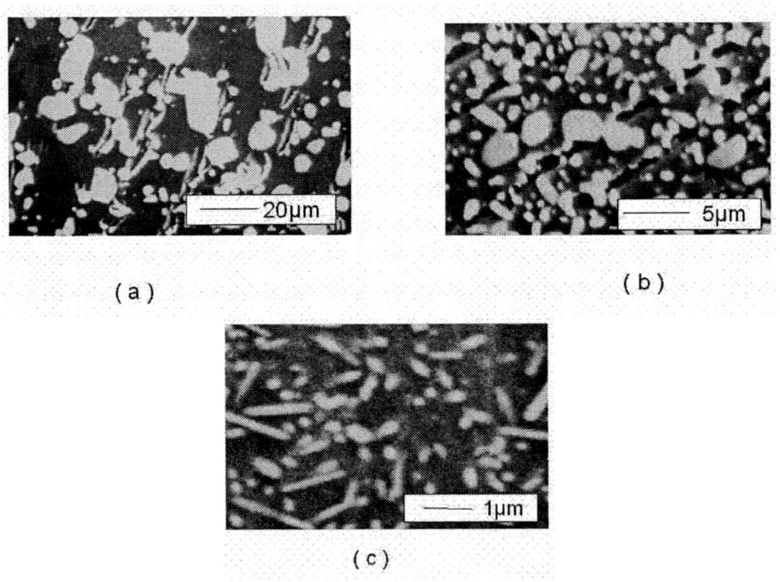

Fig. 6 Xu *et al.*: SEM image for samples fabricated in air with different size Gd211 particles: (a) commercial Gd211 powder, (b) self- made Gd211 powder, and (c) self-made Gd211 powder with ball-milling for 1 hour.

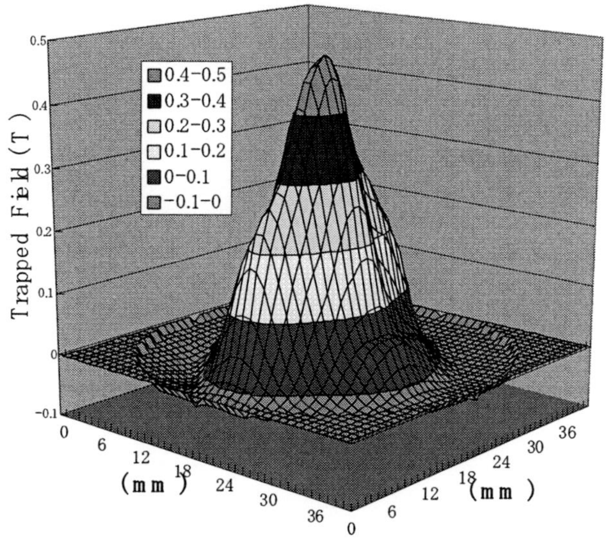

Fig. 7 Xu *et al.*: The trapped magnetic field density distribution of bulk with ball-milling 1 hour Gd211 particles addition. The maximum-trapped field density reaches 0.49 T with a 1.2 mm scanning gap between the Hall probe and the sample surface.

Fig. 7 shows the trapped magnetic field density distribution of bulk with ball-milling 1 hour Gd211 addition at 77 K. Initial magnetization field at room temperature was set as 1.0 T before field cooling. A scanning gap was 1.2 mm between the Hall probe and the sample surface. The maximum-trapped field density reaches 0.49 T.

3.4 Band Structure and Effect of Gd2411 Addition

A band structure of second phase distribution has been observed in Y123 bulks with Y2411 addition. [25]. In order to study the spatial variation of microstructure and superconducting properties of Gd123 with Gd2411 addition, we cut bulk samples along a-growth section as shown in Fig. 8 and performed the measurement of SQUID and SEM. The dimension of shadowed rectangular bar in Fig. 8 is 2mm * 2mm * 10mm. We observe the microstructure from seed position to the edge of single domain. Then, the rectangular bar was cut into five slabs marked as ss1, ss2, ss3, ss4 and ss5 with dimension as 2mm * 2mm * 2mm for the measurement of superconducting properties.

We first measured the J_c –B(T) curves with different Gd2411 addition amounts. The highest J_c at zero applied field appeared at different positions inside single domain for each composition. The J_c -B curve with highest J_c for each composition is displayed in Fig. 9. The inset accordingly shows the onset T_c. It is necessary to point out that when the additional amount $z \geq 0.02$, the J_c below the seed position was higher than that near single domain edge position, while when $z < 0.02$ the higher J_c located at the latter position. From Fig. 9, the highest J_c (at zero field) of bulk samples dramatically decreased with increasing the Gd2411

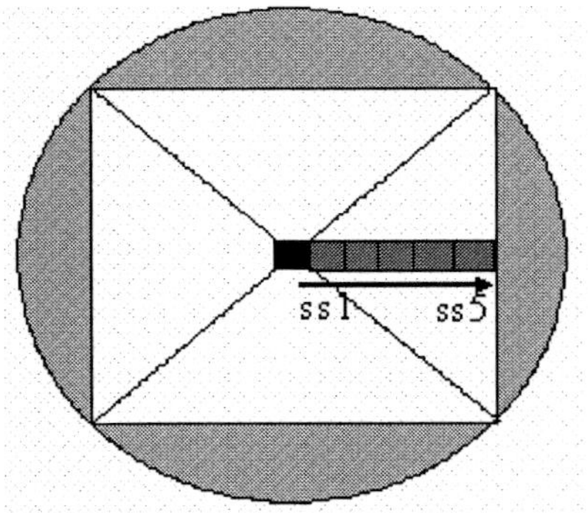

Fig. 8 Xu et al.: Illustration for cutting method of bulk samples. Microstructures were observed from ss1 position to ss5 position. SQUID was performed at each slab, ss1, ss2, ss3, ss4 and ss5.

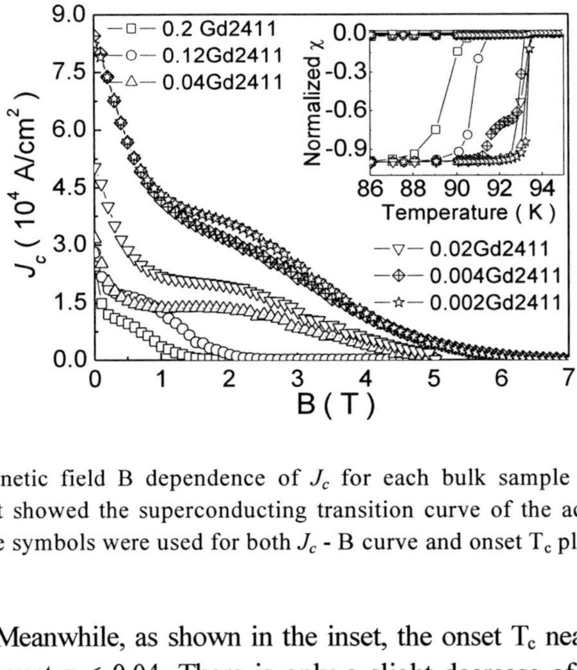

Fig. 9 Xu *et al.*: Magnetic field B dependence of J_c for each bulk sample with different Gd2411 addition amounts. Inset showed the superconducting transition curve of the according samples for J_c measurement. The same symbols were used for both J_c - B curve and onset T_c plot.

additional amounts. Meanwhile, as shown in the inset, the onset T_c nearly keeps unchanged with the addition amount $z \leq 0.04$. There is only a slight decrease of T_c with an extended transition width for a higher Gd2411 addition. Although a low T_c value may partially result in suppression of J_c, the present result evidences that Gd2411 addition does not deteriorate superconductivity with a proper amount. Thus, a depression of J_c values may arise from microstructure variation.

To clarify the spatially inhomogeneity of J_c, we focus on 0.02Gd2411. Figs. 10(a)-(e) show the microstructure of single domain bulk with 0.02Gd2411 addition observed at points A-E, from the seed position to the edge. Fig. 10(f) depicts the illustrations in scale for the band structure of 0.02Gd2411 bulk superconductor and the corresponding measured points from ss1 to ss5. There are some bands composed of second phases. Fig. 10 (a) shows the microstructure image under the seed area, A. We find there is almost no Gd211 particle but some small particles with the size much less than 1 μm are observed. They are assigned to Gd2411 from qualitative chemical analysis. Fig. 10 (b) shows a porous narrow layer with a relatively rich concentration of the second phase, which formed the first band, B. Fig. 10 (c) showed the first band for diluted second phase region, C. The second bands with relative rich and diluted second phase are shown in Fig. 10 (b) and Fig. 10 (c). As shown in Fig. 10 (d), the last band of rich second phase appears. Fig. 10 (e) shows the inner region of the last band with homogeneous distribution of the second phase, E. From Fig. 10 (b) to Fig. 10 (e), Gd2411 particles are distributed among too many Gd211 inclusions and it is thus difficult to separate them. Due to low Gd2411 addition concentration, the particles in Fig. 10 (b) to (e) should be mainly Gd211. The last band area E roughly occupies a half length of the diameter of single domain, which indicates that the band structure only locates around the seeding position. Such a band structure appears in all doping levels,[39] while it is not observed in a sample without Gd2411 addition. This shows that Gd2411 is the origin of forming inhomogeneous band structure even with a very small Gd2411 adding amount. It is worth noting that such a band structure is different from what obtained in Y123 bulk with nano-particles addition such as ZrO_2 or $BaZrO_3$ [26-27], where a relative regular band structure is reported with nearly equal band gap or width. The band structure observed in the present Gd123 system is slightly irregular with spatially variable band widths or gaps. Besides, we found the band gap or band width is a function of Gd2411 adding amount. The present structure is similar to the band structure observed in Y123 single domains with Y2411 addition grown under a large undercooling, where there is only one band [24, 25]. However, in case of Gd123 such a band structure only appears in a regime of 5 mm from each side of the seed for all Gd2411 adding levels.

This band structure can be explained by a positive difference between solid Gd123 and Gd211 and BaCuO liquid interfacial energies, $\triangle\sigma_0$, which was defined as following [26, 27, 40]:

$$\triangle\sigma_0 = \sigma_{SP} - \sigma_{LP} - \sigma_{SL} > 0$$

In the equation, σ means various interfacial energies: σ_{SP} (Gd123/Gd211), σ_{LP} (liquid/Gd211), σ_{SL} (Gd123/liquid). The critical Gd211 particle size for pushing, r*, is given by the equation:

$$R^* \propto \triangle\sigma_0 / \eta\, r^*,$$

where R* is the critical growth rate and η is the melt viscosity. From this equation, with assuming R* is constant, the increasing of interfacial energies, $\triangle\sigma_0$, will directly results in the increase of r*. Gd2411 additions could have a larger $\triangle\sigma_0$ and therefore increase the critical size of Gd211 of pushing. Thus a macrosegregation band can be formed due to accumulating the viscosity of liquid phase in growth front during pushing. A further investigation is necessary to elucidate the forming mechanism.

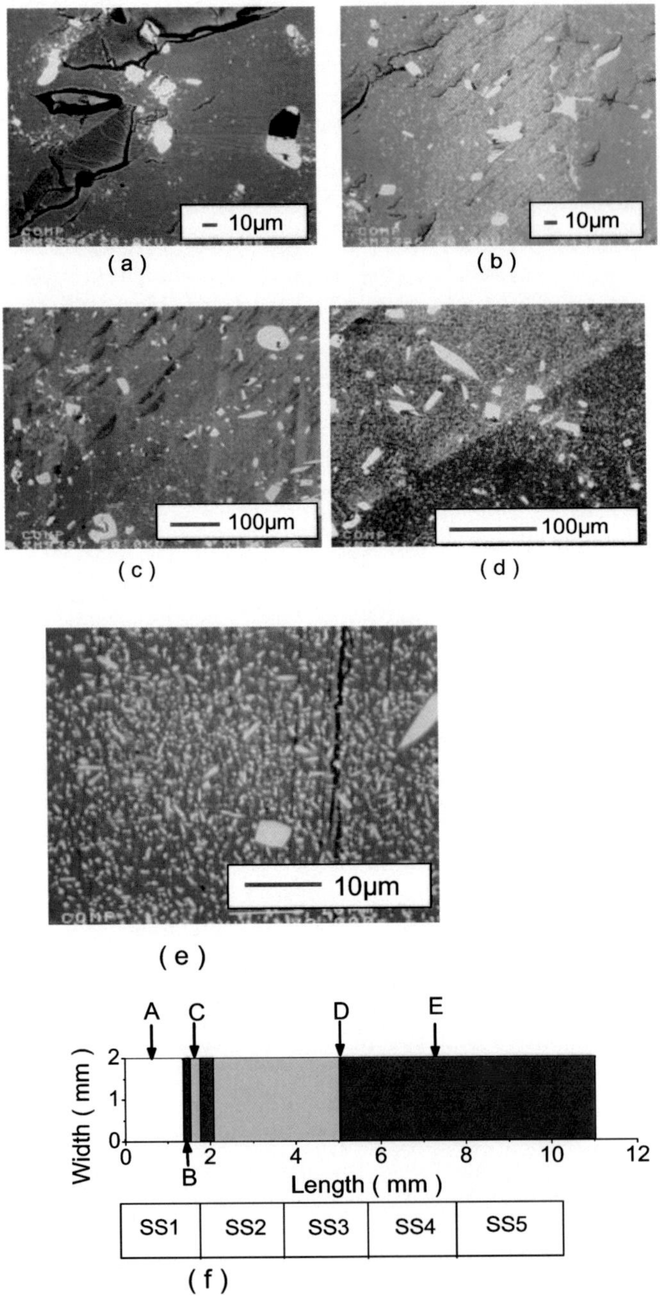

Fig. 10 Xu et al.: Microstructure for bulk with 0.02 Gd2411 in different positions from SEM. (a) locates the position under seed. (b) is the first band for rich second phase while (c) is the first band for diluted second phase. (d) shows the interface of the last diluted second phase region and the last rich second phase region. (e) shows the inside part of last rich second phase region. (f) is the illustration for band structure from seed to boundary. In the illustration of (f), stands for the low second phase concentration under the seed; while stands for the rich second phase region; means diluted second phase region. In (f), A to F are the positions accord to microstructure image (a) to (f), respectively. Meanwhile, ss1 to ss5 correspond to the positions for SQUID measurements.

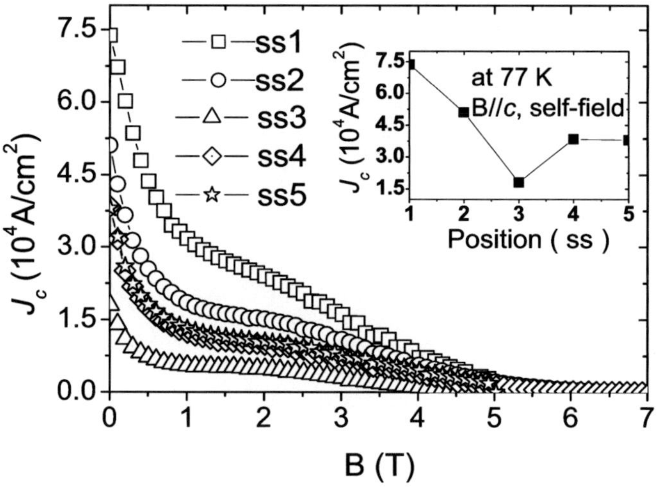

Fig. 11 Xu et al.: J_c – B curves as a function of position away from the seeding site (positions shown in Fig.8). The inset shows the tendency of self-field J_c at according positions, ss1 to ss5.

J_c as a function of position away from the seeding site is illustrated in Fig. 11 for 0.02Gd2411. ss1 to ss5 are shown in Fig. 8. The variation tendency of J_c at zero field is shown in the inset figure. The highest J_c appears below the seed position, ss1, as we discuss above while the lowest J_c locates the middle point of single domain, ss3, where should be the diluted second phase region before the last rich second phase band. Since there is almost no Gd211 inclusion below the seeding site, the present result strongly supports that Gd2411 enhances J_c values. It is notable that several previous studies have evidenced that J_c values in large RE123 single domains gradually increase from the seeding site to the edge of samples [41-44]. This is either due to the cumulative concentration of RE211 along the growth direction [35] or due to the suppression of the formation of $RE_{1+x}Ba_{2-x}Cu_3O_{7-\delta}$ solid solutions with increasing undercooling [42, 44] or even due to the contamination of seeding materials in a small size [43, 45]. The present results demonstrate a new tendency, in which J_c values follow the band structure and display a spatial variation associated with the microstructure feature of Gd211/Gd2411 distribution.

One issue still keeps unclear why a high amount of Gd2411 doping suppresses the highest J_c value shown in the inset of Fig. 9. A possible reason is the additional defects induced by Gd2411 inclusion. We note that there are a lot of macrocracks within the band as shown in Fig. 10 (b). These macrocracks certainly destroy supercurrents since they locate in the *ab* plane (parallel to CuO layer) [15]. It is known that a proper addition of either Gd211 [46] or silver [47] effectively suppress the occurring of such a crack. This suggests that the present silver amount may not be sufficient. Higher concentration of silver, such as 15% or 20% is under investigation.

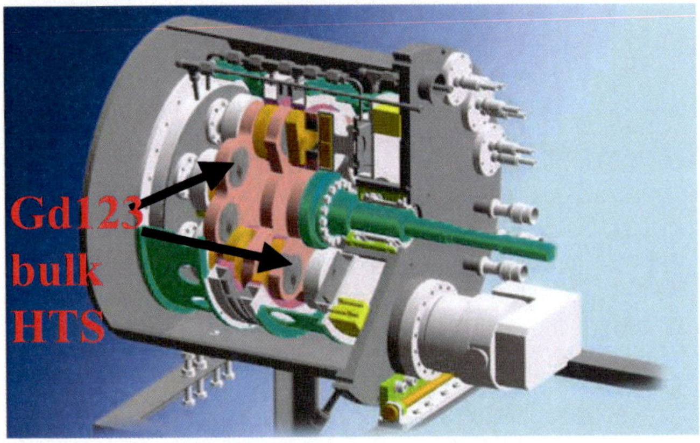

Fig. 12 Xu *et al.:* Illustration of high temperature superconductor bulk synchronous motor.

3.5 Progress of Practical Application of Large Gd123 Bulks

A variety of industrial applications have been developed based on enhanced superconducting performance of large grain high temperature bulk superconductors (HTS). H. Hayashi *et al.* reported the development of a mobile-type magnetic separator using bulk superconductors to quickly remove phytoplankton multiplying in eutrophic lakes or reservoirs [2], while T. Oka *et al.* developed a high gradient magnetic separation system using bulk superconductors to remove the particles of weak magnetization from the water slurry.[48] T. Matsuda et al. constructed a planar magnetron sputtering apparatus using a c-axis oriented single domain bulk superconductors as a very powerful magnet in place of an ordinary magnet.[49] These three cases are taken advantage of high trapped fields and field gradients of melt-processed bulks. K. Matsunaga et al. reported their studies on the radial type superconducting magnetic bearing module using bulk superconductors for a 100 k W h class flywheel system.[50] In this case, high levitation force and the self stability are used.

We have studied a high-temperature superconducting synchronous motor assembled with melt-textured Gd-Ba-Cu-O bulk field magnets. Fig. 12 showed the illustration of this kind of HTS bulk synchronous motor that was developed by Kitano Seiki Co. Ltd. and Izumi group. [1,51]The motor is an axial gap-type, brushless synchronous motor with eight bulk field magnet poles on the axial rotor plate. Liquid nitrogen is introduced and circulated inside the rotor shaft and to cool down the rotor pole bulks. Pulsed field magnetization was performed to excite the bulk field magnets by using a pair of the vortex-type armature copper windings under the zero-field cooling. The trapped peak field density on the surface of the bulk was up to to 1.0 T. Such trapped flux onto the bulk magnets provided the motor performance of 4.8 kW with 720 rpm. The optimized pulsed current waveform applied to the armature and the employing of advanced bulk crystal magnets leading to a spatially homogeneous flux trapping are promising methods for reinforcement of the field flux from the rotor and the motor torque. [51]

4 Conclusion

We fabricated the Gd-Ba-Cu-O bulk superconductors in air melt process and enhanced the superconducting performances with different methods. As a light rare earth 123 type superconductor, the Gd/Ba substitution level controls the superconducting properties, solubility and stability of $Gd_{1+x}Ba_{2-x}Cu_3O_{7-\delta}$ solid solutions. In order to suppress the solid solutions that restricted the superconducting properties, BaO_2 was added into Gd-Ba-Cu-O system as starting powders. The optimum addition amount is 0.1, molar ratio to Gd123. By refinement of second phase Gd211 particles sizes, J_c has enhanced from 20,000 A/ cm^2 to 90,000 A/ cm^2 which is the record high J_c value of Gd-Ba-Cu-O superconductors melt-processed in air. New second phase $Gd_2Ba_4CuMoO_y$ has successfully synthesized in air at 1460℃ and added into Gd-Ba-Cu-O bulks. A band structure is observed in the bulks even with a slight Gd2411 addition amount. Although this inhomogeneous distribution of second phases resulted in a spatial variation of superconducting properties, J_c in the seed positions of single domain is enhanced by the new second phase addition. This opens the way to enhance the quality of the whole bulk and accordingly offers a variety of engineering applications of bulk superconductors.

Acknowledgements

This work was partly supported by Sasakawa Scientific Research Grant from Japan Science Society. One of author (AH) was grateful for the financial support of the Alexander von Humboldt foundation, Germany, Japanese New Energy Development Organization and Japanese Society for the Promotion of Science.

References

[1] Ida T, Matsuzaki H, Akita Y, Izumi M, Sugimoto H, Hondou Y, Kimura Y, Sakai N, Nariki S, Hirabayashi I, Miki M, Murakami M, *Kitano M Physica C* 2004 Vol. 412 638-645.

[2] Hayashi H, Nagabuchi N, Saho N, Mizumori T, *Asano K Physica C* 2004 Vol. 412 766-770.

[3] Tanaka S *JSAP international* 2001 Vol. 4 17.

[4] Hu A, Sakai N, Murakami M 2001 Appl. *Phys. Lett.* Vol. 78 2539-2541.

[5] Xu C, Hu A, Sakai N, Izumi M, Hirabayashi I 2005 *Supercond. Sci.* Technol. Vol. 18 229-233.

[6] Nariki S, Sakai N, Murakami M 2005 *Supercond. Sci. Technol.* Vol. 18 S126-S130.

[7] Kramer M J, Yoo S I, McCallum R W, Yelon W B, Xie H, Allenspach P 1994, *Physica C* Vol. 219 145-155.

[8] Izumi M, Yabe T, Wada T, Maeda A, Uchinokura K, Tanaka S, Asano H 1989, *Phys. Rev. B* Vol. 40 6771-6786.

[9] Hu A M, Jia S L, Chen H, Zhao Z X 1996 *Physica C* Vol. 272 297-300.

[10] Chen S Y, Chen I G 2004 Supercond. *Sci. Technol.* Vol. 17 71-77.

[11] Xu C, Hu A, Sakai N, Izumi M, Hirabayashi I 2005 *Physica C* Vol. 417 77-84.

[12] Rogachi K 2003 *Phys. Rev. B* Vol. 68 100507 (R) 1-4.

[13] Hu A, Schätzle P, Bieger W, Verges P, Fuchs G, Krabbes G 1999 *Appl. Phys. Lett.* Vol. 75 259-261.

[14] Ikuta H, Hosokawa T, Yoshikawa M, Mizutani U 2000 *Supercond. Sci. Technol.* Vol. 13 1559-1568.

[15] Hu A, Sakai N, Zhou H, Matsui M and Murakami M 2003 *Supercond. Sci. Technol.* Vol. 16 33-38.

[16] Hu A, Zhou H, Nariki S, Murakami M and Hirabayashi I 2004 *Supercond. Sci. Technol.* Vol. 17 545-548.

[17] Dai J Q, Zhao Z X and Hu A 2004 *Physica C* Vol. 406 63-71.

[18] Haugan T, Barnes P N, Wheeler R, Meisenkothen and Sumption M 2004 *Nature,* Vol. 430 867-870.

[19] Nariki S, Matsui M, Sakai N, Murakami M, 2002 *Supercond. Sci. Technol.* Vol. 15 679-682.

[20] Kim C J, Hong G W, 1999 *Supercond. Sci. Technol.* Vol. 12 R27-R41 and references therein.

[21] Dai J Q, Zhao Z X, Xiong J W, 2003 *Supercond. Sci. Technol.* Vol. 16 815-819.

[22] Hari Babu N, Reddy E S, Cardwell D A, Campbell A M, Tarrant C D, Schneider K R 2003 *Appl. Phys. Lett.* Vol. 83 4806-4808.

[23] Hari Babu N, Sudhakar Reddy E, Cardwell D A, Campbell A M, 2003 *Supercond. Sci. Technol.* Vol. 16 L44-L45.

[24] Hari Babu N, Reddy E S, Shi Y, Iida K, Withnell T D and Cardwell D A Oct. 2004, presented in Applied Superconductor Conference (ASC2004), Florida, USA, to be published on *IEEE Trans. Appl. Supercond.*

[25] Hari Babu N, Shi Y, Iida K, Cardwell D A, Haindl S, Eisterer and Weber H W, Nov. 2004,presented in the 17th International Superconductivity Symposium (ISS2004), Niigata, Japan, to be published on *Physica C.*

[26] Iida K, Hari Babu N, Reddy E S, Shi Y-H and Cardwell D A 2005 *Supercond. Sci. Technol.* Vol. 18 249-254.

[27] Carrillo A E, Puig T, and Obradors X 2005 *Supercond. Sci. Technol.* Vol. 18 S77-s81.

[28] Izumi F, Ikeda T 2000 *Mater. Sci. Forum* Vol. 198 321-324.

[29] Hu A, Sakai N, Murakami M 2002 *Physica C* Vol. 371 19-26.

[30] Xu C, Hu A, Sakai N, Izumi M, Hirabayashi I 2005 *Physica C*, Vol. 426-431 613-617.

[31] Akinaga H, Katoh H, Takita K, Asano H, Masuda K 1988 *Jpn. J. Appl. Phys.* Vol. 27 L610-L612.

[32] Takita K, Katoh H, Akinaga H, Nishino M, Ishigaki T, Asano H, 1988 *Jpn. J. Appl. Phys.* Vol. 27 L57-L60.

[33] Hu A, Krabbes G, Schätzle P, Bieger W 1999 *Physica C* Vol. 322 129-135.

[34] Hu A, Sakai N, Murakami M 2002 *Physica C* Vol. 371 19-26.

[35] Gautier-Picard P, Beaugnon E, Chaud X Sulpice A, Tournier R 1998 *Physica C* Vol. 308 161-168.

[36] Hu A, Sakai N, Murakami M 2003 *Physica C* Vol. 386 275-278.

[37] Ubaldini A, Giovannelli F, Monot-Laffez I 2002 *Physica C* Vol. 283 107-116.

[38] Nariki S, Sakai N, Murakami M, Hirabayashi I 2004 *Supercond. Sci. Technol.* Vol. 17 S30-S35.

[39] Xu C, Hu A, Sakai N, Izumi M, Hirabayashi I 2005 *Supercond. Sci. Technol.* Vol. 18 1082-1088.

[40] Uhlmann D R, Chalmers B, Jackson K A 1964 *J. Appl. Phys.* Vol. 35 2986-2993.

[41] Mizutani U, Mase A, Tazoe K, Ikuta H, Oka T, Itoh Y, Yanagi Y, Yoshikawa M 2000 *Physica C* Vol. 335 92-96.

[42] Hu A, Sakai N, Murakami M 2003 IEEE Trans. *Appl. Supercond.* Vol. 13 3079.

[43] Dewhurst C D, Lo W, Shi Y H, Cardwell D A 1998 *Mater. Sci. Engr. B* Vol. 53 169-173.

[44] Hu A, Sakai N, Murakami M 2003 *Physica C* Vol. 571 392-396.

[45] Krabbes G, Hopfinger Th, Wende C, Diko P, Fuchs G 2002 *Supercond. Sci. Technol.* Vol. 15 665-671.

[46] Hu A M, Zhao Z X, Xiong J W, Pang C 1997 *Physica C* Vol. 288 136-140.

[47] Hu A, Zhao Z, Wu M, Wende C, Strasser T, Jung B, Bruchlos G, Gawalek W, Goernert P 1997 *Physica C* Vol. 278 43-48.

[48] Oka T, Yokoyama K, Itoh Y, Yanagi Y, Yoshikawa M, Ikuta H, Mizutani U, Okada H, Noto K 2003 *Physica C* Vol. 392 709-712.

[49] Matsuda T, Kashimoto S, Imai A, Yagagi Y, Itoh Y, Ikuta H, Mizutani U, Sakurai K, Hazama H 2003 *Physica C* Vol. 392-396 696-703.

[50] Matsunaga K, Yamachi N, Tomita M, Murakami M, Koshizuka N 2003 *Physica C* vol392-396 723-728.

[51] Matsuzaki H, Kimura Y, Ohtani I, Izumi M, Ida T, Akita Y, Sugimoto H, Miki M, Kitano M,2005 *IEEE Trans. on Appl. Supercond.*, Vol. 15 2222-2225.

INDEX

D

E

F

G

H

I

J

L

M

N

O

P